Zihai Shi, Masaaki Nakano, Yoshifumi Takahashi

Structural Analysis and Renovation Design of Ageing Sewers

Design Theories and Case Studies

Zihai Shi, Masaaki Nakano, Yoshifumi Takahashi

Structural Analysis and Renovation Design of Ageing Sewers

Design Theories and Case Studies

Managing Editor: Irmina Grzegorek

Language Editor: Adam Tod Leverton

DE GRUYTER
OPEN

ISBN 978-3-11-047173-1
e-ISBN (PDF) 978-3-11-047174-8
e-ISBN (EPUB) 978-3-11-047191-5

Published by De Gruyter Open Ltd, Warsaw/Berlin
Part of Walter de Gruyter GmbH, Berlin/Boston
The book is published with open access at www.degruyter.com.

Library of Congress Cataloging-in-Publication Data
A CIP catalog record for this book has been applied for at the Library of Congress.

Managing Editor: Irmina Grzegorek
Language Editor: Adam Tod Leverton

www.degruyteropen.com

Cover illustration: provided by Sekisui Chemical Co., Ltd.

Contents

Author Information

Zihai Shi (**1.1, 4, 5.1-5.3**)
Principal research scientist, Research & Development Centre, Nippon Koei Co., Ltd. Graduated from Civil Engineering Department, University of Tokyo in 1989. Doctor of Engineering.
Major: Fracture mechanics, mechanics of fatigue, structural engineering

Yoshifumi Takahashi (**1.2-1.3, 2**)
Senior adviser, Tokyo Sewerage Service Co., Ltd.; former Director for Technology Development, Sewerage Bureau, Tokyo Metropolitan Government. Graduated from Civil Engineering Department, Nihon University in 1971. Doctor of Engineering.
Major: Sewerage Engineering, structural engineering

Yukari Nakamura (**3, 7**)
Researcher, Research & Development Centre, Nippon Koei Co., Ltd. Graduated from Civil Engineering Department, Kagoshima University in 2008. Master of Engineering.
Major: Structural engineering

Jianhon Wang (**5.4**)
Senior researcher, Research & Development Centre, Nippon Koei Co., Ltd. Graduated from Civil Engineering Department, Waseda University in 2009. Doctor of Engineering.
Major: Buckling of thin-walled structures, tunnel engineering

Masaaki Nakano (**6**)
Senior researcher, Research & Development Centre, Nippon Koei Co., Ltd. Graduated from Science and Engineering Department (mathematics), Waseda University, Tokyo (B.S.) in1992; graduated from Environmental Science Department, Natural Resource, Carroll University, Wisconsin (B.S.) in 1994. Bachelor of Science.
Major: Fracture mechanics of concrete, structural engineering

Toru Kouchi (**8**)
Researcher, Research & Development Centre, Nippon Koei Co., Ltd. Graduated from Civil Engineering Department, Tokyo City University in 2011. Master of Engineering.
Major: Structural engineering

Preface

This book presents a systematic treatment of fracture-mechanics based structural analysis and renovation design of ageing sewers as semi-composite pipes. These have been employed for sewer renovation in Japan, especially in the Tokyo metropolitan area for more than fifteen years, with the total length of renovation construction now exceeding 700 km. The concept of a semi-composite pipe and the application of fracture mechanics of concrete in numerical modelling of the structural behaviour of a renovated sewer pipe are the two distinctive features of the established design theories that have emerged and matured with the development of a sewer renovation construction method (the SPR method, which received the top industrial technology development award in Japan, the Oukochi Memorial Prize 2013).

The general practice of sewer renovation in Japan, from site survey and structural design to renovation, is subject to the Japan Sewage Works Association's *Design and Construction Guidelines for Sewer Pipe Rehabilitation*. The Guidelines apply to two types of renovation method, namely, the independent (or stand-alone) pipe method and the composite pipe method, with the former being used mainly for non-man-entry sewers with circular cross section and the latter for both man-entry and non-man-entry sewers with arbitrary cross sections. In the composite pipe construction approach, a liner pipe is constructed inside an existing sewer by interlocking strips of surface materials made of polyvinyl chloride or polythene resin, and the annular space behind the liner is filled with cementitious grout under pressure to form a highly integrated structure. The Guidelines define a composite pipe as a composite structure requiring complete integration between the renovation layer and the existing pipe.

Test results show that the bond strength of cementitious grout is typically half the tensile strength of normal concrete. With this limited bond strength, the requirement for a rigid connection between the existing pipe and the renovation layer in the composite pipe method may not always be guaranteed. To fully explore the advantages of the composite pipe method in sewer renovation while keeping safe design in mind, a semi-composite pipe structure has been proposed based on three assumptions for the interface: a perfect bond after construction, a no-tension interface (i.e., the interface does not bear tensile force) and a perfect bond under compression. The resulting structure resembles a composite structure with mechanical connectors set in the compression zones along the interface that are intercepted by no-tension interface zones with free surfaces, and thus is called a semi-composite pipe structure.

The design approach for this semi-composite structure is based on the theory of limit state design. As a typical feature of this design approach, nonlinear structural analysis including crack analysis is performed on a semi-composite pipe under extreme load conditions to obtain its characteristic load-carrying capacity. The details of the renovation layer including the thickness and ratio of steel reinforcement are then determined using design formulas based on the ratio of member force to member strength with relevant safety factors considered. Next, a seismic performance evaluation is carried out to ensure the renovated pipe meets the code requirements for earthquake resistance.

The leading authors of this book have been involved in designing the renovation of ageing sewers since the mid 1990s, and the design theories and design practices introduced here are the crystallisation of their collaborative research efforts, which have been greatly strengthened by their younger colleagues who have joined the team in the past five or six years. The book contains eight chapters. Chapter 1 introduces the background to sewer renovation in Japan, and Chapter 2 explains the composite pipe method. Chapter 3 introduces various fracture tests on renovated pipe specimens, and structural element and material property tests. Chapter 4 presents the basic theories of the fracture mechanics of concrete, and Chapter 5 develops the structural analysis theories of composite pipes as semi-composite structures, including an analysis on the local buckling of invert lining under groundwater pressure. Chapter 6 presents renovation design theories that include limit state design and seismic performance evaluation. The last two chapters focus on applications: Chapter 7 on the development of a design-aid programme, and Chapter 8 on design case studies.

The book was written as a reference work focusing on the latest technical developments in sewer renovation in Japan for both sewer engineers worldwide and students of civil and urban engineering. The authors are both researchers and practicing engineers with extensive experience in the renovation design of ageing sewers. Each chapter was written by one or two co-authors, and was edited for content, accuracy and style.

Just like an artist in painting or music, I sometimes feel that our daily engineering work such as the issues discussed in this book is as creative as the traditional arts that we enjoy for their beauty or emotional power. This may sound farfetched, but when you face a new project for which you are aware of only some of the mechanics principles required, you really need to rely on your engineering instinct, creative skill and imagination to identify the best approach. Our work may not be enjoyed for its beauty (normally you cannot even see the sewer), but its emotional power may be associated with the safe infrastructure of a modern society, of which a sewerage system is a vital component. If the readers of this book even partially agree with me, this will mean, at least for me, 'mission accomplished'.

Z. Shi
August 10, 2015

Acknowledgement

The numerical studies in Section 5.3.2 on the load-carrying capacity tests of real-size manhole specimens, which involved a modified epoxy resin in repairing corroded manholes, were based on ongoing research by Mr. Mitsuhiro Kurozumi. We are grateful to him for kindly allowing us to use the test cases as application examples of the discrete crack modelling approach. We also thank Mr. Tsuyoshi Haibara for his assistance in the preparation of the manuscript. We are grateful to Ms. Kanae Uchiyama for her beautiful artwork and general assistance in helping us finish this huge project on time. The editorial team of De Grupter Open for this book, Drs. Elisa Capello, Agata Morka, Marek Domanski and Ms. Irmina Grzegorek deserves more than thanks than we can give for their enthusiastic approach, expert guidance and professionalism in overseeing the publication of this book.

Z. Shi
M. Nakano
Y. Takahashi
August 10, 2015

Yoshifumi Takahashi, Zihai Shi

1 Introduction

1.1 Recent Development in Sewer Renovation in Japan

In the modern world, a sewerage system increasingly serves as a vital lifeline that sustains the normal functioning of a society in which people live and work. Although sewers are normally out of sight (and thus "out of mind", as a well-known sewer engineer once lamented), a malfunction of these underground facilities can have immediate effects on those who live in the affected areas. For instance, a road-cave-in accident as shown in Photo 1.1 due to the sudden collapse of an ageing sewer can sometimes cause loss of life, not to mention inconvenience for those living in the surrounding areas, as well as the monetary cost of repairs. The inadequate discharging capacity of a network of ageing sewers for surges of rain water in stormy weather can result in floor-level inundation of large areas, which not only disrupts the normal livelihood for many but also causes enormous damage to private property. Even relatively minor problems like clogged drains and emission of foul odours due to the dysfunction of ageing sewers in a small residential area can be distressing, especially when these ordeals are prolonged and repeated.

Photo 1.1: Road cave-in caused by the collapse of a sewer main (Tokyo)

In general, as sewers approach and exceed their design service life of fifty years, the likelihood or frequency of malfunctions as discussed above increases, as a result of long-term material and structural deterioration caused by abrasion, chemical attack, excessive hydraulic flows, structural cracks and joint defects, and leaks and infiltration. Photo 1.2 shows an ageing sewer with the complete loss of cover concrete and severe corrosion of rebars in the reinforced concrete cover plate, which is a frequently-observed type of structural degradation in ageing sewers. Obviously, rehabilitation of these ageing sewers to ensure their safe operation and upgrade their functions is urgently needed and should be carried out systematically using modern renovation methods and design codes.

Photo 1.2: An ageing sewer with the complete loss of cover concrete and severe corrosion of rebars

In Japan, sewer rehabilitation began in 1987 by using the Sewerage Pipe Renewal (SPR) method. The principles of design and construction for sewer renovation are specified by the Japan Sewage Works Association's standard: *Design and Construction Guidelines for Sewer Pipe Rehabilitation* (JSWA, 2011). In this provisional code, renovation methods are classified into two categories: the so-called composite pipe method and the independent (or stand-alone) pipe method. The concept of the composite pipe method is to construct a composite structure by rigidly attaching a renovation layer, i.e., an inner lining to the existing pipe, and the renovated sewer is expected to bear the external loads using the combined resistance of the two structural components. On the other hand, in the independent pipe method a new pipe is constructed inside the existing one, ignoring the remaining strength of the ageing sewer. Note that the *Guidelines* were developed from a previous version, *Pipe Rehabilitation Handbook* (JSWA, 2001).

Since the late 1980s, the Sewerage Bureau of the Tokyo Metropolitan Government (TMG) has actively promoted the composite pipe renovation method in upgrading and reconstructing its ageing sewerage system (TMG, 2008). As a result, except when sewer replacement by open-cut construction is absolutely necessary, renovations of man-entry sewers with a wide range of cross-sectional shapes have been carried out using several certified composite pipe renovation techniques. The independent pipe method is sometimes used to renovate non-man-entry sewers with circular cross sections in the Tokyo area, but it is not used as frequently as the composite pipe method.

Photo 1.3 shows an ageing sewer during renovation construction using the SPR method, which is one of several composite pipe renovation techniques that have been approved by the Sewerage Bureau of the TMG to be employed in the metropolitan area at present. As seen, a liner-formation machine moves along the existing sewer with its pre-fabricated framework conforming to the cross-sectional shape of the sewer, creating an inner liner along the length of the existing sewer. The liner is formed by helically winding a continuous polyvinyl chloride (PVC) ribbed profile with interlocking edges. The annular space behind the liner is then filled with cementitious grout under pressure to form a highly integrated structure with the PVC liner bound to the existing sewer by grout after it has hardened naturally. For medium- to large-size man-entry sewers, the ribs of the profile are reinforced with steel to enhance the hoop strength and stiffness of the liner, and the structural strength of the renovated sewer is largely determined by the thickness and material characteristics of the renovation layer.

As seen, the ageing sewer after renovation in Photo 1.3 has been transformed into a compound structure composed of the existing sewer and the renovation layer of the reinforced liner and the grout, and is expected to function like a composite structure under external loads that include the surrounding earth pressure, groundwater pressure and the live load from traffic. Construction processes using similar renovation techniques may differ in detail from manufacturer to manufacturer. In one particular method as shown in Photo 1.4, evenly-spaced steel frames are laid against the sewer's inner walls first, and then the liner is fabricated by fixing long PVC plates longitudinally to these preset frames, which is followed by grout injection.

Photo 1.3: An ageing sewer during renovation construction by using the SPR method

In order for the renovated sewer to resist external loads as a composite structure in which the continuity of stress or force flow through the interfaces is maintained, obviously the bond strength of grout to the ageing sewer has to be strong enough to endure large interface stresses that may arise under extreme load conditions. It is known that the bond strength of various types of high-strength mortar to concrete developed by several manufacturers in Japan has a wide range of variation, changing approximately from 1.5 MPa to 3.0 MPa. In general, the bond strength of cementitious grout is less than the tensile strength of normal concrete, and on average it is one half of the latter. With this limited bond strength, the requirement for rigid connection

Photo 1.4: A composite pipe renovation method by using evenly-spaced steel frames and long PVC plates to fabricate interlinings before grout injection

between the two different structural members in the composite pipe method may not always be guaranteed. Though the required bond strength could be achieved by pre-embedding sufficient mechanical connectors in the existing pipe to rigidly connect the two parts, employing such mechanical connectors in sewer renovation is unrealistic when considering a sewer's small cross section, its long length and the sheer cost of construction.

The leading authors of this book have been involved in designing the renovation of ageing sewers since the mid 1990s, and have developed a unique structural model for the composite pipe method in which a no-tension interface model is used to define an otherwise rigid connection between the contacting surfaces of the existing pipe and the renovation layer. In other words, the presumed rigid linkage between the two surfaces following renovation construction is severed when tension develops from the interfaces. The resulting structure is referred to as *a semi-composite pipe structure*, or, *a semi-composite pipe*. The design approach for this semi-composite pipe structure is based on the theory of limit state design. As a typical feature of this design approach, nonlinear structural analysis including crack analysis is performed on a semi-composite pipe under extreme load conditions to obtain its characteristic load-carrying capacity. Details of the renovation layer including the thickness and ratio of steel reinforcement are then determined using design formulas based on the ratio of member force to member strength with relevant safety factors considered. Note that the member force is typically obtained by linear structural analysis under the design load condition, and the member strength is derived from the limit state of structural failure when the load-carrying capacity is obtained. In this book, a terminology of semi-composite pipe is coined to indicate the no-tension-interface structural model of a composite pipe. Therefore, when the term of composite pipe is used, it should be understood as such a structure.

One may question the wisdom of adopting such a seemingly complex numerical model and computational approach in the renovation design of ageing sewers: Why not choose a simpler approach using linear elastic analysis? The reason is that a good design in sewer renovation must be structurally safe and hydraulically functional. A simple but conservative model of the problem may require a thick interlining for the renovation that could greatly reduce the hydraulic capacity of the sewer. Such a design is unsuitable, even if it is structurally safe. Therefore, the purpose of employing a realistic structural model like the semi-composite pipe model in the current design approach is to obtain the minimum required thickness of the renovation layer, so that the renovated sewer is not only structurally safe but also functionally sound with its hydraulic capacity uncompromised by the renovation.

As this semi-composite structure concept and design methodology have been employed for sewer renovation in Japan, and especially in the Tokyo metropolitan area for more than fifteen years, with a total construction length exceeding 700 km, this book summarises the design theories and practices as a reference book for both sewer engineers worldwide and students of civil and urban engineering. The book contains eight chapters. Chapter 1 introduces the background to sewer renovation in Japan, and Chapter 2 explains the composite pipe renovation method. Chapter 3 introduces various capacity tests of renovated sewer pipes and structural element and material property tests. Chapter 4 presents the basic theories of fracture mechanics of concrete, and Chapter 5 develops structural analysis theories of composite pipes as semi-composite structures, including an analysis on the local buckling of invert lining. Chapter 6 presents renovation design theories that include limit state design and seismic performance evaluation. The last two chapters focus on applications with Chapter 7 on the development of a design-aid programme, and Chapter 8 on design case studies. The authors are researchers and practicing engineers with extensive experience in the renovation design of ageing sewers. Each chapter was written by one or two co-authors, and was edited by Zihai Shi for content, accuracy and style.

1.2 Past Sewer Projects and Emerging Challenges

In 1884, in the wake of a cholera outbreak, sewer construction was started in the Kanda area in Tokyo as the first project of its kind in Japan. The sewer system thus constructed is the Kanda Sewer System, the first modern sewer system in Japan (Photo 1.5).

Photo 1.5: Kanda sewer system (still in service)

For financial reasons, however, the construction had to be discontinued after a two-kilometre-long sewer system was completed. In 1894, Osaka began transforming its open-channel, masonry sewerage system into culverts, with necessary reinforcement and modifications. During the decade from 1887 to 1897, sewer plans for other major cities such as Nagoya, Fukuoka, Hiroshima and Kobe were drawn up, and the construction of sewerage systems gradually began. As so, sewerage construction in Japan in the major cities began.

In the subsequent years, the functions and purposes of sewer systems were expanded to include improvement of the living environment, flood control, and water contamination prevention. Thus, sewerage systems became widely recognised throughout the country as part of urban infrastructure. As a result, the percentage of sewered population in Japan has reached 97% in the ordinance-designated cities. The percentage in other cities, however, remains about 64%, and the national average is 73% (as of 2008).

As these figures indicate, the task for sewerage projects in non-ordinance-designated cities is to raise the percentage of the sewered population. In the ordinance-designated cities, where sewerage projects began early, it is to rehabilitate and functionally upgrade the ageing facilities. This section outlines the sewerage projects that have been carried out thus far in Tokyo and explains the challenges to be addressed and the measures being taken.

1.2.1 Modern sewerage projects

In the 1880s, there was an outbreak of cholera in Tokyo. The number of patients exceeded 750,000, and more than 260,000 died. The main cause was thought to be the urban structure, particularly the drinking water supply and sewerage systems in those days, and there were increasing calls for sewer construction to prevent cholera. In response to this public demand, the Kanda Sewer System, which was a pioneering sewerage project in modern Japan, was constructed in 1884 and 1885 prior to the construction of drinking water supply systems. Its construction, however, was halted two years later after just two kilometres of sewers had been completed. This was partly because the necessity and benefits of sewerage systems were not widely recognised. The direct reason was the lack of public funding because the use of government subsidies was not permitted.

One reason why the Kanda Sewer System project was discontinued was the Government's poor understanding of its importance, as well as the fact that the project was not being carried out systematically as part of urban reform. In addition, the construction technology was immature, and it was difficult to procure construction materials. Also, releasing untreated sewage into rivers was a problem.

After the project was stopped, in 1890 the Government decided to give priority to constructing drinking water supply systems; construction began in 1892, and water supply commenced in some areas in 1898. Later, the construction of sewerage systems resumed, and in 1908 the Tokyo City Sewer Design, which formed the basis of sewerage plans for the present-day Tokyo, was adopted.

The plan called for the construction of combined sewerage systems, which meant that sewers and sewage treatment facilities were to be constructed in an integrated manner. Under the plan, sewer construction began in 1913 in Ryusenji in Shitaya Ward, and construction of the Mikawajima Sewage Treatment Plant began in 1914. Figure 1.1 shows the length of sewers constructed each year and how the unit sewage volume, which is the basis for sewerage planning, and the storm-water runoff coefficient have changed over the years in Tokyo.

As shown, the sewerage systems constructed in Tokyo can be classified, according to quality and quantity, into two groups: those constructed during the pre-war period and those constructed from 1955. The sewerage systems constructed during the pre-war period have small unit sewage volumes and storm-water runoff coefficients. The total length of those sewers, which were constructed to serve the former Tokyo City area, is about 1,950 km. Most of the sewerage systems constructed from 1955 were built rapidly from around 1960, when Tokyo was selected as the host of the 1964 Olympic Games. From the mid-1960s to 1994, when the percentage of sewered population reached 100%, sewers were constructed at a rate of about 400 km/year. During that period, the unit sewage volume increased by a factor of 4 and the storm-water runoff coefficient by a factor of about 1.5 compared with the pre-war levels. These trends in sewer construction reflect changing lifestyles and urbanisation.

1.2.2 Emerging challenges related to urban sewerage

1.2.2.1 Functional degradation

Changes in the volume of incoming sewage directly affect the drainage capacity of sewers. As shown in Fig. 1.1, storm-water runoff increased with the progress of urbanisation, and the drainage capacity of most of the sewers constructed during the pre-war years and up to around 1980 is smaller than that of the later sewers by about 40% (the runoff coefficient increased from 53–59% to 75–80%). This factor has contributed to the aggravation of flood damage in recent years. Table 1.1 shows the number of flooded houses in Tokyo from 1989 to 2000. As shown, many houses were flooded in those years (4,136 houses in 1989, 2,907 in 1991, 5,719 in 1993, 3,542 in 1999) although such flooding was caused by frequent localised heavy rains.

Sewage volume has also increased. The volume of sewage released into sewers per person per day has quadrupled from 167 to 680 litres, and population has also increased. To cope with the situation, various storm-water control measures have been taken to enlarge and strengthen infrastructure facilities such as augmenting sewer mains, increasing pump stations, increasing pump drainage capacity and the capacity of treatment plants, and projects for improving combined sewer systems have also been carried out. It has not been possible, however, to replace all old sewers mainly because of financial constraints.

Table 1.1: Number of stormwater-flooded houses in central Tokyo and frequency of heavy rains

Year	1989	1990	1991	1992	1993	1994	1995	1996	1997	1998	1999	2000
Number of flooded houses	4,136	508	2,907	13	5,719	515	94	65	233	163	3,542	619
Number of more-than-50-mm rainfall events*	2	2	0	0	4	4	1	0	1	2	4	6

* The design rainfall adopted by the Tokyo Metropolitan Government is 50 mm/hr.

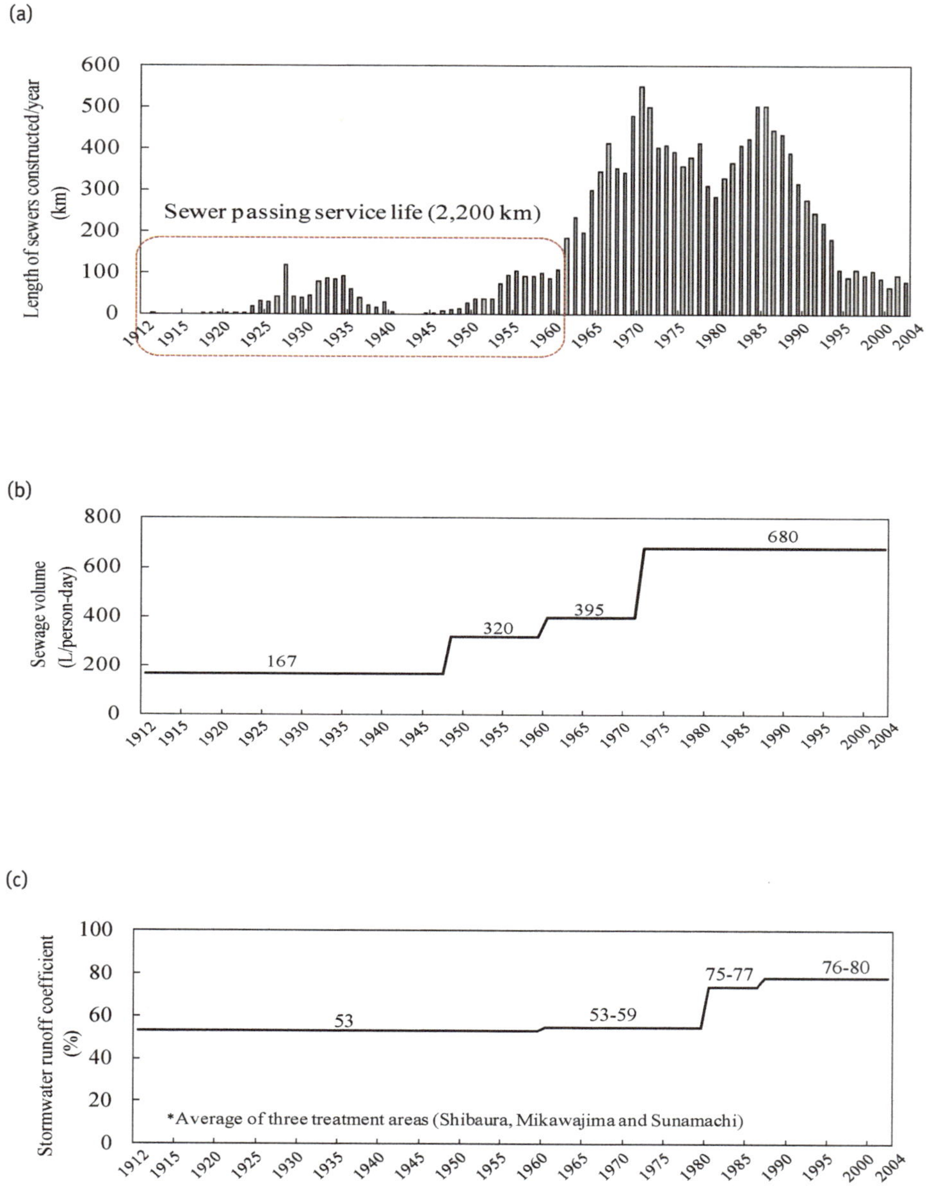

Figure 1.1: Variations of sewer length, volume and runoff coefficient with time in central Tokyo: (a) the length of sewers constructed; (b) sewage volume; and (c) stormwater runoff coefficient

1.2.2.2 Structural degradation

The sewer environment is corrosive and degrading, and sewers can also be adversely affected by increasing traffic load and land subsidence. Therefore, phenomena such as breakage of sewer pipes and disconnection of a branch pipe from a sewer main can occur, allowing sediment to flow into the sewer with groundwater and cause the road to cave in. Figure 1.2 shows the damage counts per span of sewers constructed during different periods. Needless to say, old sewers tend to show more damage than new sewers.

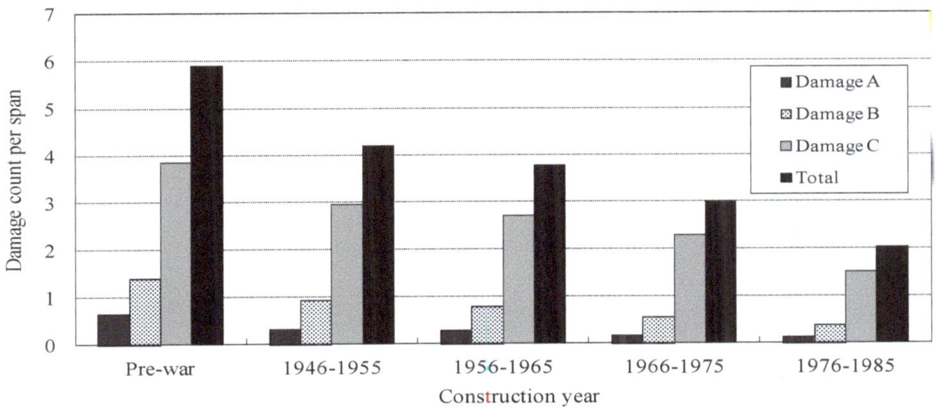

Note: 1) About 20% of total length was surveyed; 1 span ≒ 30 m.
2) Damage refers to sewer breakage, cracking or joint displacement.
3) Damage counts are calculated based on survey results of 1993.

Figure 1.2: Damage counts per span in centra Tokyo

Figure 1.3 shows the relationship between the number of road cave-ins (per 10 km) caused by sewer damage in Tokyo and the average age (the number of years elapsed after construction) of sewers based on 1998 data. As is evident from the figure, there are differences in the number of cave-in incidents between the wards where sewers were constructed in recent years such as Adachi, Edogawa, Katsushika and Setagaya and those where sewers were constructed earlier such as Chuo, Bunkyo, Minato and Toshima. In wards located on the soft soil layers of the Yurakucho formation, some of which have suffered regional land subsidence, such as Arakawa, Sumida and Kita, the number of road cave-ins tends to be large relative to the average age of sewers. These data clearly indicate that degradation in structural strength is correlated with not only age-related factors but also other factors such as ground conditions and past regional land subsidence events.

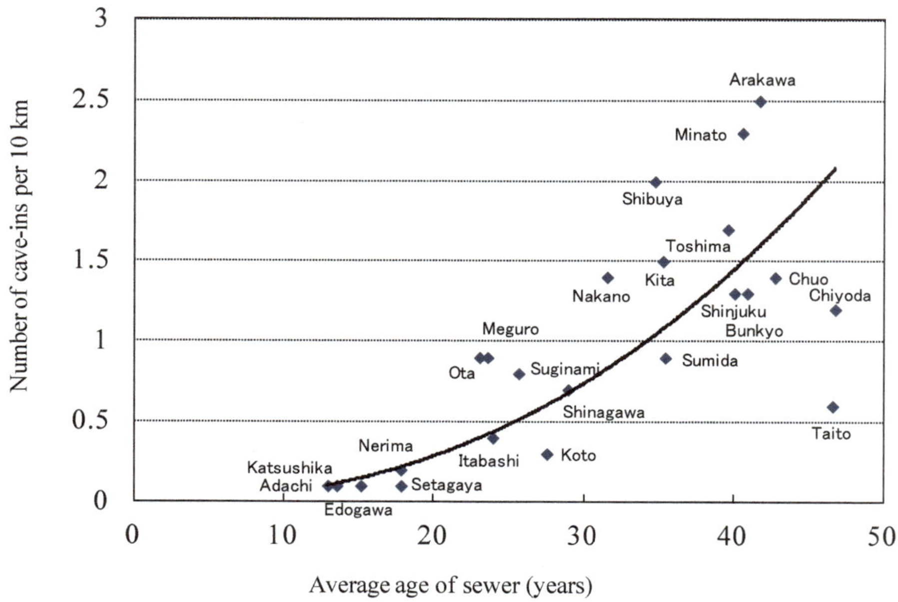

Figure 1.3: Number of road cave-ins and average age of sewers in central Tokyo

On the basis of the sewer damage and road cave-in data shown above, an attempt has been made to predict the future situation of the sewers, assuming the current level of maintenance. Figure 1.4 shows predicted sewer damage counts based on the data shown in Fig. 1.2, obtained by taking into consideration possible future increases in the length of old and deteriorating sewers. For the purpose of calculation, it is assumed that in the 23 wards of Tokyo, a single sewer span is 30 m. As seen, the predicted damage count per span was 2.9 in 1998, and it will increase by a factor of 1.7 in 2018, and by a factor of 2.4 in 2038. Thus, the number of damage sites is likely to increase as the total length of old sewers increases.

Similarly, the occurrence of road cave-ins in the coming years can be estimated from the fact that the average number of cave-in incidents per year in the 10 years up to 1998 was 1,600. The number of cave-ins is expected to increase by a factor of 2.3 in 20 years and by a factor of 5.2 in 40 years. These grim predictions clearly indicate the importance of taking measures to prevent the ageing of sewerage systems, also from the viewpoint of maintaining road traffic and normal urban functions.

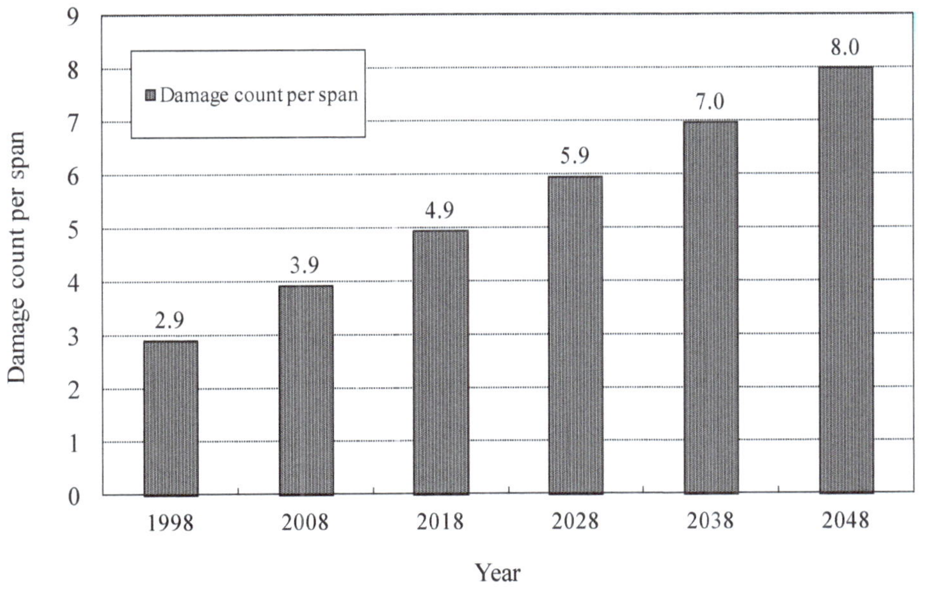

Note: 1 span ≒ 30 m

Figure 1.4: Predicted damage counts per span (average in central Tokyo, 1998)

1.2.3 Necessity of reconstruction projects

In order to cope with the problems mentioned above, the TMG has been working since 1994 to switch from merely responding to the symptoms, to a preventive maintenance approach by not only replacing or repairing parts of old sewers but also carrying out reconstruction projects to increase capacity as well as rehabilitate its ageing facilities. Note that reconstruction projects imply the construction of new sewers and the rehabilitation of ageing sewers, with the latter including sewer renovation and repair.

Figure 1.5 shows the number of road cave-in incidents in a designated region (273 ha) covered by the priority plan (a plan specifying the priorities of areas and measures drawn up to accelerate the effects of reconstruction projects adopted in 2003) and the coverage of sewer reconstruction between 1999 and 2003 (PFIJ, 2012). As shown, the number of road cave-in incidents has decreased with the progress of reconstruction projects, demonstrating the effectiveness of the projects.

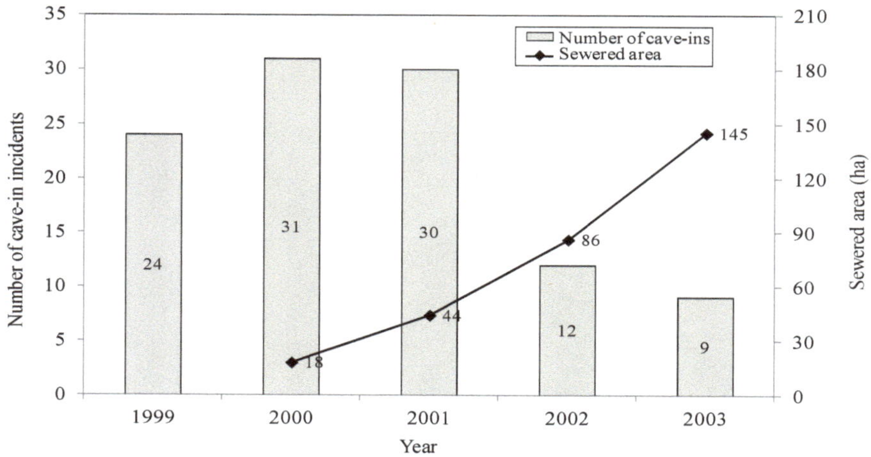

Figure 1.5: Relationship between the number of road cave-in incidents and the coverage area of reconstructed sewerage systems (TMG, 2008)

1.3 Sewer Renovation in Reconstruction Projects

The use of sewer renovation methods is indispensable in urban sewer reconstruction projects. Today, there are various methods with a growing track record. This section explains why sewer renovation is necessary in reconstruction projects, and shows the classification of renovation methods and their track records in Japan. Also, the basic methodology for selecting a renovation method by the Sewerage Bureau of the TMG to be used in its sewer reconstruction projects is discussed.

1.3.1 Background: Why rehabilitation?

In order to renew or reconstruct ageing sewers and meet the required functionality and durability, it had been common practice to dig out old sewer pipes and install new ones by the cut-and-cover method. In urban areas, however, the underground space is increasingly crowded with many utility pipes and buried structures, and it is also necessary to carry out sewer reconstruction economically without affecting road traffic and the surrounding environment as much as possible.

This is why various sewer renovation methods have been developed. A common feature of these methods is that construction materials and equipment can be brought in through existing manholes and, without removing the existing pipe, renovation construction can be carried out from inside the existing structure by manufacturing strong inner linings that are rigidly attached to the walls of the ageing sewer and greatly strengthen it.

1.3.2 Classification of renovation methods and their track records

Figure 1.6 shows the classification of pipe renovation methods based on the *Guidelines* and ISO standards (JSWA, 2011; ISC, 1986). The *Guidelines* deal with investigation, design and construction management related to sewer renovation methods. The *Guidelines* classify renovation methods according to the method of construction and the structural type after renovation. The former classification corresponds to pipe-forming methods, and the latter corresponds to design methods. The *Guidelines* include tables that classify renovation methods based on the type of material and the type of pipe-forming method, and introduces 23 independent pipe methods and 5 composite pipe methods.

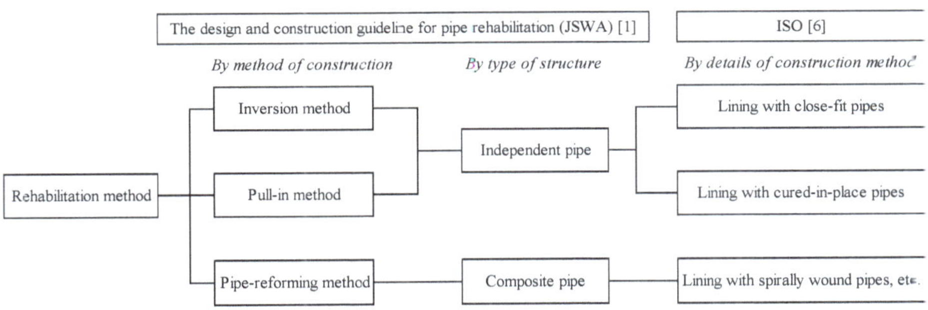

Figure 1.6: Classification of pipe renovation methods

As of 2011, the total length of sewers rehabilitated by renovation methods in Japan was 5,617 km (PFIJ, 2012), and the total length of renovation construction per year has been on the increase. Even so, the length of sewers rehabilitated thus far accounts for only 1.4% of the total sewer length of 400,000 km. Hence, it is obvious that rehabilitation of ageing sewers will continue to be the major category of sewerage projects for a long time. The renovation methods classified by the method of construction as described in the *Guidelines* are explained below.

1. Inversion method: A liner pipe is formed by inserting a cylindrical liner made of a base material (glass fibre, organic fibre, etc.) impregnated with thermosetting resin into an existing pipe by the inversion method, then the resin is hardened by applying hot water, steam or other heat source while keeping the liner in close contact with the inner surface of the existing pipe by applying air or water pressure from inside the liner.

2. Pull-in method: A liner pipe is formed by softening and deforming a thermoplastic resin (polyvinyl chloride or high-density polyethylene) pipe and inserting it into an existing pipe by the pull-in method. The new pipe is then left to expand while still hot and then cool down and harden, with the liner kept in close contact with the inner surface of the existing pipe by applying air or water pressure from inside the liner. Also, a liner pipe can be formed by the pull-in method by following the same construction procedure as the inversion method, except for changing the method of construction from inversion to pull-in.
3. Pipe-reforming method: In this method, a renovated pipe is constructed by forming interlinings along the length of the existing pipe by interlocking strips of surface materials (polyvinyl chloride, polyethylene, etc.) and grouting the annular gap between the lining and the existing pipe with cementitious material to achieve structural integrity of the composite pipe thus formed. As explained in Section 1.1, due to the limited bond strength of the cementitious grout, such a composite pipe as defined by the *Guidelines* is structurally treated as a semi-composite pipe in this book. There are several pipe-reforming methods in Japan, which mainly differ in how the interlining is produced. The three widely-used methods for lining fabrication are the profile winding interlocking, the assembled panel interlocking and the liner segment interlocking method.

According to the *Guidelines*, pipe structures after renovation can be classified into two types: an independent pipe and a composite pipe, as defined below.
1. Independent pipe: A pipe that is strong enough to act as an independent structure and that has the same or higher load-carrying capacity and durability as a newly installed pipe
2. Composite pipe: A composite structure that consists of an old pipe and a renovation layer rigidly bonded together as a single integrated structure, and that has the same or higher load-carrying capacity and durability as a newly installed pipe

The classification into independent pipe and composite pipe categories is reflected in their respective design methods. The main difference is that while the former does not take the existence of the old pipe into consideration, the latter treats not only the liner but also the old pipe as its structural members. The design theory for an independent pipe is based on the flexible pipe assumption and is used for the inversion method and the pull-in method, and the design theory for a semi-composite pipe as discussed in this book is applied to the pipe-reforming method. This semi-composite pipe concept and the corresponding design theory are unique to Japan and were first adopted by the Sewerage Bureau of the TMG in its sewer reconstruction projects.

1.3.3 Tokyo's approach to reconstruction

Of the total length of sewer pipes of about 15,000 km (as of the end of 2011) in central Tokyo, the length of those exceeding the design service life of 50 years is approximately 2,200 km (about 14%). As exemplified by the Kanda Sewer System which is still in service 120 years after its construction, some of the ageing sewers in this category are still functional. Therefore, the TMG has drawn up a policy of extending the service life of those still in sound condition as much as possible. To improve the entire sewer network, priority in selecting ageing sewers for reconstruction is determined by considering the degree of deterioration, drainage capacity, construction conditions and cost. Criteria for actions to be taken according to the degree of pipe damage are as follows:

1. Pipe without damage: To be kept in service
2. Pipe with minor damage: To be renovated and kept in service
3. Pipe with major damage: To be replaced

The existing pipe is checked as follows (the survey method is described in detail in Chapter 2): in the case of a man-entry pipe with a diameter of 800 mm or more, the degree of soundness of the pipe is evaluated through visual inspection and laboratory tests on sample specimens taken from the site; in the case of a non-man-entry pipe with a diameter of less than 800 mm (80% of all sewer pipes fall into this category), a self-propelled video camera is used to inspect the pipe.

Before rehabilitation methods were developed, it was standard practice to use the cut-and-cover method for reconstruction projects even when damage was minor. Today, the sewer rehabilitation approach is adopted in many projects in order to reduce the influence on road traffic and local residents. In general, the cost of no-dig sewer rehabilitation is only half that of cut-and-cover construction. Another advantage of no-dig rehabilitation is that its environmental impact is small because construction by-products from excavation are not generated. No-dig sewer rehabilitation, therefore, is essential for reconstruction projects.

The TMG approves the use of a no-dig pipe rehabilitation method in the metropolitan area according to criteria such as design conditions, strength and performance requirements, and verification test methods. So far, three pipe-reforming methods and eight inversion or pull-in methods have been approved.

Figure 1.7 shows Tokyo's methodology for selecting a pipe rehabilitation method. As shown, the kinds of damage to which the no-dig rehabilitation method cannot be applied are "breakage (A: chipping)," "joint displacement A," "slack or meander A/B," and "step displacement" shown in Table 1.2. In principle, the pipe-reforming method is to be used in other cases. If the required drainage capacity cannot be obtained when the pipe-reforming method is used for a small-diameter pipe, the independent pipe method, which employs thinner liners, is to be used. As discussed in Section 2.2 of Chapter 2, the composite pipe method is superior to the independent pipe method in terms of reliability and economy.

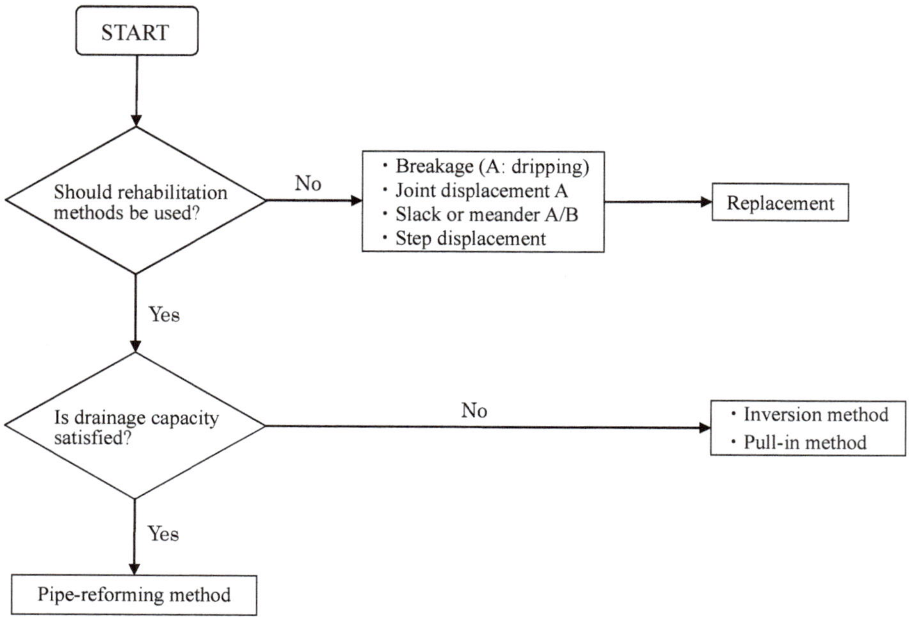

Figure 1.7: Flowchart illustrating Tokyo's approach for selecting methods for sewer reconstruction

Table 1.2: Soundness evaluation criteria for video-camera-based and eyeball investigation (Tokyo)

Item		Damage level A	B	C
Pipe breakage	Reinforced concrete pipe	Chipping Longitudinal crack with a width of 5 mm or more	Longitudinal crack with a width of 2 mm or more	Longitudinal crack with a width of less than 2 mm
	Earthenware pipe	Chipping Longitudinal crack with a length equal to or greater than 1/2 of pipe length	Longitudinal crack with a length of less than 1/2 of pipe length	
Pipe cracking	Reinforced concrete pipe	Circumferential crack with a width of 5 mm or more	Circumferential crack with a width of 2 mm or more	Circumferential crack with a width of less than 2 mm
	Earthenware pipe	Circumferential crack with a length equal to or greater than 2/3 of circumference	Circumferential crack with a length of less than 2/3 of circumference	
Pipe joint displacement		Disjointing	Earthenware pipe: 50 mm or more Reinforced concrete pipe: 70 mm or more	Earthenware pipe: less than 50 mm Reinforced concrete pipe: less than 70 mm
Pipe corrosion		Reinforcing bar exposure	Aggregate exposure	Roughened surface
Slack or meander of pipe		Equal to or greater than inside diameter	Equal to or greater than 1/2 of inside diameter	Smaller than 1/2 of inside diameter
Water intrusion		Gushing	Flowing	Seeping
Lateral pipe protrusion		Equal to or greater than 1/2 of inside diameter of lateral pipe	Equal to or greater than 1/10 of inside diameter of lateral pipe	Smaller than 1/10 of inside diameter of lateral pipe
Lard deposition or tree root intrusion		1/2 or more of inside diameter blocked	Less than 1/2 of inside diameter blocked	

References

ISO (1986). Techniques for rehabilitation of pipeline systems by the use of plastic pipes and fittings. In: Sewerage Rehabilitation Manual: Volume III-Sewer Renovation. ISO TR 11295.
JSWA (2001). Pipe Rehabilitation Handbook. Tokyo, Japan Sewage Works Association.
JSWA (2011). Design and Construction Guidelines for Sewer Pipe Rehabilitation. Tokyo, Japan Sewage Works Association.
PFIJ (2012). Survey on Sewer Rehabilitation in Sewer Reconstruction. Tokyo, Public Fund Investment Journal.
TMG (2008). The Quick Plan for Sewer Reconstruction. Tokyo, The Sewerage Bureau of the Tokyo Metropolitan Government.

Yoshifumi Takahashi

2 The Composite Pipe Construction Method

2.1 What Is a Composite Pipe?

The term "composite pipe" was first coined in the *Pipe Rehabilitation Handbook* published by the Japan Sewage Works Association in 2001 (JSWA, 2001). Based on its definition, the existing sewer and the renovation layer in a composite pipe are designed to resist external loads as an integrated structure, and the load-carrying capacity and durability are required to be comparable or superior to a newly-constructed sewer. In order for the existing pipe to function like a permanent member, concrete strength needs to be retained for about 50 years after completion of rehabilitation work.

To evaluate the strength of old concrete pipes in preparation for reconstruction projects, the Tokyo Metropolitan Government (TMG) dug up a total of 130 ageing pipes from 26 sewer lines over a period of two years from 1999 to 2000. A series of tests including the JIS external pressure test, uniaxial compression test and carbonation depth test were carried out on specimens taken from the dug-up pipes. Figure 2.1 shows the relationships of the cracking and fracture loads of each pipe divided by the code-specified value at the time of construction and the age of the pipe. Data points for the 60-year-old or older pipes indicate the hand-cast reinforced concrete pipes buried in the pre-war years, and those for the less-than-60-year-old pipes indicate the centrifugally-cast reinforced concrete pipes (Hume pipes: HP) buried in the post-war years. As shown, there is no clear correlation between age and strength ratio, and even if pipe deterioration occurs during a period of 70 years of service, the strength of installed pipes still satisfies the code requirements.

Figure 2.2 shows the relationship between the uniaxial compressive strength and the age of HP specimens by type of damage. As shown, the strength of buried concrete does not decrease for about 40 years. A comparison between damaged pipes and undamaged pipes reveals no significant difference in average strength, indicating that the damage was not caused by pipe deterioration.

From these test results, it can be concluded that because the initial strength of buried concrete pipes is retained even in cases of chemically induced deterioration, the concept of the composite pipe for sewer renovation is valid.

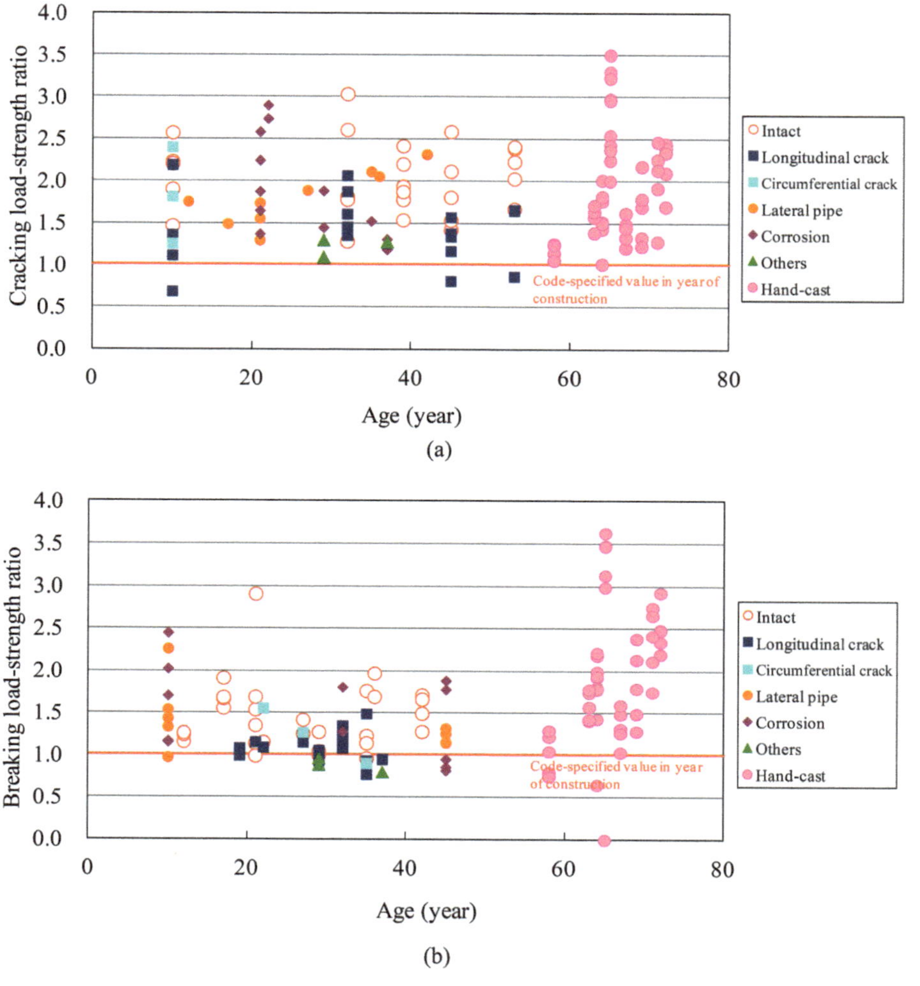

Figure 2.1: Relationship between the strength of dug-up pipes and the age: (a) cracking load; and (b) fracture load

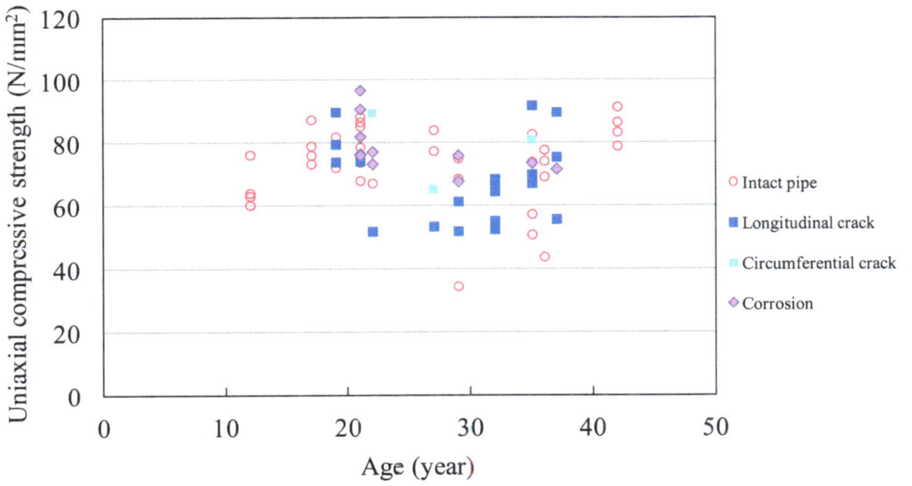

Figure 2.2: Uniaxial compressive strength of dug-up Hume pipes versus pipe age

2.2 Classification of Pipe-Reforming Methods

In Japan, five pipe-reforming methods for sewer renovation have been developed and employed in sewer reconstruction projects. These methods are summarised in Table 2.1. In all the methods, materials and equipment can be brought in through a manhole 60 cm in diameter, the existing pipe can be renovated by the trenchless method, and the gap between the liner pipe and the existing pipe is grouted with mortar so as to achieve structural integrity. In the SPR method and the Danby method, a PVC profile strip is spirally deployed by a winding machine to form a liner pipe. In the 3S segment method, the PALTEM Flow-Ring method and the PFL method, plastic and high-density polyethylene pipe materials are manually assembled in the existing pipe. These five methods can be used to form not only circular pipes but also pipes with a rectangular or horseshoe-shaped cross section. Among them, the SPR method, the Danby method and the PALTEM Flow-Ring method allow the interlining to be fabricated while keeping the sewer in service.

Table 2.1: Pipe-reforming methods (JIWET, 2009a–c, 2011, 2012a)

Method	**SPR Method**	**3S Segment Method**
Method illustration	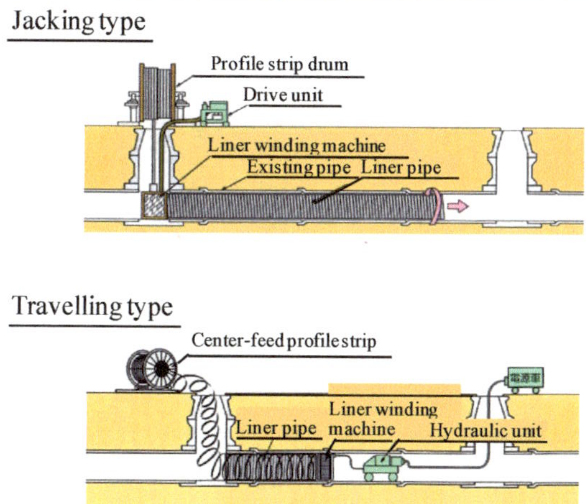	 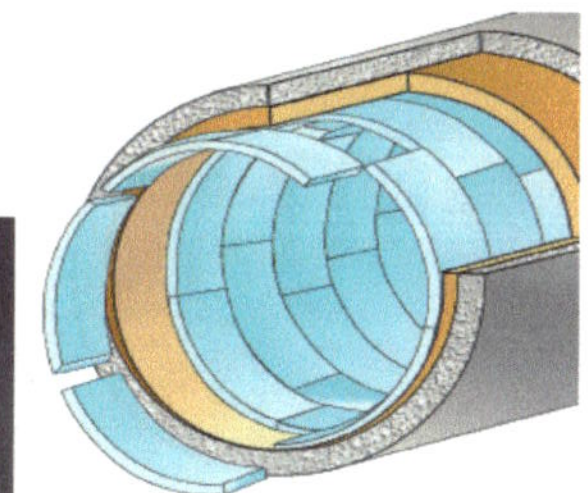
Description	A PVC profile pipeline is formed in the existing pipe, and the annular gap between the existing pipe and the profile liner is grouted to form a strong composite pipe. (1) Circular pipe: Pipes smaller than 2,200 mm in diameter are constructed by the jacking method, and pipes 800–4,750 mm in diameter are constructed by the traveling winder method. (2) Non-circular pipe: Pipes with a shorter side length of 800 mm or more and a longer side length of 6,000 mm or less are constructed by the traveling winder method.	Transparent and lightweight plastic liner segments are manually brought into the existing pipe and assembled with nuts and bolts. Then, the gap between the liner and the existing pipe is filled with 3S grout to ensure structural integrity of the composite pipe thus formed.

Lining method	Winding a profile strip and inserting a wound liner	Manually bringing in and assembling segments
Materials	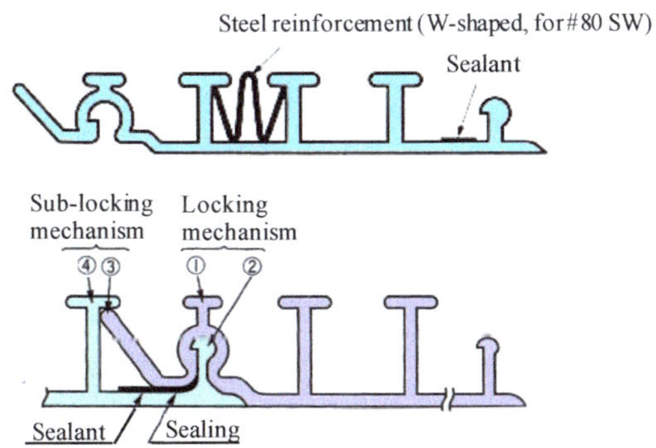	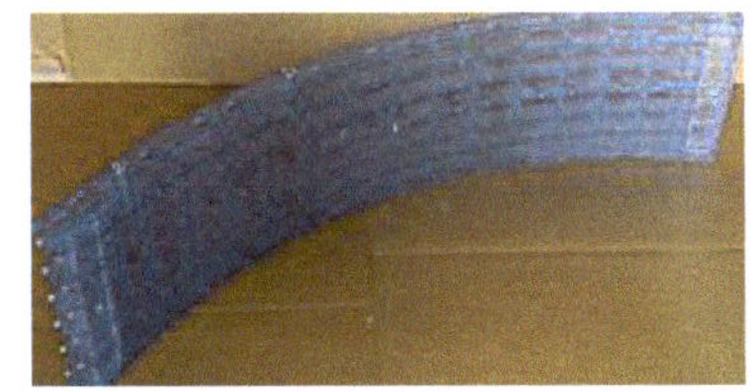 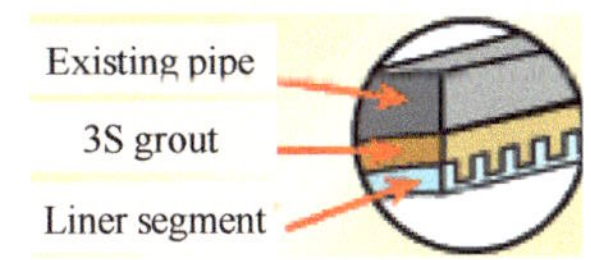
Main materials	PVC (profile strip), mortar	Plastic segments, 3S grout
Applicable diameter range (mm)	250–4,750 mm dia., non-circular pipe: shorter side 800 mm or more, longer side 6,000 mm or less	1,000–4,000 mm dia., non-circular pipe 1,000–5,000 mm on side
Characteristics	• Constructibility: The jacking method and the traveling winder method can be used. Applicable to various pipes such as non-circular cross section pipes and sharply curved (5D) pipes. Slope adjustments can be made according to pipe diameter. • Quantity: 18–460 m (pipe fabrication only) • Sewer service during rehabilitation: Possible (water depth 60cm) • Work area: The equipment and work area required are small (length: 25 m, number of vehicles: 3).	• Constructibility: There is no need for large equipment for bringing in liner segments through a manhole. The degree of completeness of grouting can be visually observed easily. Versatility is high so as to meet diverse construction needs such as curved pipe construction and concurrent liner construction in two directions. • Quantity: Circular pipe (800 mm dia.): about 6.5 m/day (typical) Non-circular pipe (2,200 × 2,200 mm): about 3.1 m/day (typical) • Sewer service during rehabilitation: Not possible • Work area: Small work area (length: 25 m, number of vehicles: 3)
Trade association	Japan SPR Method Association	3SICP Technology Association
Date of certification	May, 1993	March, 2004

Table 2.1: Pipe-reforming methods (JIWET, 2009a–c, 2011, 2012a) (Continued)

Method	**PALTEM Flow-Ring Method**	**Danby Method**
Method illustration	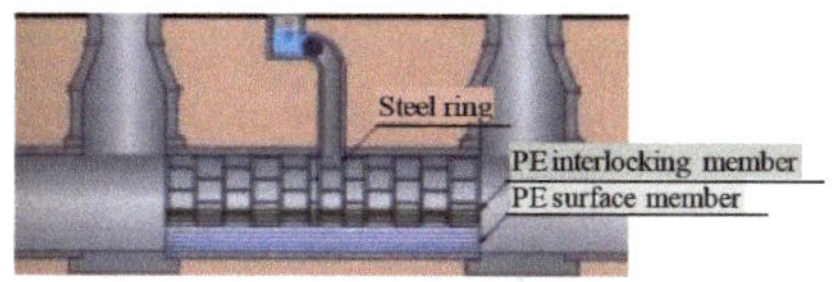  	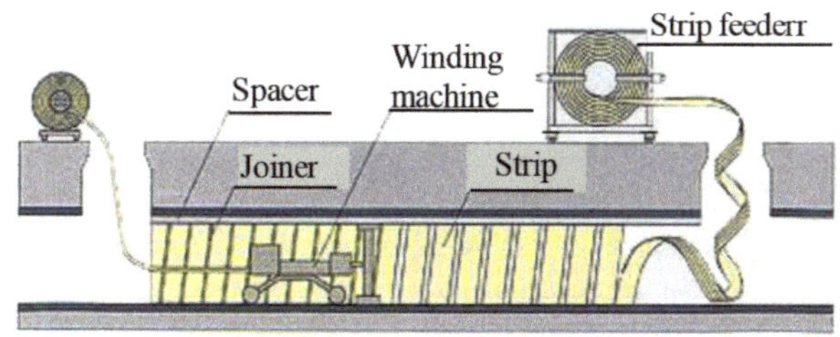
Description	Steel rings brought into the existing pipe through a manhole are assembled with bolts, and interlocking members and surface members are installed to the steel rings. Then, the gap between the existing pipe and the surface members is grouted with high-fluidity high-strength mortar to form a composite pipe.	A PVC strip is pulled into the existing pipe, wound spirally and joined together by using a winding machine to form a continuous spiral-wound pipe. Then, the gap between the existing pipe and the spiral-wound pipe is grouted.
Lining method	Manually bringing in and assembling steel rings	Manually bringing in and winding a strip
Materials	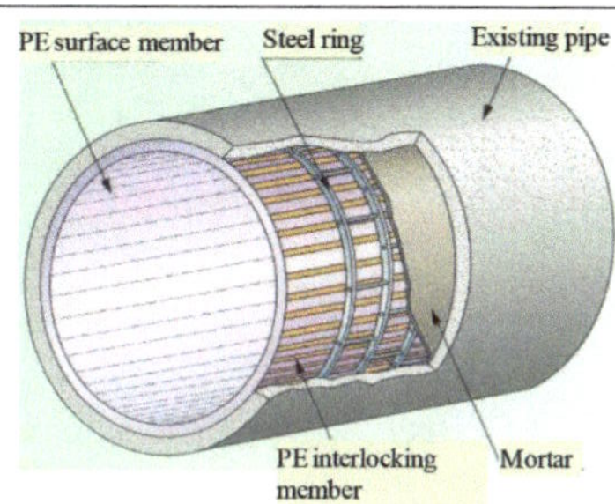	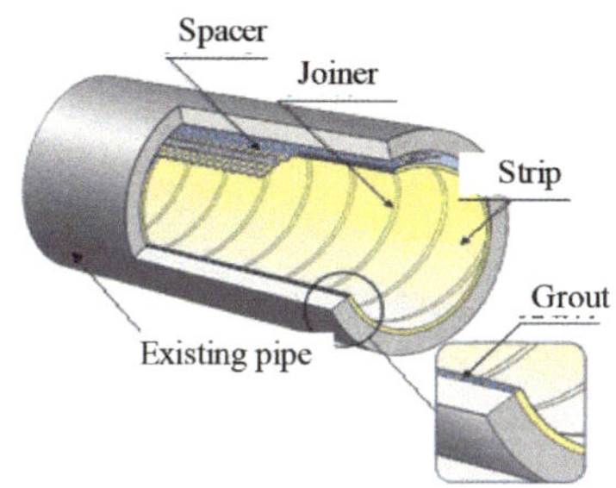

Main materials	High-density polyethylene resin, mortar, steel rings	PVC, mortar, spacer
Applicable diameter range (mm)	800–3,000 mm dia., non-circular pipe 800–3,000 mm	800–3,000 mm
Characteristics	• Constructibility: KBM segments and PFL panels are brought in through a manhole and assembled in the existing pipe, and grouting is carried out. PFL panels are highly wear resistant (about 30 times higher than PVC), and high-strength, no-shrink grout is used. • Quantity: An advance rate of 700 m in three months has been achieved. • Sewer service during rehabilitation: Possible (water depth: 15% or less of existing pipe diameter) • Work area: Relatively small work area (length: 25 m, number of vehicles: 3)	• Constructibility: All equipment and materials can be brought in and taken out through a manhole 60 cm in diameter. • Quantity: About 8–10 m/day (1,200 mm dia., 50 m: 6 days, 1,350 mm dia., 100 m: 10 days) • Sewer service during rehabilitation: Possible (water depth: 30 cm) • Work area: Relatively large equipment and work area (length 30 m, number of vehicles: 5)
Trade association	PALTEM Systems Association	EX. Danby Association
Date of certification	February, 2002	March, 1996

Table 2.1: Pipe-reforming methods (JIWET, 2009a–c, 2011, 2012a) (Continued)

Method	**PFL Method**
Method illustration	 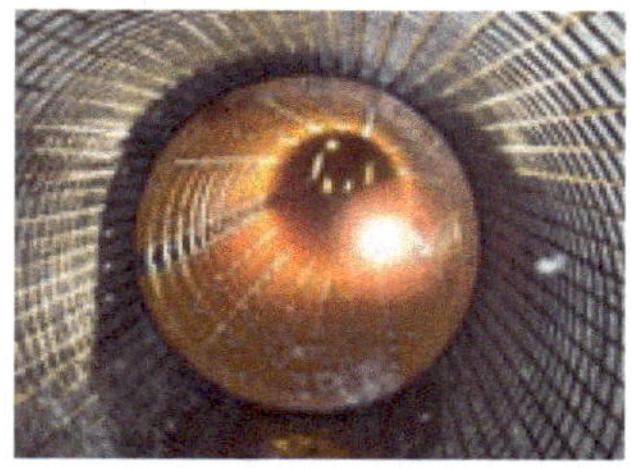
Description	A high-tensile-strength carbon fibre grid (KBM) is installed on the inside surface of the existing pipe, and high-density polyethylene PFL panels with surface protrusions are installed. Then, the gap between the panelling and the existing pipe is grouted to rehabilitate the old pipe.
Lining method	Manually bringing in and assembling materials
Materials	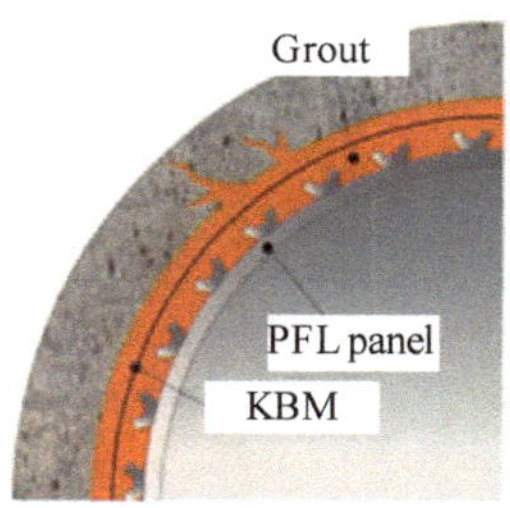 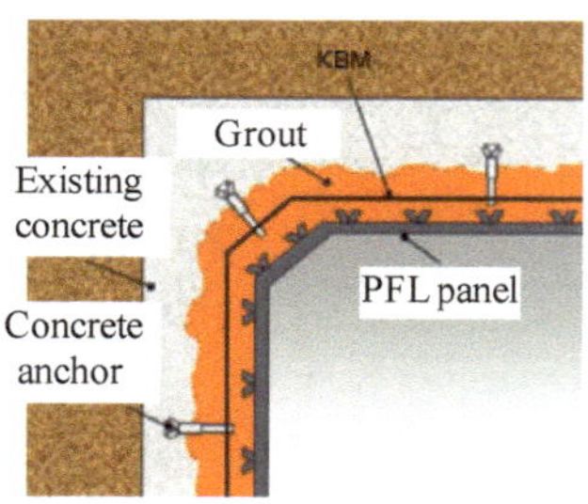
Main materials	Carbon fibre reinforcing grid (KBM), high-density polyethylene PFL panels
Applicable diameter range (mm)	Circular pipe 800 mm or more in diameter; applicable also to non-circular pipes
Characteristics	• Constructibility: All equipment and materials can be brought in and taken out through a manhole 60 cm in diameter. • Sewer service during rehabilitation: Not possible
Trade association	Polyethylene-Lining Method-of-Construction Association
Date of certification	March, 2011

2.3 Superiority of Composite Pipes

Compared with independent pipes, composite pipes have the following advantages:

1. A composite pipe uses the existing pipe as part of its structure by reinforcing the existing pipe by annulus grouting. As a result of renovation, therefore, it is possible to construct a pipe structure that is stronger than an independent pipe in which the existing pipe is not expected to contribute to strength.
2. A composite pipe does not allow soil intrusion, while an independent pipe may allow soil to enter the annular gap between the old pipe and the new pipe.
3. The interlining of a composite pipe is more resistant to external chemical attack than an independent pipe because the former is covered by the existing pipe and grout.

In the case of an independent pipe placed in an old sewer pipe by the inversion method or the pull-in method, the load carrying ratio of the independent pipe increases with the progress of deterioration of the old pipe, as shown in Fig. 2.3. Eventually, the new pipe alone has to carry all loads.

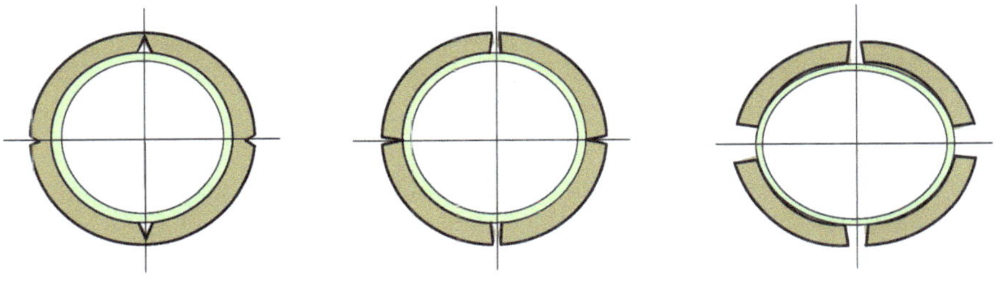

(a) (b) (c)

Figure 2.3: Independent pipe: (a) upon completion of rehabilitation; (b) with passage of time; (c) final stage

When designing an independent pipe, it is common practice to permit a deflection of up to 5%. If, for example, the diameter of the pipe is 300 mm, a deflection of up to 1.5 cm is permissible; if the diameter is 1,000 mm, a deflection of up to 5.0 cm is permissible. If such deflection occurs in a sewer pipeline, the road surface may cave in or pipelines of other utilities buried over the sewer pipeline may be adversely affected. In the case of a composite pipe, in contrast, the progress of deterioration of the old pipe can be prevented because annulus grouting will likely fill the cracks in the old pipe, compensate for the decrease in pipe wall thickness due to corrosion and strengthen the pipe structure as shown in Fig. 2.4. Since the existing pipe, the liner pipe and the grout filling the gap between them are integrated as a composite structure (more exactly, a semi-composite structure as explained in Chapter 1), composite pipes have higher stiffness and undergo very little deflection under the applied loads. Unlike independent pipes, therefore, composite pipes are relatively free from risks associated with land subsidence.

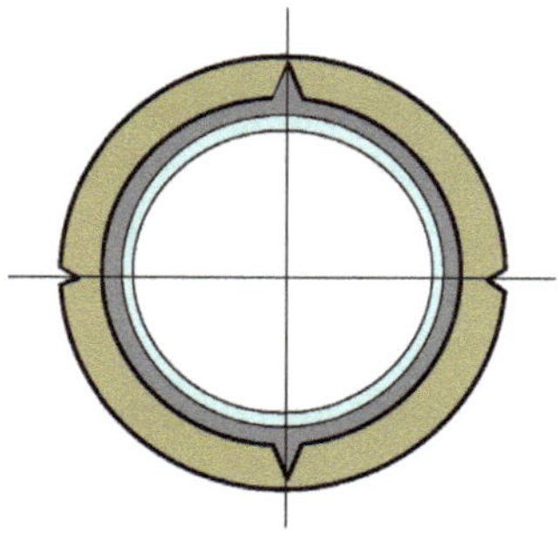

Figure 2.4: Composite pipe

If independent pipes are used for pipe rehabilitation, in many cases a gap develops between the existing pipe and the new pipe (independent pipe) because the new pipe shrinks in the existing pipe after hardening. Since no remedial or preventive measures are taken to prevent deterioration of the existing pipe surrounding the independent pipe, soil and water around the existing pipe may flow into the gap through cracks or other openings. As a result, cavities form around the existing pipe, increasing the possibility of road cave-ins. In contrast, in the case of a composite pipe, cracks in the existing pipe are filled with grout, and the gap between the existing pipe and the liner pipe is also filled with grout to ensure structural integrity. Consequently, composite pipes are basically not prone to the intrusion of groundwater and soil into the gap between the existing pipe and the liner pipe.

As shown in Fig. 2.3(c), an independent pipe installed for pipe rehabilitation comes into direct contact with the ground soil. Therefore, if there are corrosive substances in the ground, they accelerate the deterioration of the independent pipe. Composite pipes, however, are free from such risk because the liner pipe is doubly protected by the existing pipe and the grout.

2.4 The SPR Method

To illustrate technical developments in sewer renovation using the composite pipe methodology in Japan, this section briefly describes the SPR method developed by a joint team of three companies including the authors over a period of three decades.

The development of the SPR method dates back to 1984, when Adachi Construction and Industry Co., Ltd., one of the developers of the SPR method, proposed a technical plan for rehabilitating an old pipeline by spirally winding a steel strip. At around the same time, Sekisui Chemical Co., Ltd., another developer, entered into a licensing agreement with Rib Loc, an Australian company, for its PVC profile strip products, and improved their quality and shapes to enable them to be used as liner pipes for the rehabilitation of old sewer pipes. The first field test was carried out in Yokohama City in a sewer rehabilitation project. In 1985, the Tokyo Metropolitan Sewerage Service Corporation (TGS) joined the development team, since then the three companies have been jointly developing the technologies concerned.

Initially, a method called the jacking method was used for lining construction. As shown in Fig. 2.5, in the jacking method, a jack-type winder is installed in a manhole, and a profile strip fed from a profile strip drum placed on the ground surface is pushed into the existing pipe while the winder performs spiral winding and interlocking operations (refer to Photos 2.1 and 2.2). At first, this method was applied to small-diameter pipe rehabilitations. In 1987, it was used for renovating pipes of 250 to 800 mm in diameter, and in 1988 for pipes of 900 to 1,200 mm in diameter.

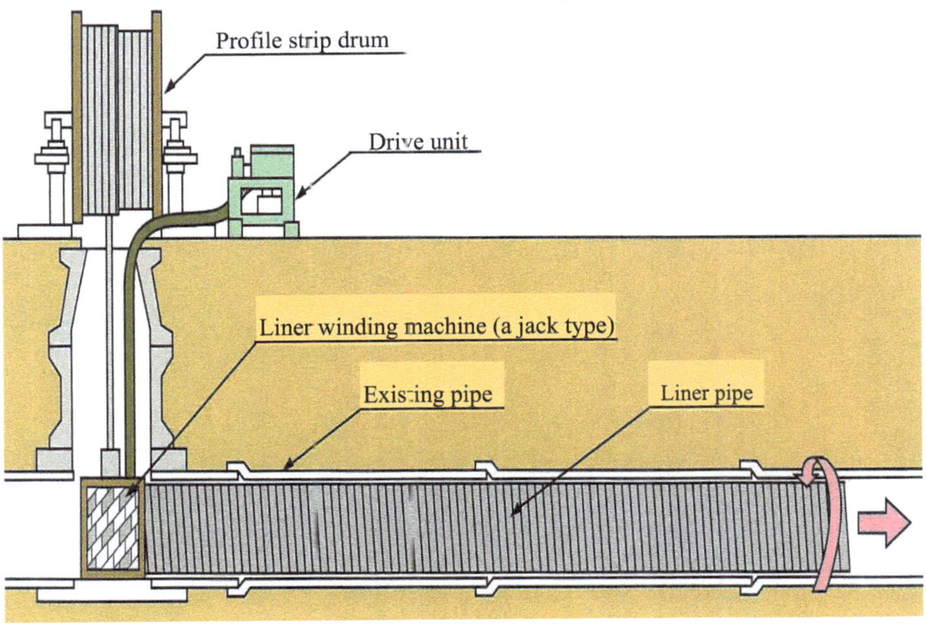

Figure 2.5: Construction of liner pipe by spiral winding using a jacking method

Photo 2.1: A jack type winding machine

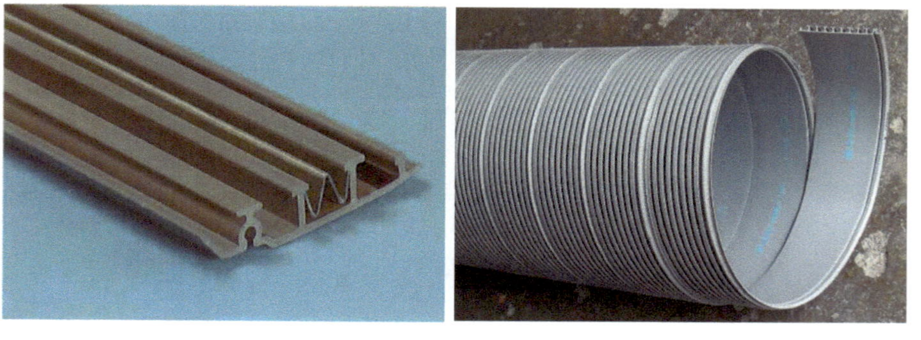

Photo 2.2: PVC profile strips

Later, as experience was accumulated in rehabilitating medium- and small-diameter pipes, demand grew for large-diameter-pipe rehabilitation technology. In order to meet this demand, the Super SPR method was developed for rehabilitating large-diameter pipes ranging in diameter from 1,650 mm to 2,200 mm. The Super SPR method was made possible by a new technology for installing steel reinforcements to conventional profile strips. The Super SPR method was approved for use in state-subsidized projects for large-diameter sewer restoration following the Great Hanshin Earthquake and was used in many projects in the subsequent years.

In the jacking method of liner making, a spirally wound liner pipe is rotated and pushed into an existing pipe. Naturally, there is a limit to the pipe length and diameter to which the method can be applied, due to the limitations in both the driving power and stiffness of the liner. Therefore, the development team began to develop a liner-making method using a travelling winder capable of advancing in the existing pipe while winding a profile strip, as shown in Fig. 2.6 and Photo 2.3. In 1996, the newly developed method was used to renovate an ageing sewer of 1,200 mm in diameter.

Old sewer mains include various non-circular sewers such as rectangular, horseshoe-type and covered sewers, and the team began to develop a method for renovating such sewers. In 1998, a method called the Free Cross Section SPR method began to be used in the Tokyo area, and has since been used to rehabilitate various sewers including a rectangular sewer 6 m wide and 3 m high. As shown in Photo 2.4, it is now possible to rehabilitate pipes of any shape. Major factors that have made this possible include the availability of prefabricated frames designed to fit into existing pipes and the shape-retaining effect of the reinforcing steel members installed in the profile strips.

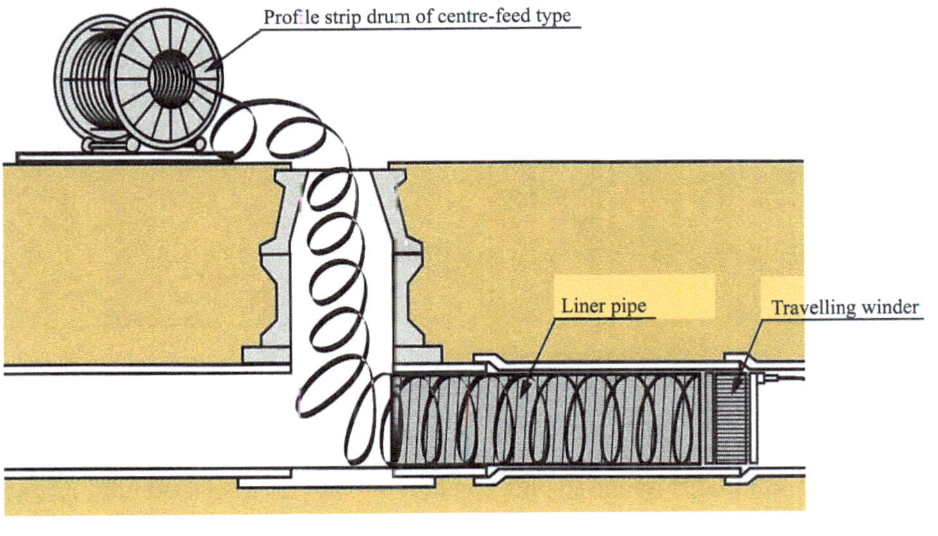

Figure 2.6: The travelling winder method of liner making

Photo 2.3: A travelling type winding machine

Photo 2.4: Free cross section SPR method

Concurrently with the development of the winders and liner pipe materials mentioned above, work continued on developing a special mortar to be injected into the gap between the existing sewer and the liner pipe. The requirements for such grouting mortar included not only high strength but also fluidity, segregation resistance, shrinkage resistance and the bond strength with existing concrete. Improvements were made through many mix design tests, laboratory element tests, laboratory full-scale model tests and in situ tests, and currently four types of grout materials are used to meet project-specific needs. The strongest mortar, No. 4, has a compressive strength of 55 N/mm^2.

As shown in Table 2.1, with the SPR method, it is possible to keep the host pipe in service, allowing a water depth of up to 60 cm. The construction cost of the SPR method is about half that of cut-and-cover construction. Because of these advantages, the SPR method is the most widely used pipe-reforming method of sewer renovation in Japan. Figure 2.7 shows the length of sewers rehabilitated by the SPR method by fiscal year: the total length rehabilitated by the method by 2011 was 762 km.

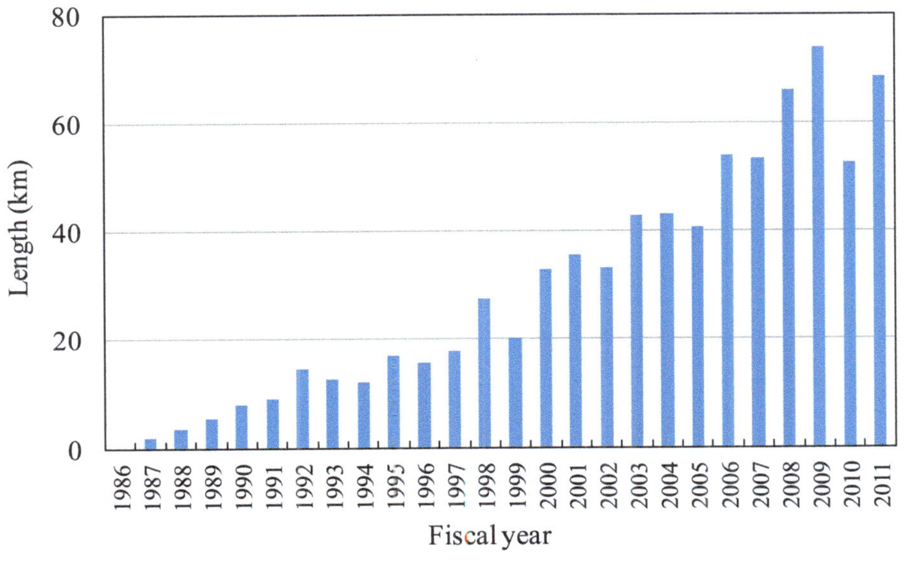

Figure 2.7: The length of ageing sewers renovated by the SPR method each fiscal year since 1986

At present, the development team is studying ways to solve challenges such as dealing with sharp curves and improving economy and safety. For sharp curves, profile strips capable of expansion and contraction in the width direction have been developed. As a result, it is now possible to deal with a radius of curvature of $R = 5D$ (D: existing pipe diameter), as shown in Fig. 2.8 and Photo 2.5.

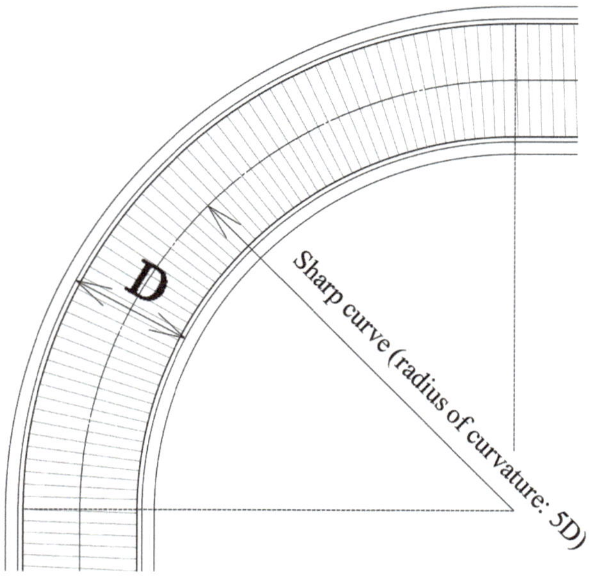

Figure 2.8: Dealing with curve construction

Photo 2.5: Curve construction by the SPR Method

2.5 Renovation Construction of Composite Pipes: From Investigation to Construction

This section briefly explains the stages of investigation, design and construction in a sewer rehabilitation project with examples of the SPR method.

2.5.1 Investigations necessary for composite pipe design

2.5.1.1 Medium- to large-diameter pipe

In a project involving a medium- to large-diameter pipeline of man-entry size (inside diameter: 800 mm or more), four types of investigations are conducted:

1. Drainage area investigation
2. Surveying
3. In-pipe investigation
4. Environmental investigation

The drainage area investigation involves collecting reconstruction-related information, specifically information on the conditions of and future plans for the drainage area and sewer mains concerned. Field reconnaissance is also conducted to investigate the conditions around the sewer. Surveying involves collecting information on the sewer specifications, ground elevation, earth cover, etc. in the area over the sewer line. The in-pipe investigation includes visual observation, internal cross-sectional measurement, wall flatness investigation and existing pipe structure investigation. The environmental investigation is conducted to collect information on the flow rate, flow velocity and other site conditions that may affect pipe winding operation such as step displacement, heaving and differential settlement and decide on actions to take during construction.

The structural investigation conducted on the existing pipe as part of the in-pipe investigation necessary for structural analysis and renovation design is briefly explained below, with the two ageing sewers shown in Photo 2.6 as examples under investigation.

1) Concrete strength investigation

Core samples are taken from the existing sewer, and a compressive strength test is conducted. Prior to core sampling, the locations of reinforcing bars and the degree of damage are investigated through visual inspection and reinforcing steel detection, and core sampling locations are determined, taking into consideration the minimisation of adverse effects on safety and the number of samples needed to achieve the required level of accuracy. Specimens for compressive strength testing are taken from non-carbonated regions.

The number of core samples taken should be limited for the safety of the existing pipe. As a complementary test, therefore, a rebound test using a Schmidt hammer is also conducted. Photos 2.7 to 2.10 show how the core sampling, Schmidt hammer test and compressive strength test are conducted.

(a)

(b)

Photo 2.6: Ageing sewers: (a) rectangular sewer with exposed reinforcing bars; and (b) covered stone masonry sewer with accumulated sediment

Photo 2.7: Core boring

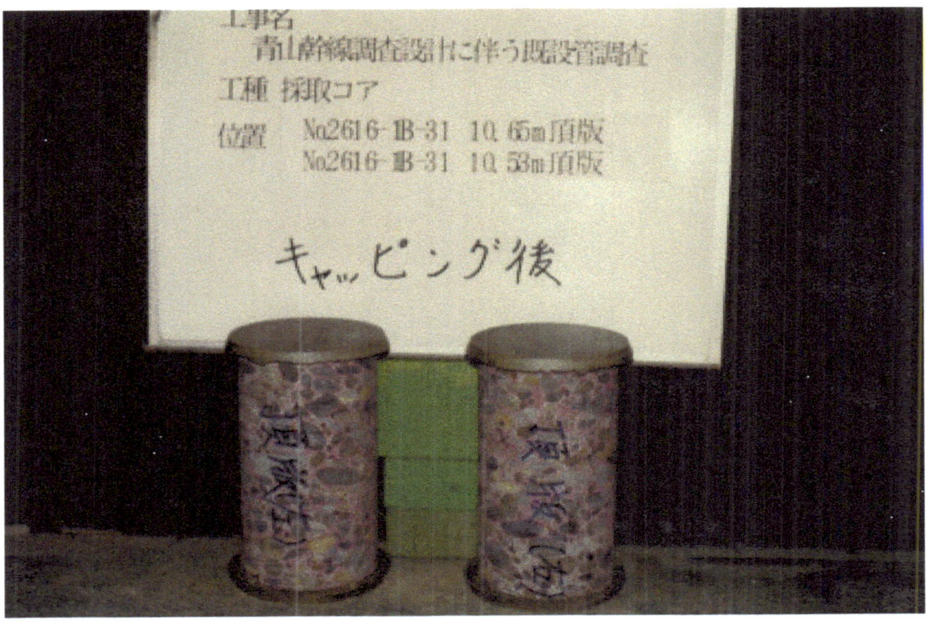

Photo 2.8: Core samples

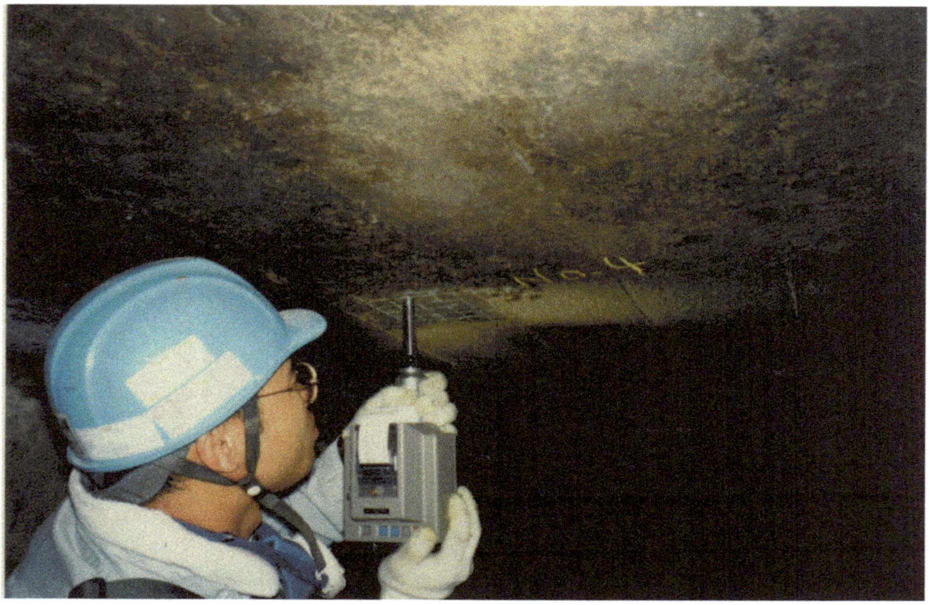

Photo 2.9: Rebound test

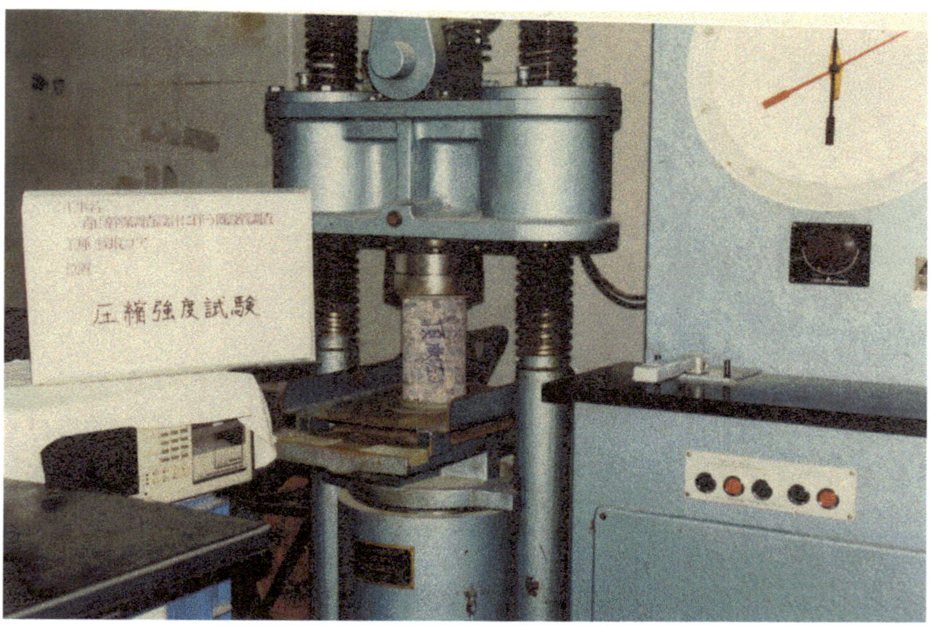

Photo 2.10: Compressive strength test

2) Carbonation test

A carbonation test is conducted for the purpose of investigating the carbonation depth of the existing concrete. The test can be conducted by the scraping method or the coring method. In the scraping method, concrete in an area measuring about 200 mm by 200 mm is scraped off from the surface until reinforcing bars are exposed, and a 1% phenolphthalein solution is sprayed to the scraped area. Since the depth at which a red-purple colour does not appear indicates the carbonation depth, the distance from the concrete surface to that point is measured. Photos 2.11 and 2.12 show how the scraping and measurement are carried out.

In the coring method, each core is immediately washed with water after it is extracted. After the core is drained, a 1% phenolphthalein solution is sprayed. The distance from the concrete surface to the depth at which the colour of the concrete turned red-purple is measured with a ruler. Photos 2.13 and 2.14 show how this investigation is carried out.

Photo 2.11: Scraping

3) Rebar arrangement investigation

The reinforcement arrangement is investigated by using a combination of two methods: the electromagnetic wave method and the electromagnetic induction method.

Photo 2.12: Measuring the carbonation depth

Photo 2.13: Cores for carbonation test

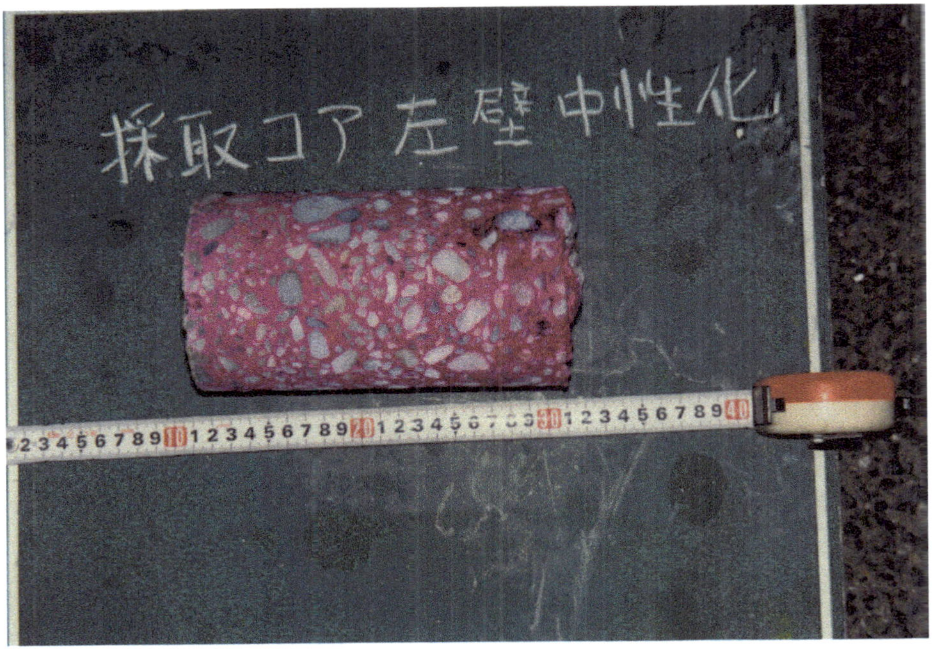

Photo 2.14: Measurement of carbonation depth

(a) Reinforcing bar detection by the electromagnetic wave method

From the concrete surface, the reinforcing patterns in the upper slab and the side walls are investigated, and reinforcement drawings are prepared. As a measuring instrument, a reinforcing bar detector, RC radar, is used.

To achieve high resolution at shallow depths (about 30 cm or less) in the concrete, the reinforcing bar detector uses pulse waves with a pulse width of about one nanosecond (1/1,000,000,000 second) for transmission. Equipped with an integrated transmit/receive antenna and capable of moving along the concrete surface, the detector can also be used to measure the spacing of reinforcing bar. Photo 2.15 shows reinforcing bar detection by the electromagnetic wave method.

(b) Reinforcing bar detection by the electromagnetic induction method

To check whether there are reinforcing bars on the outer side of an ageing pipe, reinforcing bar detection is carried out by the electromagnetic induction method at the holes from which cores were extracted. This method utilises the large difference in magnetic permeability between concrete and steel. Compared with the relative magnetic permeability of concrete which is approximately 1, reinforcing bars are ferromagnetic and have a relative permeability as high as several dozen to several hundred.

Photo 2.15: Reinforcing bar detection by the electromagnetic wave method

The scanner sends an alternating current of relatively low frequency through the coil in the probe and, by moving along the concrete surface while producing an alternating magnetic field, measures changes in magnetism and displays the results as the distance between the probe (concrete cover) and reinforcing bars. Photo 2.16 shows how reinforcing bar detection is performed by the electromagnetic induction method.

Photo 2.16: Rebar detection by the electromagnetic induction method

4) Rebar corrosion investigation

Concrete is scraped off (about 200 × 200 mm) until the reinforcing bars are exposed, and the concrete cover and the reinforcing bar diameter are measured. Photos 2.17 and 2.18 show how this investigation is carried out.

5) Rebar tension test

Exposed and corroded portions of reinforcing bars are cut off, and the samples thus obtained are used to measure the corrosion loss ratio and tensile strength. The reinforcing bar samples are immersed in a 10 w/v% solution of diammonium hydrogen citrate containing 2-mercaptobenzothiazole (150 ppm) to remove corrosion products. Photos 2.19 and 2.20 show examples of reinforcing bar samples. Following the measurement of the corrosion loss ratio, the yield point and tensile strength are measured in accordance with JIS Z 2241; refer to Photos 2.21 and 2.22.

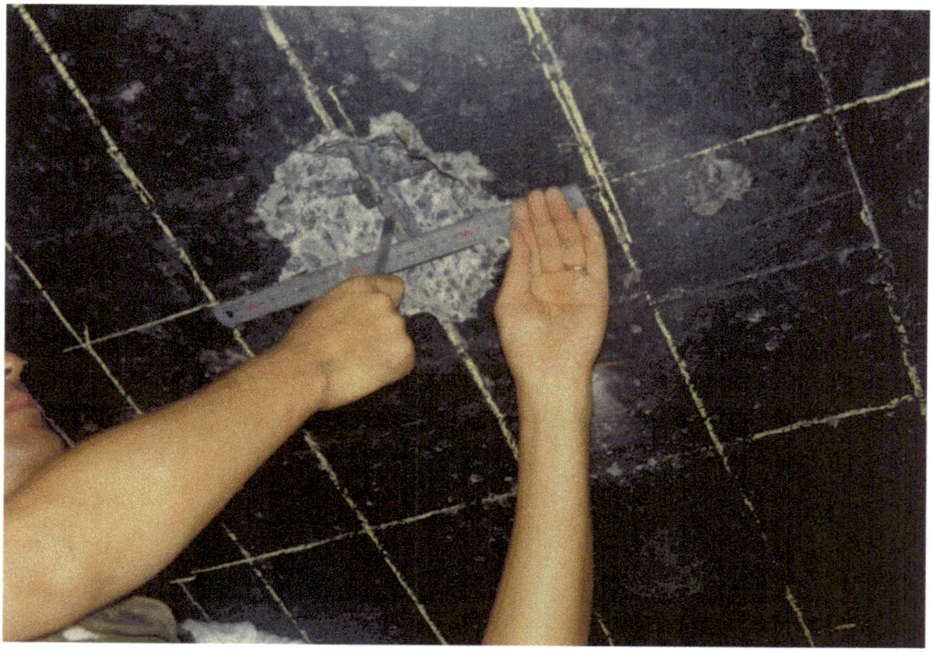

Photo 2.17: Concrete cover measurement

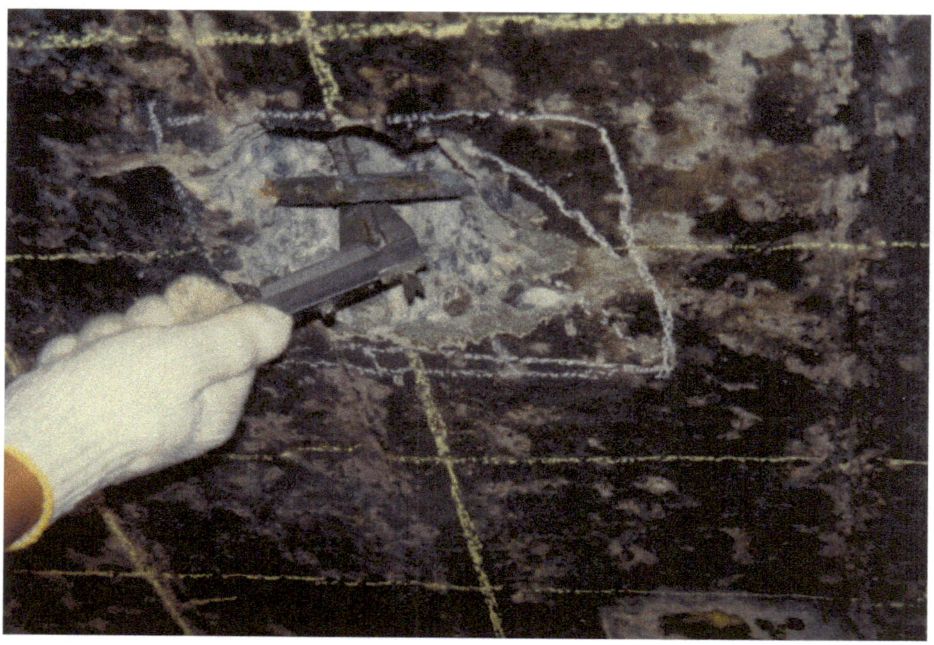

Photo 2.18: Rebar diameter measurement

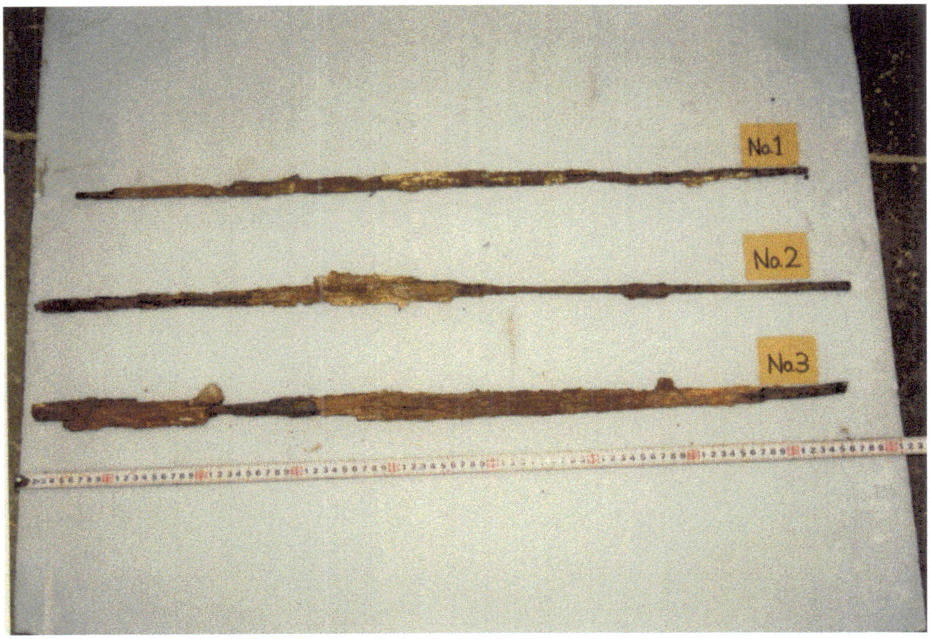

Photo 2.19: Corroded rebar samples

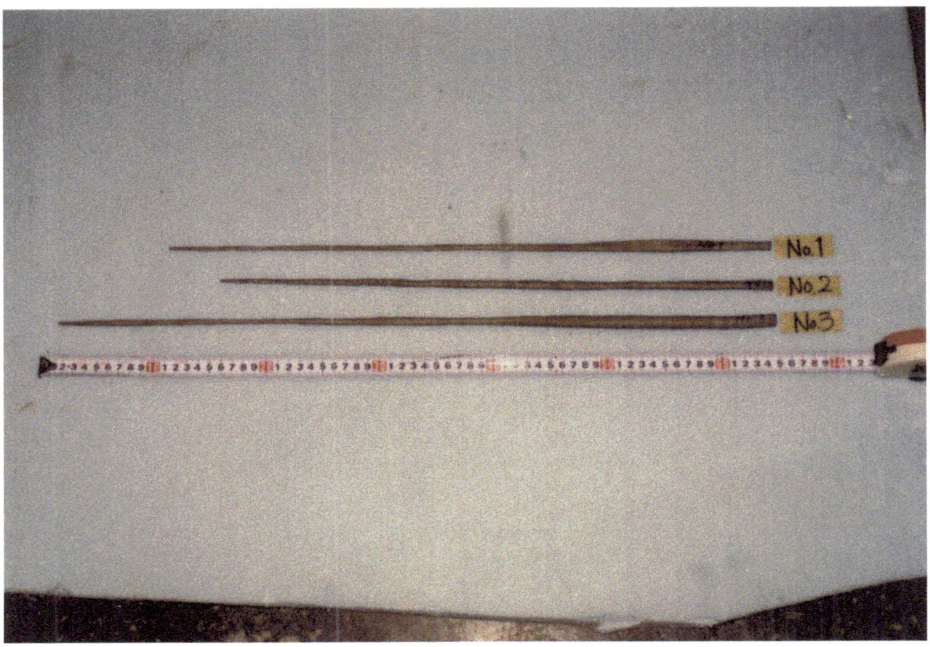

Photo 2.20: Rebar samples after removing corrosion products

Photo 2.21: Tension test of a corroded rebar

Photo 2.22: Fracture at the tensile strength of a corroded rebar

6) Structural member thickness investigation

If the design and completion documents of an ageing sewer do not exist, the thickness of concrete members is measured by the ultrasonic method, as shown in Fig. 2.9. The ultrasonic method utilises the fact that ultrasonic waves (sound waves having a frequency of 20 kHz or higher) are elastic waves capable of propagating more easily through a solid than electromagnetic waves, and are reflected at a boundary surface. The thickness of a structural member is calculated as

$$L = \frac{V}{2f} \tag{2.1}$$

where L: thickness of concrete member (mm); V: propagation velocity (sound velocity) (about 4,000 m/s in undamaged concrete); f: frequency of reflection (kHz).

As shown in Fig. 2.9, two sensors, one for excitation and another for detection, are attached to the same concrete surface with grease or other similar materials to keep complete contact between the sensors and the surface. Longitudinal ultrasonic waves are transmitted and the received waves are Fourier-analysed with the FFT analyser. The reciprocal of the return time of the reflected wave is shown as the frequency of reflection on the spectrum displayed on the FFT analyser, and the thickness is estimated based on Eq. (2.1). Photos 2.23 and 2.24 show how the thickness measurements are conducted.

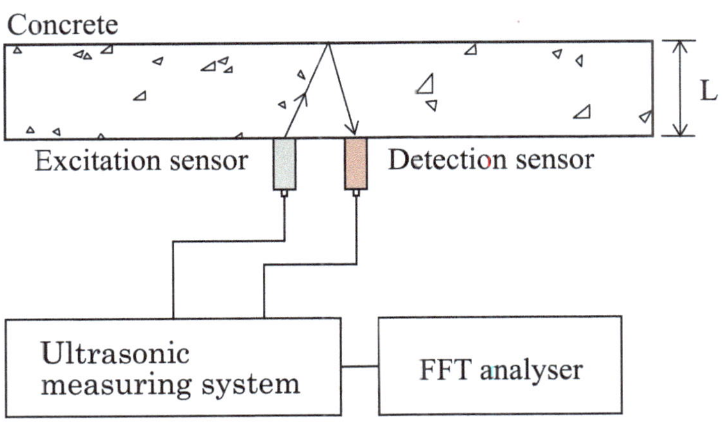

Figure 2.9: Thickness measurement of structural members by the ultrasonic method

Photo 2.23: Thickness measurement of structural members by the ultrasonic method (1)

Photo 2.24: Thickness measurement of structural members by the ultrasonic method (2)

2.5.1.2 Small-diameter pipe

Ageing pipes smaller than 800 mm in diameter cannot be surveyed from inside by inspectors directly to check on the pipe strength and rebar corrosion. Therefore, for non-man-entry sewers the current design practice for the composite pipe method is to prepare a damaged specimen representing an ageing sewer, renovate it and conduct an external pressure test on the renovated pipe to verify the effectiveness of the renovation method. If the ultimate strength of the renovated pipe exceeds the standardised value of a new pipe, the renovation method is certified as effective and is adopted.

Figure 2.10 presents the comparison results between the fracture strengths of renovated pipes and the standardised strengths and actual test results of new pipes. These fracture tests were performed first on virgin Hume pipes 250 to 3,000 mm in diameter under concentrated line loads. After renovating these damaged pipes by the SPR method, the same fracture tests were carried out again on the renovated pipes (JSWA, 2001). Figure 2.10(a) shows the comparison with the standardised strength of a Hume pipe, and Fig. 2.10(b) shows the comparison with the actual strength of the Hume pipe obtained from the fracture tests.

As shown in Fig. 2.10, the fracture strength of the renovated pipes exceeds both the standardised values and the actual strength values of virgin Hume pipes, and this tendency is particularly noticeable for small-diameter pipes. These results clearly indicate that an ageing pipe with cracks and exposed reinforcing bars can be rehabilitated into a composite pipe stronger than the original one, thus verifying the effectiveness of the renovation method in rehabilitating small-diameter ageing pipes.

The damaged pipe specimens before renovation in Fig. 2.10 had reached the stage of reinforcement yielding, but the rebars had not broken. This damage state should serve as the limit state of damage when applying the renovation method to rehabilitate small-diameter ageing sewers. To evaluate the degree of damage of existing pipes, a travelling mirror-type TV camera as shown in Photo 2.25 is used by the TMG. This TV camera can be used for pipe diameters of 150 to 700 mm and, when moving forward, is capable of simultaneous shooting in the forward and lateral directions. By adding a pipe interior imaging system and a pipe inspection and diagnosis auxiliary system to the mirror-type TV camera system, the TMG has developed a systematic procedure for producing an expanded drawing of the pipe interior based on the image data and making a quick judgement on the condition of the pipe.

Figure 2.11 shows an example of automatic diagnosis based on the expanded drawing generated from image data. The entries shown in red indicate Level A damage as shown in Table 1.2, and the entries shown in black indicate Level B or Level C damage as specified in the same table. The entries shown in green indicate information concerning a lateral pipe. As shown in Table 1.2, Level A damage includes longitudinal cracks with a width of 5 mm or more, disjointing of pipes and a sag in a pipe equal to or greater than the inside diameter of the pipe. In principle, pipes in such conditions must be replaced rather than rehabilitated. Thus, the mirror-type

TV camera system and automatic diagnosis system can accurately clarify the interior conditions of a small-diameter pipe, enabling a rational decision on the method of pipe reconstruction to be made.

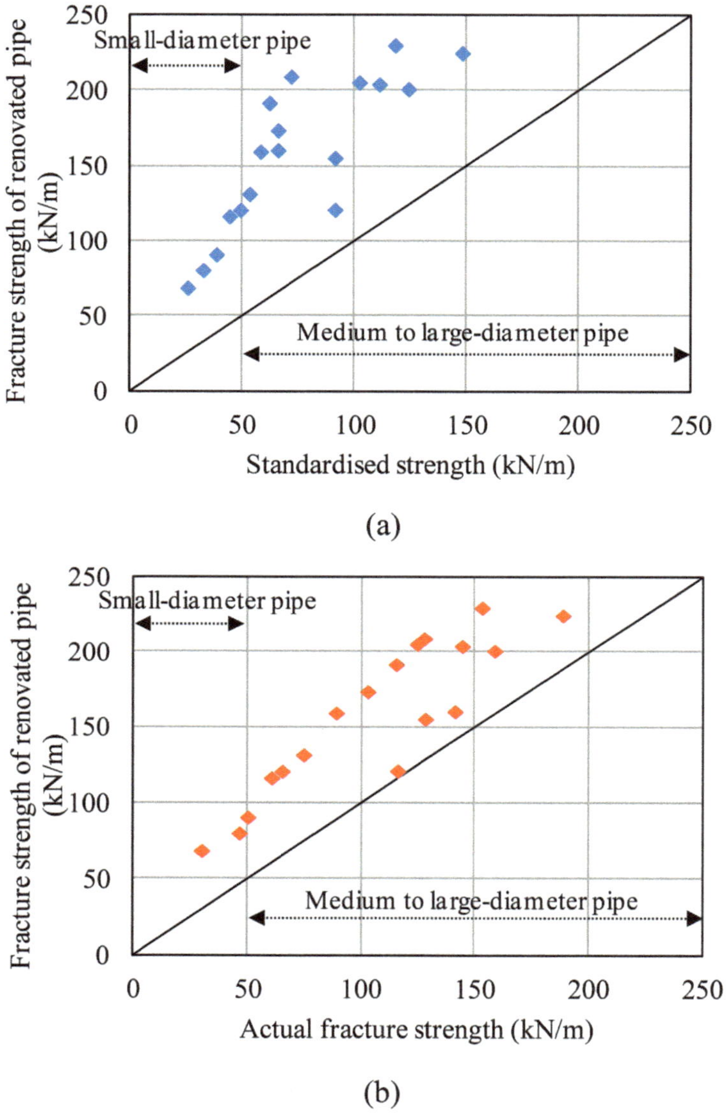

Figure 2.10: Restoration of fracture strength by pipe renovation: (a) comparison with standardised values; and (b) comparison with actual fracture strength of virgin pipes

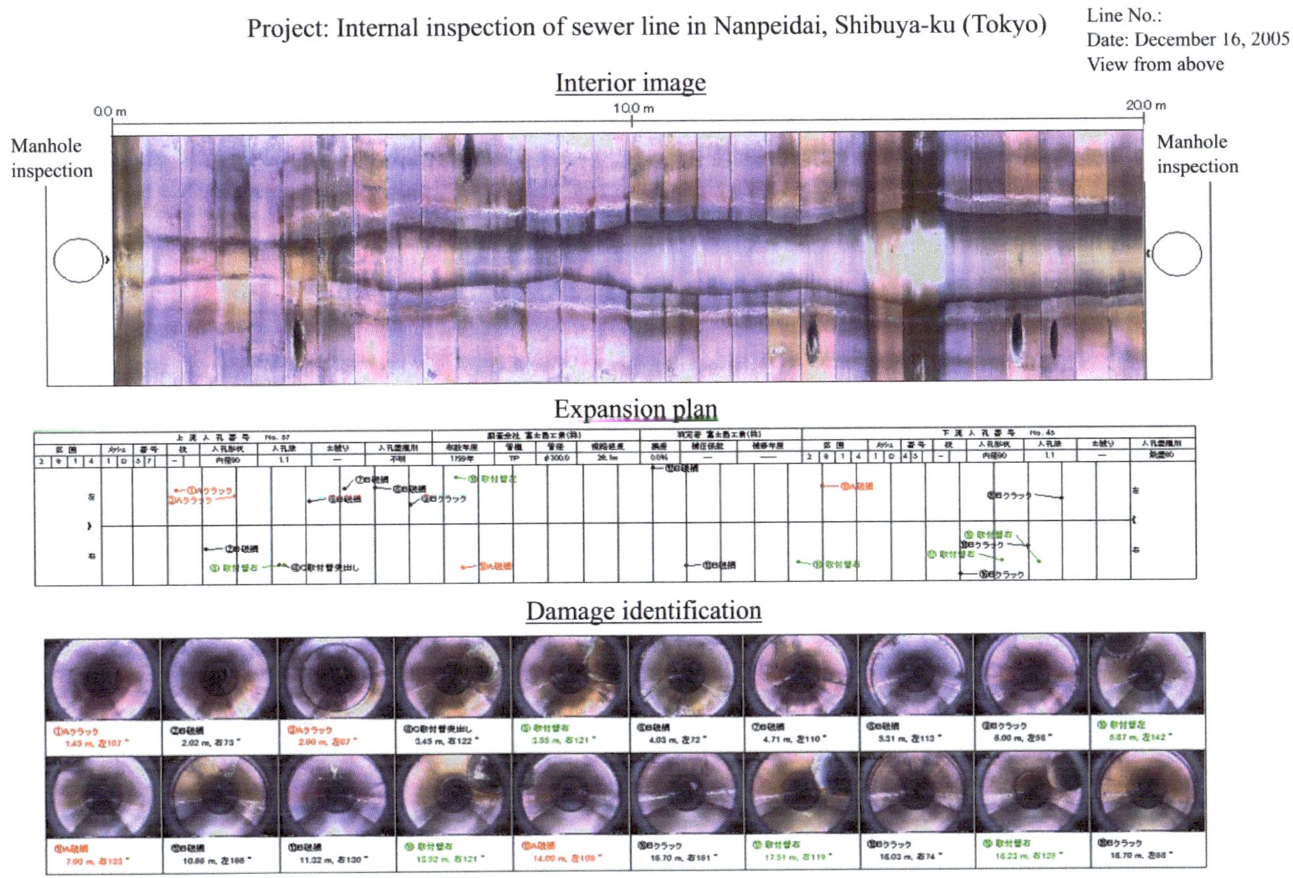

Figure 2.11: Automatic evaluation results of a small-diameter pipe

Photo 2.25: A mirror-type TV camera for pipe damage inspection

A number of studies have been carried out to find effective means for evaluating the load-carrying capacity of small-diameter non-man-entry-size pipes. As a promising method in this category, the impact elastic wave method introduced below has been certified in Japan and is now being used in practice (JIWET, 2012b). In this method, an impact is applied to the inside surface of a pipe from one end as shown in Fig. 2.12, and the elastic wave is detected at the opposite end of the pipe. As shown in Fig. 2.13, the frequency spectrum is examined to determine whether high-frequency components have been reduced (reduced high-frequency components indicate the existence of cracks or a reduced wall thickness) and whether the relationship between the wall thickness and the high-frequency component ratio is linear. A system for evaluating the load-carrying capacity of ageing pipes by applying this method has been developed. Photo 2.26 shows an inspection robot consisting of an impactor unit and a receiver unit. At present, the accuracy of the method is still being improved by accumulating inspection data.

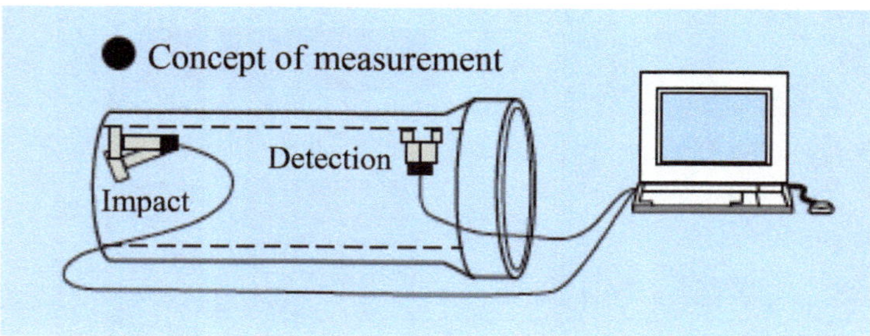

Figure 2.12: The impact wave method

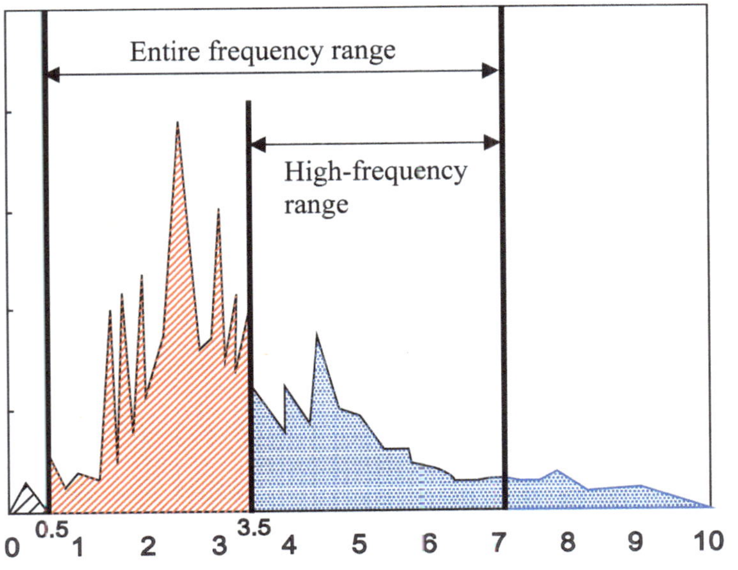

Figure 2.13: Frequency characteristics

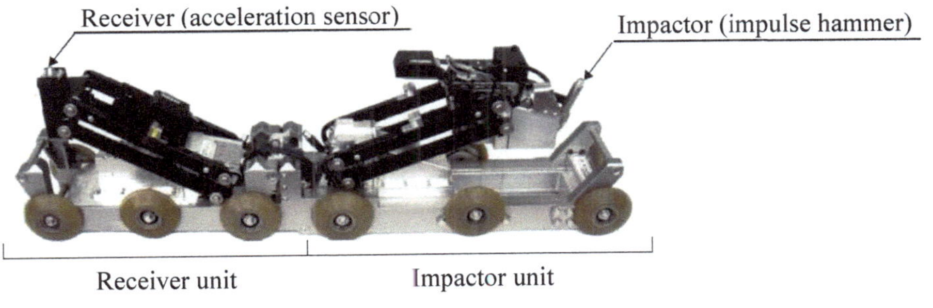

Photo 2.26: An inspection robot of the impact wave method

2.5.2 Design flow

Upon completion of the field investigation on an ageing sewer, the cross section of the renovation layer is determined by nonlinear structural analysis based on the limit state design concept, which will be discussed in Chapters 5 and 6. This section presents a brief outline of the design flow, as shown in Fig. 2.14.

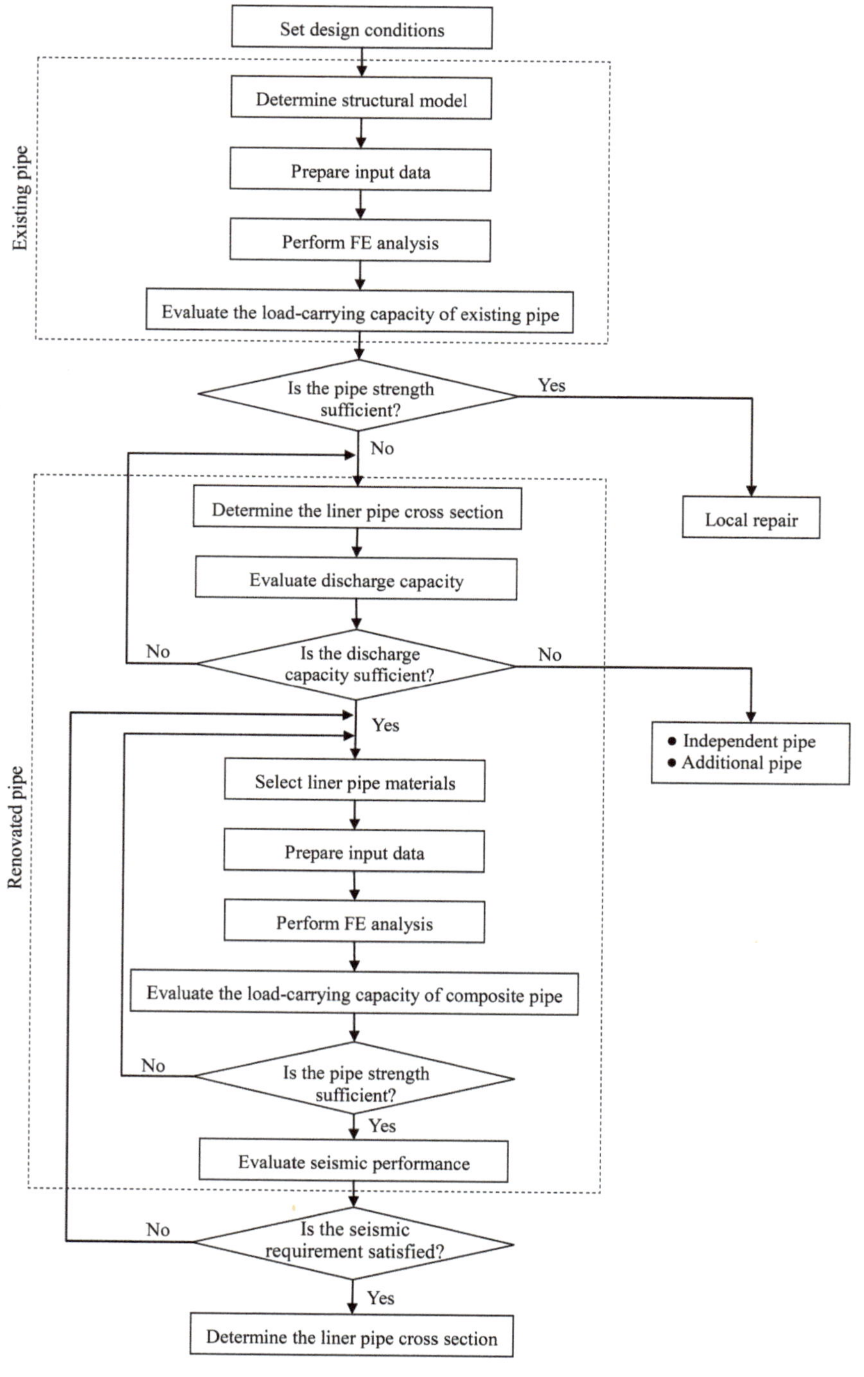

Figure 2.14: Outline of the design flow

First, the load-carrying capacity of the existing pipe is assessed, and if the strength of the pipe is judged to be insufficient, the possibility of pipe renovation is then evaluated through a computational procedure that takes into account several important factors. If the load-carrying capacity of the existing pipe meets the code requirement, the damaged portions should be repaired locally. Note that there are cases when the insufficient load-carrying capacity of an existing pipe is not caused by pipe deterioration or damage, but by a change in the design standards or safety factors. In such cases the discharge capacity of the pipe is likely to be insufficient, too.

Second, the cross section of the liner pipe is determined so as to meet discharge capacity requirements. If the discharge capacity is insufficient because of an increase in the flow rate, an independent pipe whose liner pipe minimises the cross-sectional reduction will be adopted. If the discharge problem persists, the construction of additional pipes should be considered. After the internal cross section of the renovation layer is determined, specifications for liner pipe materials (in the SPR method, for example, profile strip, reinforcing steel and mortar) are set, and a seismic performance evaluation is carried out to ensure the renovated pipe meets the code requirements for earthquake resistance.

2.5.3 Construction flow of the SPR method

Figure 2.15 illustrates the construction flow of the SPR method as an example for the pipe-reforming method; the main procedures are explained below (JSPRA, 2009).

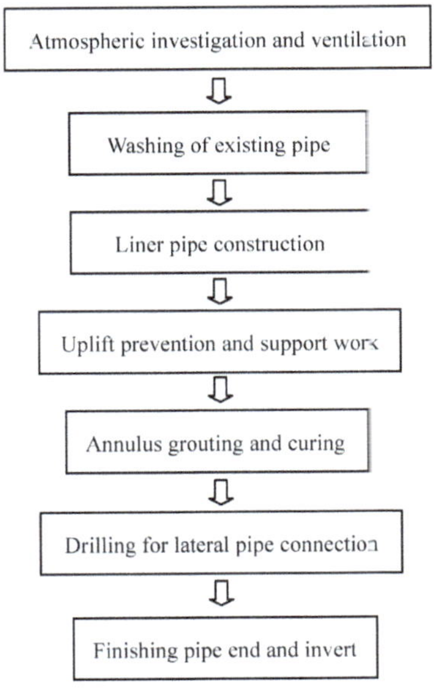

Figure 2.15: Construction flow of the SPR method

(1) Washing the existing pipe

After measuring the concentration of oxygen and toxic gases in the pipeline and ventilating it, the inside of the pipe is washed to ensure structural integrity between the grout and the existing pipe. In the case of a small-diameter pipe, a spray nozzle attached to the front opening of a high-pressure hose is advanced in the pipe while the nozzle sprays water so as to stir the sludge in the sewer as shown in Fig. 2.16(a). Then, as shown in Fig. 2.16(b), the nozzle is slowly pulled backward, collecting and removing the sludge. In the case of a medium- or large-diameter pipe, as shown in Fig. 2.17, workers enter the pipe and wash its inside by using a spray-gun type washer, brushes, etc.

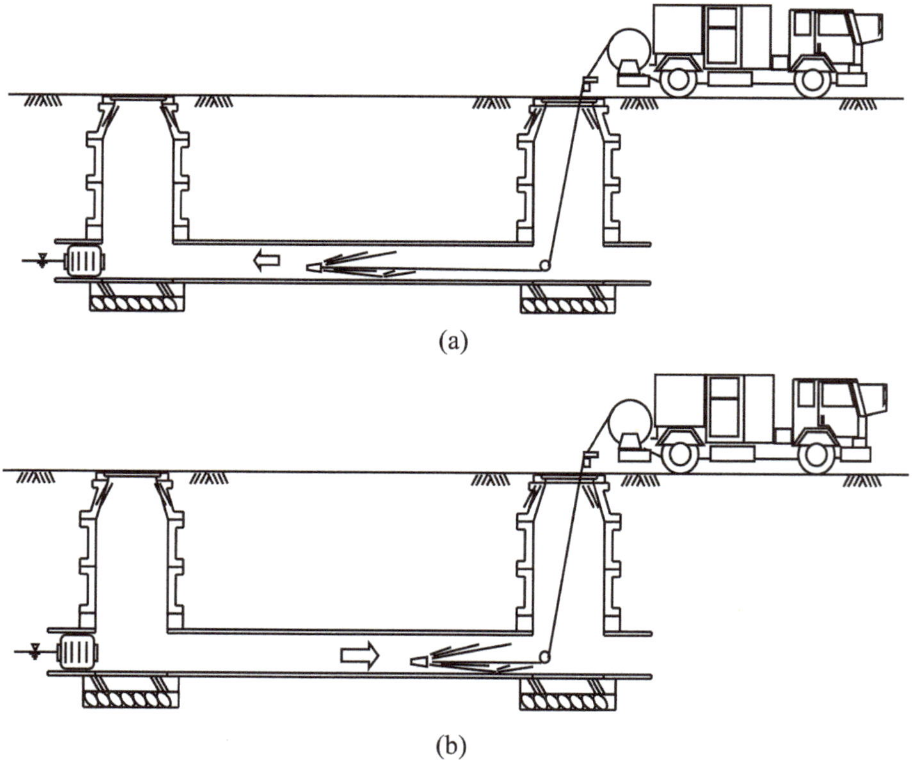

Figure 2.16: Washing a small-diameter pipe: (a) moving the nozzle forward; and (b) pulling the nozzle back

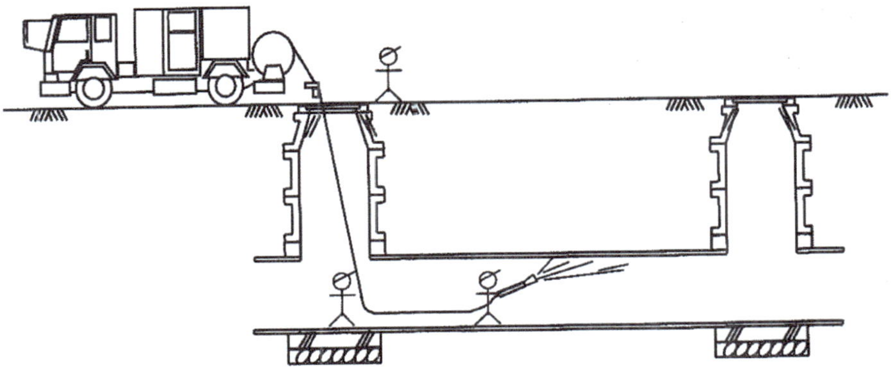

Figure 2.17: Washing a medium- and large-diameter pipe

(2) Liner pipe construction

Figure 2.18 illustrates the jacking method of liner pipe construction used for small-diameter pipes and the travelling winder method used for medium- and large-diameter pipes. In both methods, the winder is disassembled and brought into a 600-mm-diameter manhole and assembled in the manhole. The profile strip is fed from an outer winding drum in the jacking method and from an inner winding drum in the travelling winder method.

In the jacking method, a winding roller is operated in the manhole, and winding is continued while performing strip interlocking and pushing the newly-formed interlining forward until the front tip of the liner pipe arrives at the manhole used as the receiver shaft.

In the travelling winder method, a hydraulic unit is set up in the pipe and a winding roller moves along the pipeline, forming the liner pipe by winding and interlocking the profile strip, which is fed to the winding machine spirally from the profile drum above the ground. Winding is continued until the winder arrives at the manhole used as the receiver shaft.

(3) Uplift prevention and support work

Prior to annulus grouting, uplift prevention and support work is carried out to prevent uplift due to the uplift pressure acting on the liner pipe. Figure 2.19 illustrates how to carry out uplift prevention work for a small-diameter pipe while keeping the sewer in service. Weights such as metal chains are spread in the liner pipe and water is poured into the pipe to prevent uplift.

For a medium- to large-diameter pipe, braces as supports are installed to prevent uplift and the buckling of the liner pipe due to the grouting pressure. Figure 2.20 shows examples of 6-point and 8-point bracings. In order to prevent the uplift of the liner pipe, holes are drilled at the top of the liner pipe so that the braces are supported by the existing pipe.

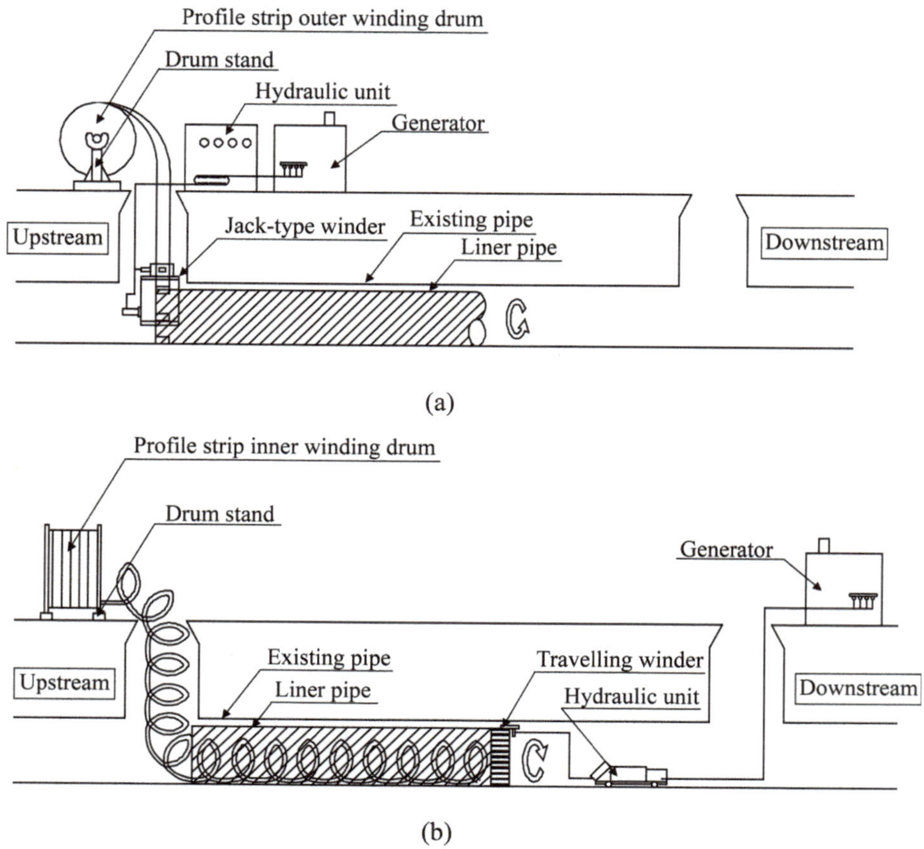

Figure 2.18: Liner pipe construction systems: (a) jacking method of pipe-making; and (b) travelling winder method of pipe-making

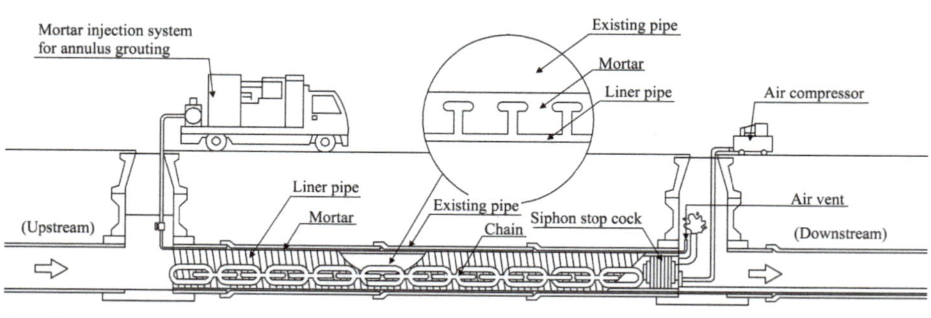

Figure 2.19: Uplift prevention work for small-diameter pipe

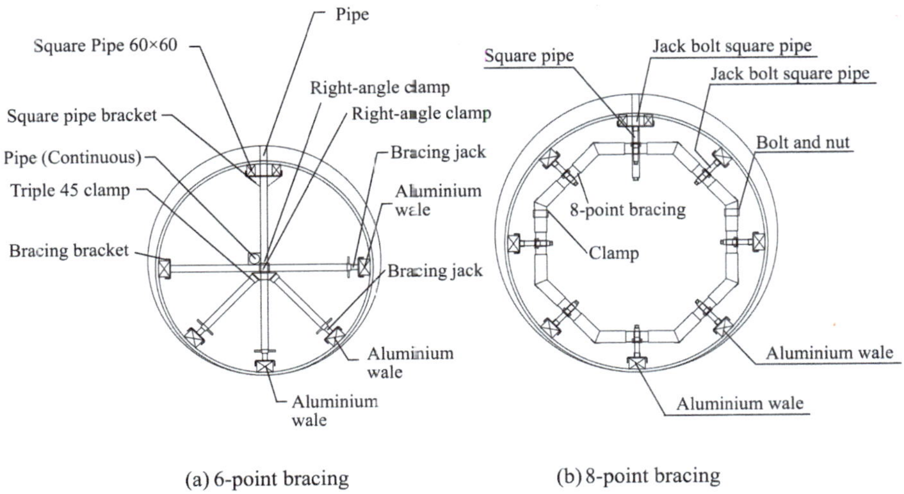

Figure 2.20: Uplift prevention framework for medium- and large-diameter pipes

(4) Annulus grouting

In order to prevent grout from flowing into the lateral pipe, an air plug is installed in the lateral pipe inlet from inside the catch basin as shown in Fig. 21. Next, the annular gap between the existing pipe and the liner pipe at the upstream and downstream ends of the pipeline is sealed with a 50 mm-thick layer of clay cement. This sealing is provided with a grout inlet and an air vent. Special mortar is injected from the upstream end, and when the grout spills out of the air vent-cum-overflow valve at the downstream end, the valve is closed. Then, after an overpressure of 0.02 MPa is applied, the inlet valve is closed, and curing is carried out while the overpressure is maintained (Fig. 2.22).

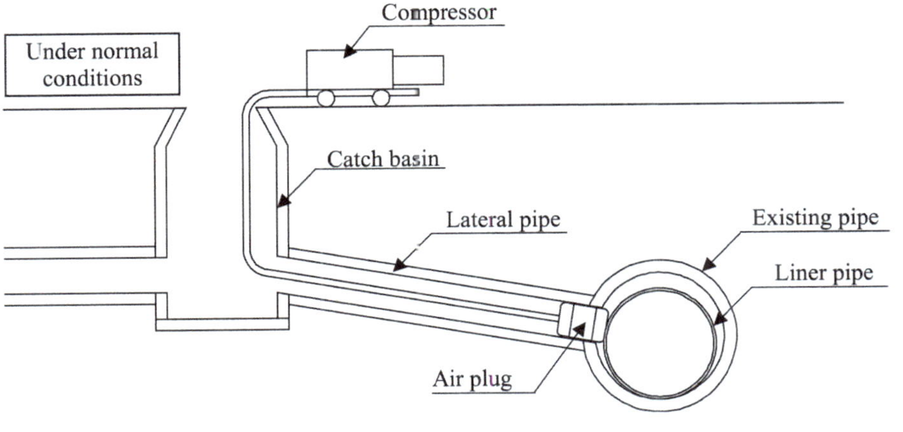

Figure 2.21: Air plugging of lateral pipe

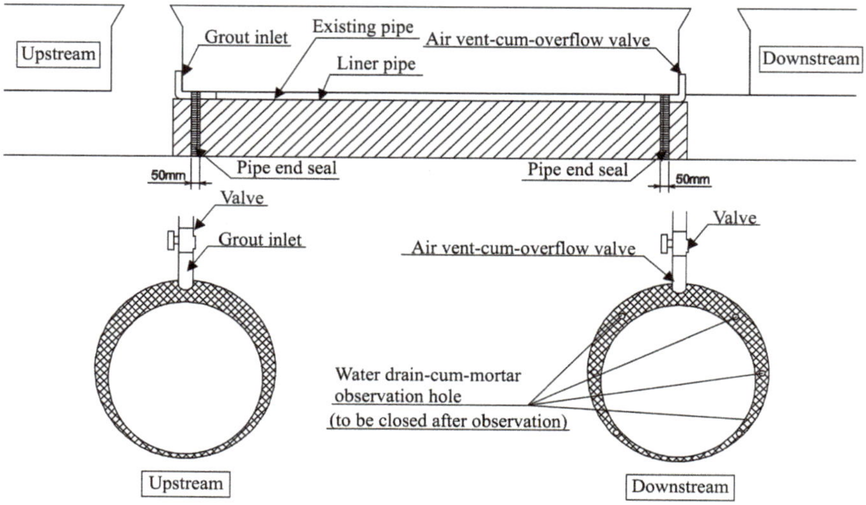

Figure 2.22: Pipe end sealing and grout inlet installation methods

In the case of a small-diameter pipeline, a single span is about 30 m and the full cross section is grouted in one operation. When dealing with a large-diameter pipeline, however, there may be cases where it is not possible to grout the full cross section in one operation. In such cases as shown in Fig. 2.23, piping (steel pipes) for grouting and grout inlets are provided in the liner pipe, and grouting is carried out in two or more stages.

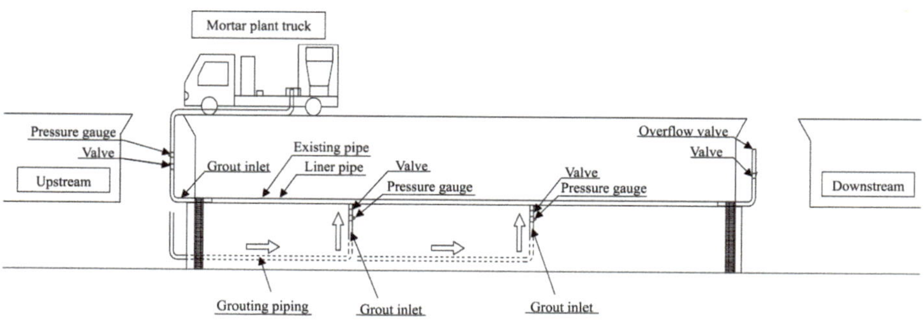

Figure 2.23: Phased grouting in the longitudinal direction

(5) Drilling for lateral pipe connection
After the completion of annulus grouting, drilling for lateral pipe connection is carried out. First, preliminary drilling is performed to determine the drilling location by inserting a drilling machine into the pipe from the catch basin as shown in Fig. 2.24. After locating the pipe position, the main drilling is carried out from inside the

liner pipe. For small-diameter pipes a robot is used while the drilling operation is monitored through a TV camera as shown in Fig. 2.25; for man-entry pipes it is done manually.

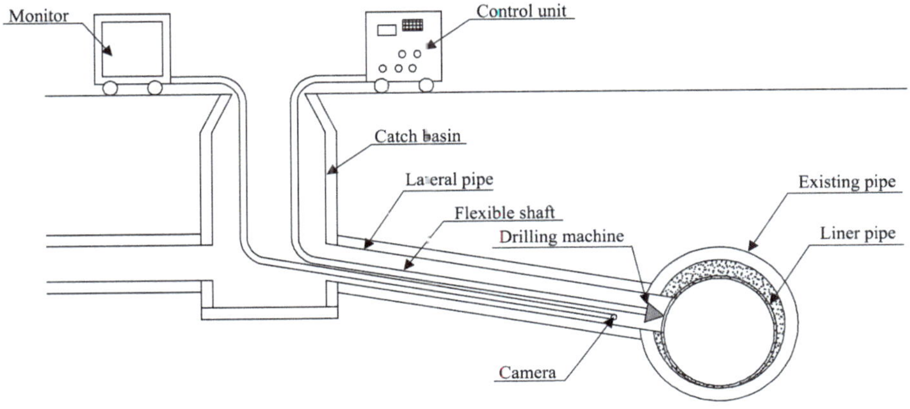

Figure 2.24: Preliminary drilling for lateral pipe connection

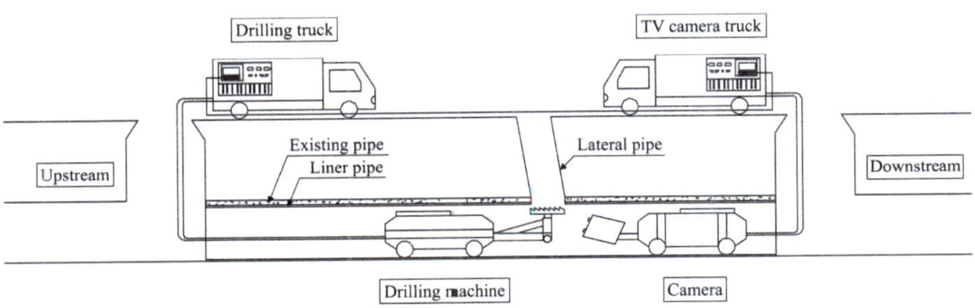

Figure 2.25: Finish drilling for lateral pipe connection

References

JSWA (2001). Pipe Rehabilitation Handbook. Tokyo, Japan Sewage Works Association.

JIWET (2009a). Construction Technology Certification Report: 3S Segment Method. Tokyo, Japan Institute of Wastewater Engineering Technology.

JIWET (2009b). Construction Technology Certification Report: PALTEM Flow-Ring Method. Tokyo, Japan Institute of Wastewater Engineering Technology.

JIWET (2009c). Construction Technology Certification Report: Danby's Method. Tokyo, Japan Institute of Wastewater Engineering Technology.

JIWET (2011). Construction Technology Certification Report: PFL Method. Tokyo, Japan Institute of Wastewater Engineering Technology.
JIWET (2012a). Construction Technology Certification Report: SPR Method. Tokyo, Japan Institute of Wastewater Engineering Technology.
JIWET (2012b). Technical Data Book for Pipeline Diagnosis by Impact Elastic Wave Method. Tokyo, Japan Institute of Wastewater Engineering Technology.
JSPRA (2009). SPR Method Construction Manual: Sewer Rehabilitation. Tokyo, Japan SPR Method Association.

Yukari Nakamura

3 Fracture Tests of Full-Scale Pipe Specimens and Various Structural Element and Material Property Tests

In developing and improving the SPR method, external pressure tests on many full-scale specimens have been conducted to investigate the load-carrying capacity of renovated pipes. Various structural element and material property tests have also been conducted to obtain basic data. This chapter briefly describes the tests conducted to investigate the effects and performance attributes listed in Table 3.1 and presents the results of those tests.

Table 3.1: List of conducted tests and the aims of the tests

	Aim of test	
Type of test	**Test**	**Effect/performance confirmation**
Load-carrying capacity verification test	External pressure test	Renovation effect
	Preload test	Renovation effect
	Verification test on earthquake resistance	Renovation effect, seismic performance
Element test	Autogenous shrinkage test	Shrinkage properties of SPR mortar
	Bending test of beam specimen	Integrity of concrete and liner material
	Direct tension test	Bond strength of SPR mortar
	Compressive shear test	Bond strength of SPR mortar
	Double shear test	Shear strength of SPR mortar
	Profile pullout test	Seismic performance of renovated pipe (longitudinal direction)
	Verification test on effectiveness of additional rebar	Effectiveness of additional rebar
Basic properties test	Compressive strength test	Compressive strength, modulus of elasticity, Poisson's ratio, unit weight
	Splitting tensile strength test	Tensile strength
	Fracture energy test	Fracture energy
	Single shear test	Shear strength

3.1 Fracture Tests on Load-Carrying Capacity

To evaluate the load-carrying capacity of renovated pipes, a series of fracture tests were conducted on full-scale specimens of circular and rectangular pipes of different sizes, as shown in Fig. 3.1. The renovated pipe specimens reflected different states of sewer damage or deterioration, the initial stress state in a sewer for which renovation was performed, and the state of seismic loading endured during strong earthquakes.

(a)

(b)

Figure 3.1: External pressure test: (a) original RC pipes; (b) renovated RC pipes

3.1.1 External pressure test

Fracture tests were conducted on full-scale pipe specimens, which contained various degrees of artificial damage and were renovated by the SPR method.

(1) Test cases

Table 3.2 lists seven test cases for rectangular pipes, and Table 3.3 shows five test cases for circular pipes. The tables specify the original pipe conditions and types of renovated pipe, which will be explained below.

Table 3.2: Test cases for 1500 × 1500 mm rectangular pipes

Test case	Original pipe condition	Type of specimen	Quantity
1	Standard doubly-reinforced cross section	Original pipe	3
2	Standard doubly-reinforced cross section	Standard renovated pipe	3
3	Standard doubly-reinforced cross section	Renovated pipe as double-layered structure	3
4	Cross section without inner concrete cover	Original pipe	3
5	Cross section without inner concrete cover	Standard renovated pipe	3
6	Cross section without inner concrete cover and tension rebar	Original pipe	3
7	Cross section without inner concrete cover and tension rebar	Standard renovated pipe	3

Table 3.3: Test cases for ϕ1000 mm circular pipes

Test case	Original pipe condition	Type of specimen	Quantity
1	Standard doubly-reinforced cross section	Original pipe	3
2	Standard doubly-reinforced cross section	Renovated pipe with steel reinforcement in profile	3
3	Standard doubly-reinforced cross section	Renovated pipe as double-layered structure with steel reinforcement in profile	3
4	Standard doubly-reinforced cross section	Renovated pipe without steel reinforcement in profile	3
5	Standard doubly-reinforced cross section	Renovated pipe without steel reinforcement in profile (in contact with bottom slab)	3

(2) Material properties of test specimens
The material properties of the test specimens are shown in Tables 3.4 and 3.5 for the rectangular and circular pipes, respectively.

Table 3.4: Material properties of 1500 × 1500 mm rectangular pipe specimens

	Material type			
Material property	**Concrete**	**Mortar**	**Reinforcing bar**	**Profile-reinforcing steel (79SW)**
Compressive strength (N/mm^2)	56.25	42.04		
Tensile strength (N/mm^2)	3.75	4.17		
Poisson's ratio	0.20	0.21		
Unit weight (kN/m^3)	23.0	23.0		
Modulus of elasticity (kN/mm^2)	31.85	20.09	210	170
Yield strength (N/mm^2)			295	210

Table 3.5: Material properties of ϕ1000 mm diameter circular pipes

	Material type			
Material property	**Concrete**	**Mortar**	**Reinforcing bar**	**Profile-reinforcing steel (79SW)**
Compressive strength (N/mm^2)	57.62	14.11		
Tensile strength (N/mm^2)	4.41	1.23		
Poisson's ratio	0.19	0.17		
Unit weight (kN/m^3)	23.0	12.0		
Modulus of elasticity (kN/mm^2)	31.95	6.17	210	170
Yield strength (N/mm^2)			593	210

(3) Dimensions and steel reinforcement arrangement of test specimens
Figures 3.2 and 3.3 show the dimensions and rebar arrangements of the test specimens for rectangular and circular pipes, respectively. The following types of 1500×1500 mm rectangular specimens were used: (a) an original pipe having the same standard doubly-reinforced cross section as a new pipe and a renovated pipe obtained by using such a pipe, (b) an original pipe with a reduced wall thickness, prepared by removing the inner concrete cover to simulate an ageing pipe, and a renovated pipe obtained by using such a pipe and (c) an original pipe whose inner concrete cover was removed

and the tension rebar in the top slab was removed to simulate an ageing, severely deteriorated pipe, and a renovated pipe obtained by using such a pipe. To compare the load-carrying capacities of a standard renovated pipe and a double-layered pipe, which has no bond strength between the original pipe and the liner, a double-layered pipe was prepared by placing a thin film on the inner surface of the original pipe (with a standard doubly-reinforced cross section) before renovation.

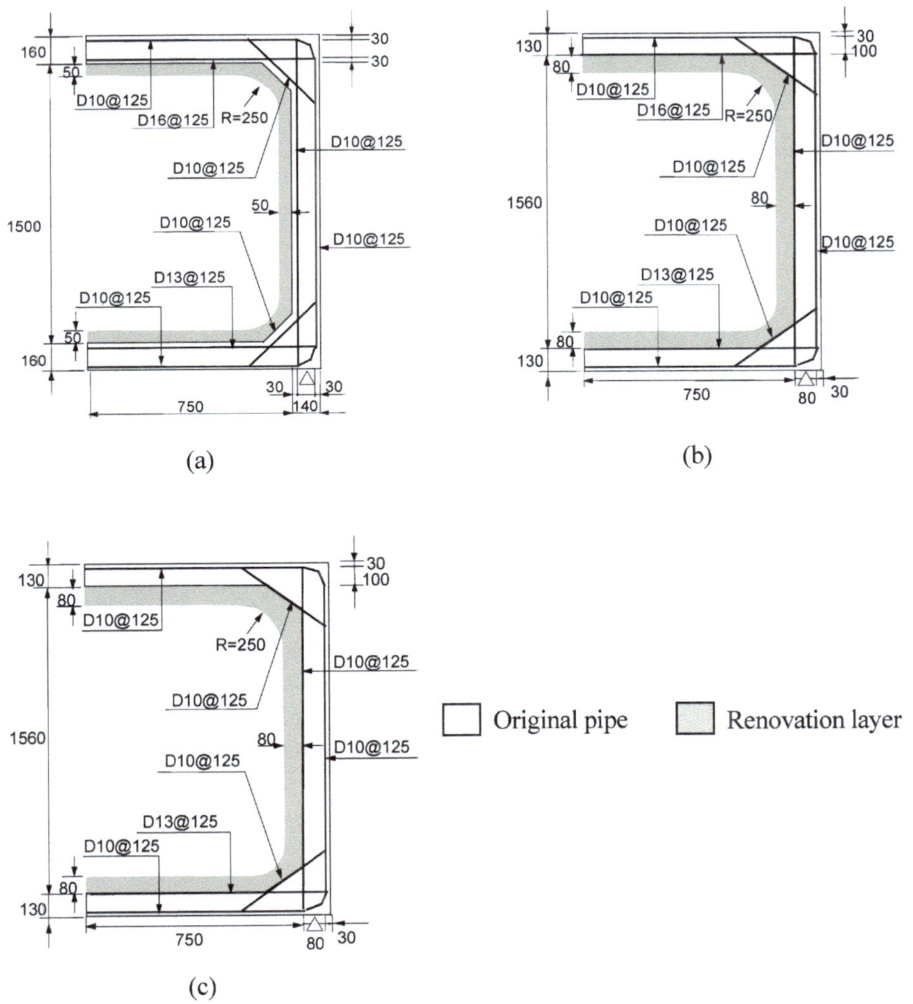

Figure 3.2: Structural dimensions and rebar arrangement of 1500 × 1500 mm rectangular pipe: (a) standard doubly-reinforced cross section (original pipe, renovated pipe); (b) cross section without inner concrete cover (original pipe, renovated pipe); (c) cross section without inner concrete cover and tension rebar (original pipe, renovated pipe)

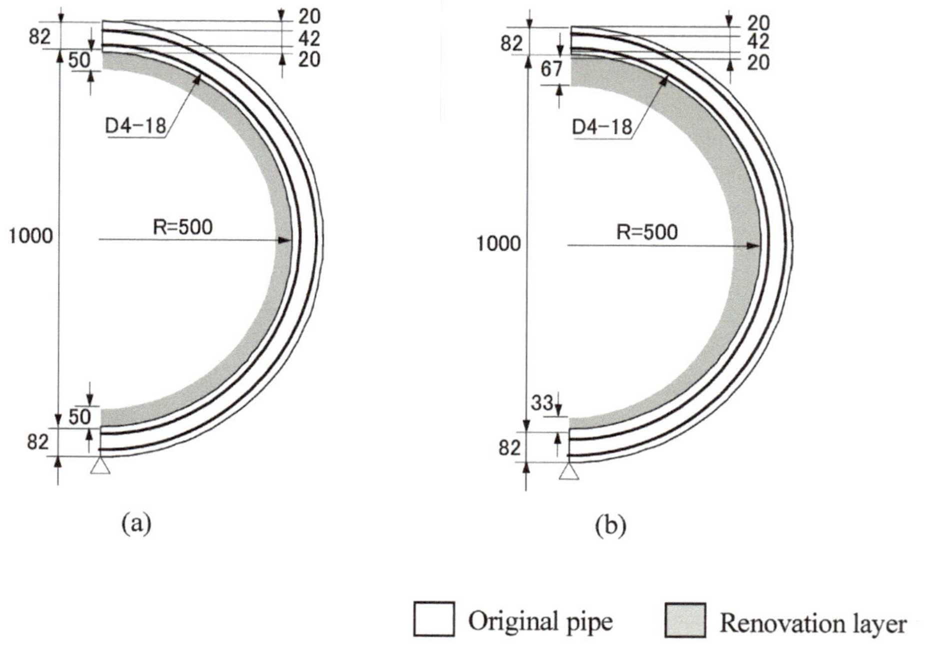

Figure 3.3: Structural dimensions and rebar arrangement of ϕ1000 mm circular pipe: (a) standard doubly-reinforced cross section (original pipe, renovated pipe); (b) standard doubly-reinforced cross section (renovated pipe in contact with bottom slab)

The following types of ϕ1000 mm circular specimens were used: (a) an original pipe having the same standard doubly-reinforced cross section as a new pipe and a renovated pipe (with or without profile-reinforcing steel) obtained by using such a pipe and (b) a renovated pipe obtained by using an original pipe of a standard doubly-reinforced cross section with its liner profile in contact with the bottom slab. To compare the load-carrying capacities of a standard renovated pipe and a double-layered pipe, a double-layered pipe was prepared by placing a thin film on the inner surface of the original pipe (with a standard doubly-reinforced cross section) before renovation.

(4) Test method

The external pressure tests were conducted in accordance with JIS A5363 (2010) and JIS A5372 (2010). Displacement gauges were installed to measure deformation in the vertical and horizontal directions. Figure 3.4 illustrates the test method.

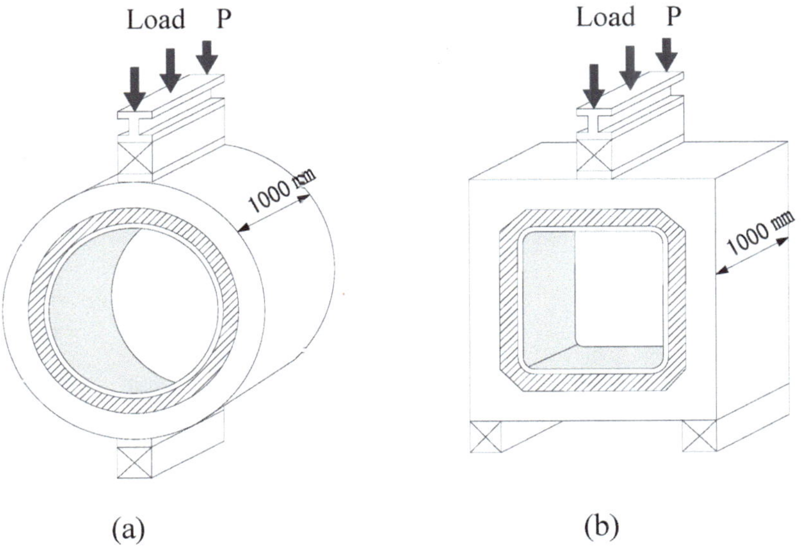

Figure 3.4: External pressure test of SPR renovated pipe: (a) circular pipe specimen; (b) rectangular pipe specimen

(5) Results of external pressure test
Tables 3.6 and 3.7 show the results of fracture tests, and Figs. 3.5 and 3.6 present the load–displacement relationships of each type of test specimen. It was confirmed that the load-carrying capacity of the 1500×1500 mm standard renovated rectangular pipe was greater than that of the original pipe with various degrees of simulated damage or structural deterioration by a factor of 1.8 to 3.4. It was also confirmed that the load-carrying capacity of the ϕ1000 mm standard renovated circular pipe was greater than that of the original pipe by a factor of 1.6 in the profile-with-reinforcement case and 1.3 in the profile-without-reinforcement case.

Figures 3.7 to 3.9 compare the maximum loads of rectangular pipe specimens obtained for cases 1 to 7 of Table 3.6. As shown in Fig. 3.7, the loss of inner concrete cover alone resulted in a decrease in load-carrying capacity of about 30% compared with the original pipe, and the loss of inner concrete cover and tension rebar resulted in a decrease of 70%. As shown in Fig. 3.8, a renovated pipe is at least as strong as the original pipe in the standard cases of renovation, including the case in which the pipe has lost both the inner concrete cover and the tension rebar.

In Fig. 3.9, the maximum load of the double-layered pipe is compared with that of the standard renovated pipe for the circular and rectangular test specimens. As shown, although the load-carrying capacity of the double-layered pipe is at least comparable with that of a new pipe, it is still about 30% smaller than that of the standard renovated pipe.

Table 3.6: Results of external pressure test of 1500 × 1500 mm rectangular pipes

Test case	Type of specimen	Maximum load (kN/m)	
		Test results	Average
1	Original pipe	366.3	364.3
		362.5	
		364.0	
2	Standard renovated pipe	673.7	643.4
		622.1	
		634.4	
3	Renovated pipe as double-layered structure	487.1	477.5
		473.1	
		472.4	
4	Original pipe without inner concrete cover	246.9	251.2
		262.3	
		244.5	
5	Standard renovated pipe without inner concrete cover	625.2	545.1
		455.8	
		554.4	
6	Original pipe without inner concrete cover and tension rebar	106.6	104.5
		110.6	
		96.3	
7	Standard renovated pipe without inner concrete cover and tension rebar	(249.3)	359.4
		372.0	
		346.8	

* The value in parentheses was excluded from average calculation.

Table 3.7: Results of external pressure test of ϕ1000 mm circular pipes

Test case	Type of specimen	Maximum load (kN/m)	
		Test results	Average
1	Original pipe	88.7	92.6
		94.8	
		94.2	
2	Renovated pipe with steel reinforcement in profile	145.9	145.5
		145.0	
		(112.2)	
3	Renovated pipe as double-layered structure with steel reinforcement in profile	99.4	99.8
		99.1	
		100.9	
4	Renovated pipe without steel reinforcement in profile	105.1	118.1
		117.2	
		131.9	
5	Renovated pipe without steel reinforcement in profile (in contact with bottom slab)	120.1	118.6
		114.0	
		121.6	

* The value in parentheses was excluded from the average calculation.

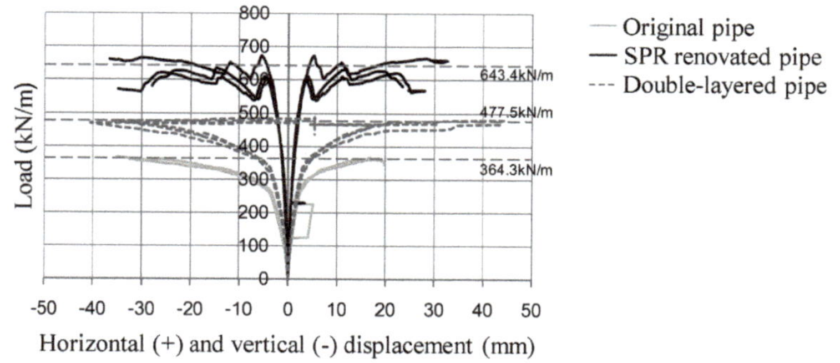

(a)

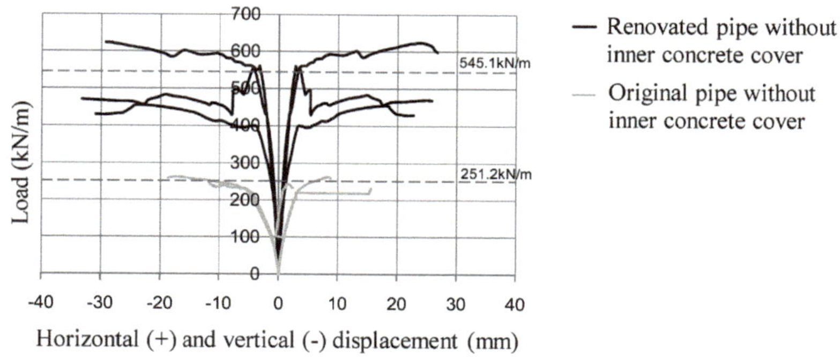

(b)

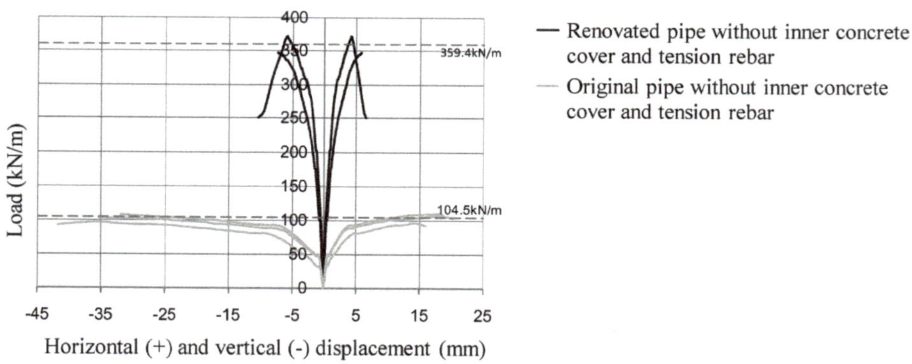

(c)

Figure 3.5: Load–displacement relationships of 1500 × 1500 mm rectangular test specimens: (a) original pipe, standard renovated pipe and double-layered pipe; (b) original pipe and standard renovated pipe without inner concrete cover; (c) original pipe and standard renovated pipe without inner concrete cover and tension rebar

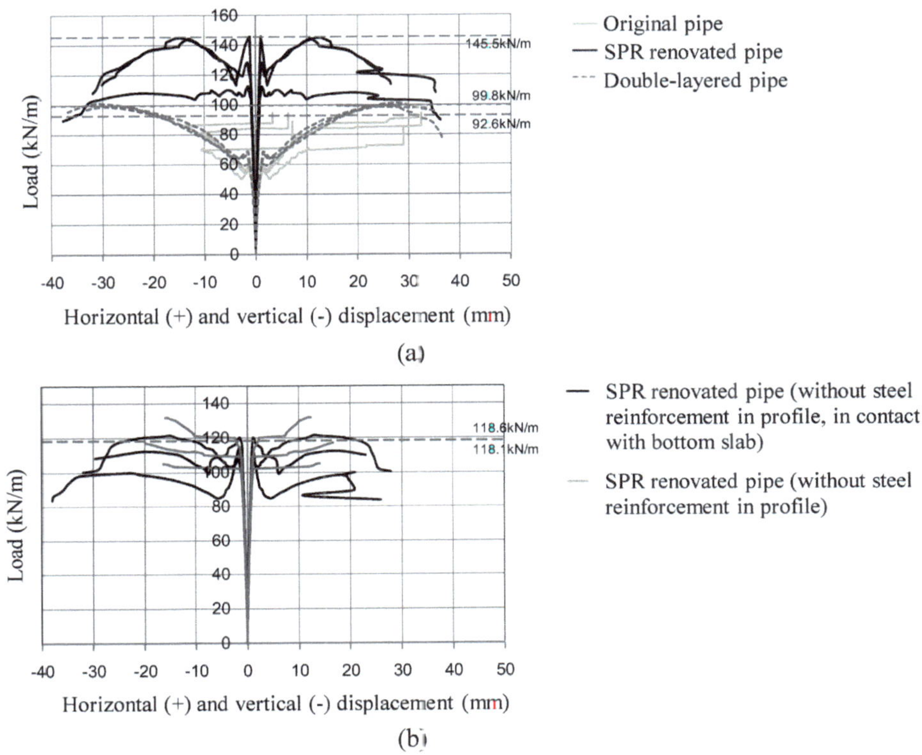

Figure 3.6: Load–displacement relationships of ϕ1000 mm circular test specimens: (a) original pipe, standard renovated pipe and double-layered pipe; (b) standard renovated pipe and renovated pipe with profile in contact with bottom slab: without steel reinforcement in profile

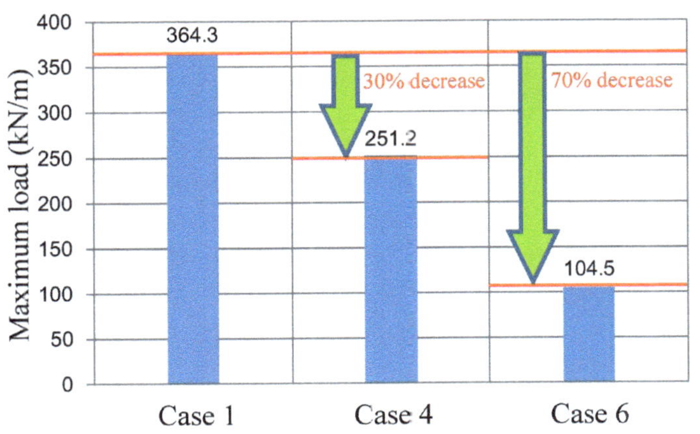

Figure 3.7: Influence of degree of damage on strength (1500 × 1500 mm)

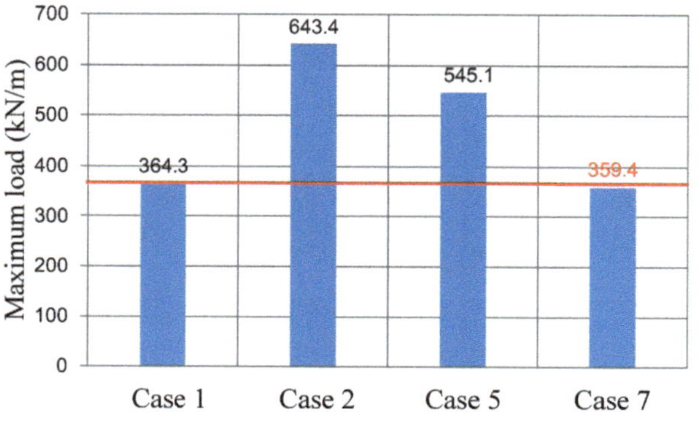

Figure 3.8: Recovery of strength of standard renovated pipe (1500 × 1500 mm)

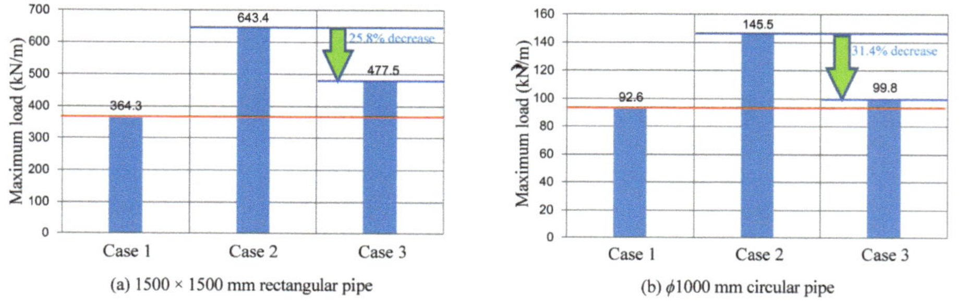

Figure 3.9: Comparison of strength of standard renovated pipe and double-layered pipe

(6) Load-strain relationships near the interfaces of the standard renovated pipe and the double-layered pipe

To investigate the continuity of the interfacial strains between the original pipe and the renovation layer, strain gauges were installed across the interface between the original pipe concrete of the top slab and the mortar of the renovation layer at the two free surfaces of the specimen. Figure 3.10 shows the test results of load-displacement relations in the vertical and horizontal directions for the rectangular and circular test specimens, and Table 3.8 compares the maximum loads of the two types of renovation. As shown, in the case of the double-layered structure, both the rectangular and circular pipe specimens exhibit smaller initial stiffness, and their load-carrying capacities are approximately 70% of those of the standard renovated pipe specimens.

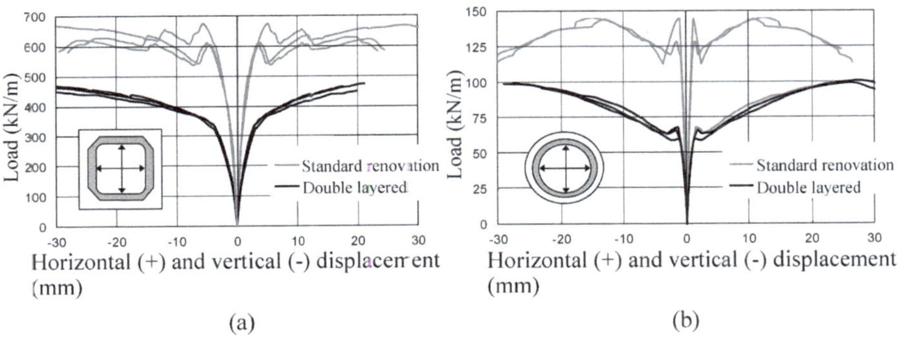

Figure 3.10: Load–displacement relationships of the standard renovated pipe and the double-layered pipe: (a) box culvert (1500 mm × 1500 mm); (b) circular cross section (ϕ1000 mm)

Table 3.8: Results of external pressure test

Type of pipe	Type of specimen	Average maximum load (kN/m)	Ratio of maximum loads
			(b) / (a)
Rectangular pipe 1500 × 1500 mm	Standard renovated pipe (a)	643.4	0.74
	Double-layered pipe (b)	477.5	
Circular pipe ϕ1000 mm	Standard renovated pipe (a)	145.5	0.69
	Double-layered pipe (b)	99.8	

For the rectangular test specimens, the measured strain behaviours near the interface are shown in Fig. 3.11. In the case of the standard renovated pipe, the strain at point C of the mortar surface is greater than the strain at point B of the concrete surface of the original pipe. This indicates the continuity of strain flow across the interface of the original pipe and the renovation layer. In the double-layered pipe, the strain in the concrete surface of the original pipe is greater than the strain in the mortar surface from the beginning of loading, indicating the discontinuity of strain flow across the interface.

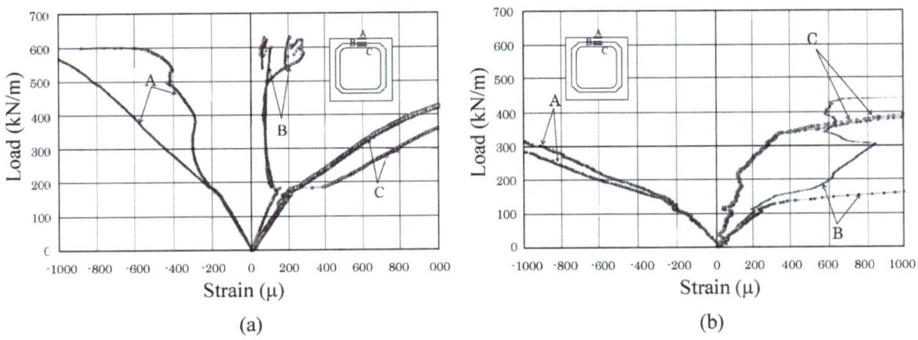

Figure 3.11: Measured load-strain relationships of rectangular specimens: (a) standard renovated pipe; (b) double-layered pipe

3.1.2 Preload test

Existing pipelines are constantly subjected to various loads such as earth pressure, traffic loads, groundwater pressure, etc. To investigate how the initial stress state affects pipe renovation, two types of renovated specimen were prepared: one with the original pipe specimen being subjected to preloading, and the other without subjecting the original pipe specimen to preloading. Fracture tests were carried out on these two types of renovated specimen.

(1) Test cases
Table 3.9 shows the test cases and the preload conditions. A circular pipe of ϕ1000 mm was chosen for testing because this is a typical size of circular pipes used for integrated renovation and has been used in many tests. In order to simulate a severe preload condition, the state of yielding of rebar (the maximum steel strain of 2700 μ) was set as the target of preloading, which is also within the applicable damage limit of the SPR method. In view of the variability of preload and renovation conditions, three test specimens were prepared in each case.

Table 3.9: Preload test cases

Test case	Type of pipe	State of original pipe before renovation	Degree of preloading	Quantity
1	ϕ1000 mm circular pipe (ϕ900 mm after renovation)	With preload	Yielding of rebar	3
2		Without preload		3

(2) Material properties of test specimens
Table 3.10 shows the material properties of the test specimens. These property values were obtained by preparing standard test specimens for strength testing when the test specimens for preloading were prepared and by conducting a strength test with the fracture test.

Table 3.10: Material properties of preloaded test specimens

	Material type		
Material property	**Concrete**	**Mortar**	**Rebar**
Compressive strength (N/mm^2)	79.2	28.35	
Tensile strength (N/mm^2)	4.20	2.93	
Poisson's ratio	0.225	0.25*	
Unit weight (kN/m^3)	24.1	13.0*	
Modulus of elasticity (kN/mm^2)	40.0	8.9	200
Yield strength (N/mm^2)			540

* Adopted design value

(3) Structural details of test specimens
Figure 3.12 shows the dimensions and reinforcement arrangements of the test specimens. The length of the specimens was 1500 mm. For renovation, type #2 SPR mortar and type #79S profile were used.

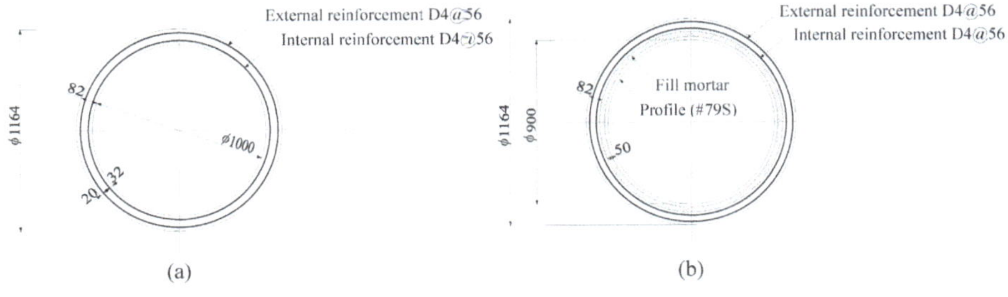

Figure 3.12: Structural details of test specimens: (a) original pipe (inner diameter 1000 mm); (b) renovated pipe (inner diameter 900 mm)

(4) Preparation of test specimens
To determine the level of preload, loading was continued until the yield strain of rebar in both the upper and lower parts of the original pipe was reached or exceeded 2700 μ. The corresponding load was recorded as within the range of 56.0 to 66.3 kN/m. This load was kept constant while pipe renovation was carried out to prepare the test specimens, as shown in Photo 3.1.

Photo 3.1: Pipe renovation under preloading

(5) Results of fracture tests

Details of the test results at the first and second peak loads are summarised in Table 3.11, and Fig. 3.13 shows the obtained load–displacement relations in each case. As seen, for Case 1 the obtained loads ranged from 121.7 to 145.9 kN/m at the first peak and from 120.5 to 149.0 kN/m at the second peak. The displacement of the preloaded specimens tended to increase after the first peak while the maximum load remained unchanged. In contrast, in Case 2, the load-carrying capacity of the un-preloaded specimens slightly increased after the first peak. The loads and displacements of the two cases at the first peak load were similar under the same renovation conditions, regardless of the presence or absence of preload. Since the maximum load-carrying capacity of a renovated pipe is determined based on the first peak load, the test results show that the initial stress state of an existing pipe has an insignificant effect on renovation performance and can be ignored in renovation design.

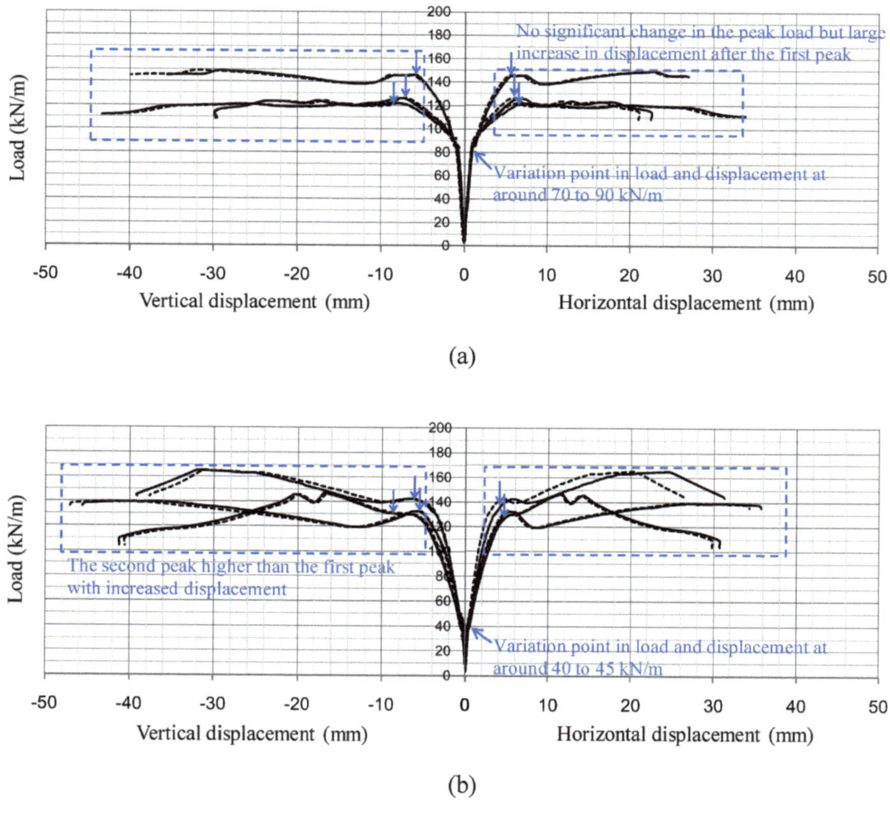

Figure 3.13: Load–displacement relationships of preloaded and un-preloaded test specimens: (a) with preload; (b) without preload

Table 3.11: Results of fracture tests on renovated test specimens for preloading test

Test case	State of original pipe before renovation	Type of data	First peak			Second peak		
			Load (kN/m)	Displacement (mm)		Load (kN/m)	Displacement (mm)	
				Vertical	Horizontal		Vertical	Horizontal
1	With preload	Measured value	145.9	6.15	5.53	149.0	30.95	22.99
			126.5	7.22	6.02	123.3	22.36	16.22
			121.7	8.06	6.83	120.5	26.52	19.88
		Average	136.7	7.08	5.94	144.9	29.73	21.81
2	Without preload	Measured value	142.5	7.03	5.45	165.1	31.71	22.57
			130.6	7.92	6.05	147.2	16.73	11.54
			132.2	5.35	5.16	139.4	38.14	26.64
		Average	129.8	6.83	5.74	136.6	25.74	18.13

3.1.3 Verification test on earthquake resistance

According to the *Design and Construction Guidelines for Sewer Pipe Rehabilitation* (JSWA, 2011), renovated sewer pipes are required to retain discharge capacity even after suffering major structural damage caused by a Level 2 strong earthquake. Therefore, horizontal reversed loading tests on full-scale, renovated pipe specimens were conducted to check if discharge capacity can be retained by preventing the joint disconnection of profile from occurring, and if the load-carrying capacity of the damaged pipe is still comparable to that of a new pipe.

(1) Test cases
Table 3.12 shows three test cases: an original pipe with its inner concrete cover removed, a new pipe, and an SPR renovated pipe based on the original pipe. To verify the discharge capacity and the load-carrying capacity of a renovated pipe during and after an earthquake, reversed loading tests were conducted on these three test specimens. The test methods are explained below.

Table 3.12: Test cases for seismic performance verification

Test case	Original pipe condition	Type of specimen	Quantity
1	Cross section without inner concrete cover	Original pipe	1
2	Doubly-reinforced cross section	New pipe	1
3	Cross section without inner concrete cover	Renovated pipe	1

(2) Material properties of test specimens

Table 3.13 shows the material properties of the test specimens. These property values were obtained by preparing standard test specimens for strength testing when the test specimens for the seismic loading test were prepared.

Table 3.13: Material properties of test specimens for verifying seismic performance

	Material type			
Material property	**Concrete**	**Mortar**	**Reinforcing bar**	**Profile-reinforcing steel (79SW)**
Compressive strength (N/mm^2)	47.4	26.1		
Tensile strength (N/mm^2)	4.39	1.45		
Poisson's ratio	0.214	0.25*		
Unit weight (kN/m^3)	23.0	13.0*		
Modulus of elasticity (kN/mm^2)	31.8	8.9	200	165
Yield strength (N/mm^2)			295	205

* Adopted design value

(3) Structural details of test specimens

Figure 3.14 shows the dimensions of the test specimens, and Fig. 3.15 illustrates their reinforcement arrangements. Photo 3.2 shows core samples taken from the bottom slabs of renovated pipe specimens, revealing good adhesion between the pipe concrete and the renovation layer. Note that the length of the concrete core includes the base concrete, as shown in Fig. 3.14.

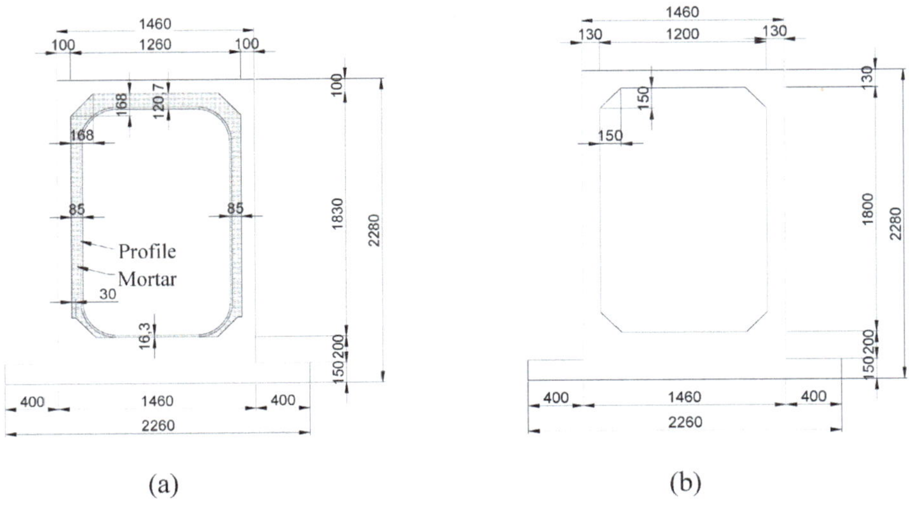

Figure 3.14: Dimensions of test specimens (mm) for seismic performance test: (a) cross section without inner concrete cover (original pipe, renovated pipe); (b) doubly-reinforced cross section (new pipe)

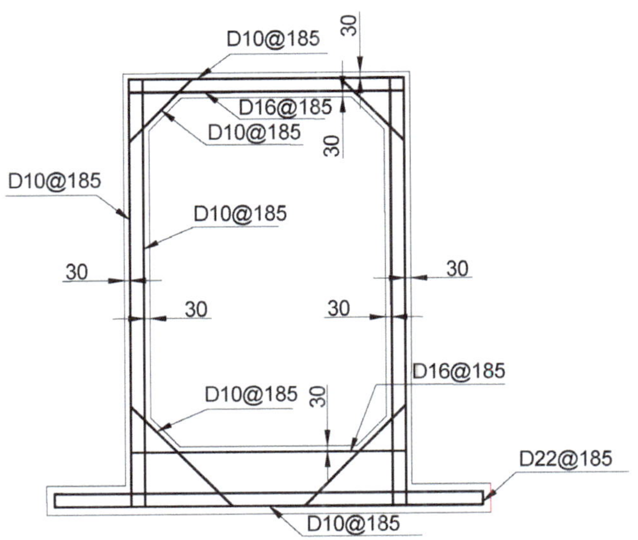

Figure 3.15: Reinforcement arrangement of test specimen

Photo 3.2: Good adhesion between the pipe concrete and the renovation layer revealed by core samples taken from the bottom slabs of renovated pipe specimens during earthquake resistance verification test

(4) Test methods

A two-stage loading test was conducted to evaluate seismic performance for the renovated test specimen. Fig. 3.16 shows the test flow.

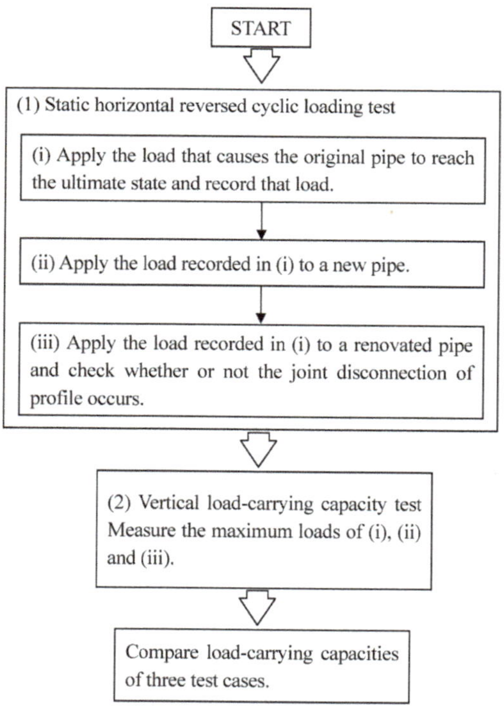

Figure 3.16: Test flow for seismic performance verification of renovated pipe specimens

1) Horizontal reversed loading test
A horizontal reversed cyclic loading test was conducted, assuming the case in which a sewer pipe suffers severe structural damage due to large ground motion caused by a Level 2 strong earthquake. Photo 3.3 shows a test specimen set up on the testing apparatus. A weight of about 5.0 tons was placed on top of the test specimen to simulate a two-metre-thick overburden.

Photo 3.3: Reversed cyclic loading test

In this section, the load corresponding to a Level 2 earthquake ground motion is defined as "a load that causes the original pipe to reach the ultimate failure state." The loads shown in Table 3.14 were applied to the original pipe first, and then they were applied to an SPR renovated pipe and to a new pipe, respectively. Finally, a comparison of the three cases was made.

Table 3.14: Loads applied in static horizontal reversed cyclic loading test

Cycle	Applied load
1st cycle	Initial crack loads for right and left sides of original pipe
2nd cycle	Reinforcement yield loads for right and left sides of original pipe
3rd cycle	Peak loads for right and left sides of original pipe
4th cycle	Peak loads for right and left sides of original pipe
5th cycle*	Peak loads for right and left sides of original pipe

* Upon completion of the 5th cycle, horizontal displacement is returned to zero for load-carrying capacity testing.

2) Load-carrying capacity test

As shown in Photo 3.4, a load-carrying capacity test was conducted by applying vertical loads to specimens that had undergone severe shear deformation due to a strong earthquake motion simulated by the horizontal reversed loading test. In the capacity test, the maximum loads withstood by the original, new and SPR renovated pipes were measured to evaluate the degree of strength degradation.

Photo 3.4: Load-carrying capacity test

(5) Results of reversed loading tests

Figure 3.17 shows the results of the reversed loading tests. The main observations are summarised below:

- During unloading carried out after completion of the third cycle in which the peak load of the original pipe was reached, buckling of the steel reinforcement in both right and left walls was confirmed. Therefore, the third cycle became the last cycle of loading.
- The peak load of the original pipe (approximately 50 kN) was applied to the new and renovated pipe specimens in the third to fifth cycles, but the reinforcement of the two pipes did not yield. The maximum strain of rebar was 1800 μ in the new pipe and 188 μ in the SRP renovated pipe.
- From the test results, it was concluded that failure of the interlocking connection of the profile strips did not happen, and the water-tightness and discharge capacity of the pipe could be retained even when subjected to a major seismic force that could cause the original pipe to reach an ultimate failure state.

- As shown in Fig. 3.17, when subjected to the peak load of the original pipe, the maximum deformation of the SPR renovated pipe was as small as that caused by the cracking load for the original pipe, and was only about one-sixth of the deformation of the new pipe. There was little residual displacement, too. Thus, it has been confirmed that sewer renovation is effective for reducing major structural deformations of sewers due to a strong earthquake.

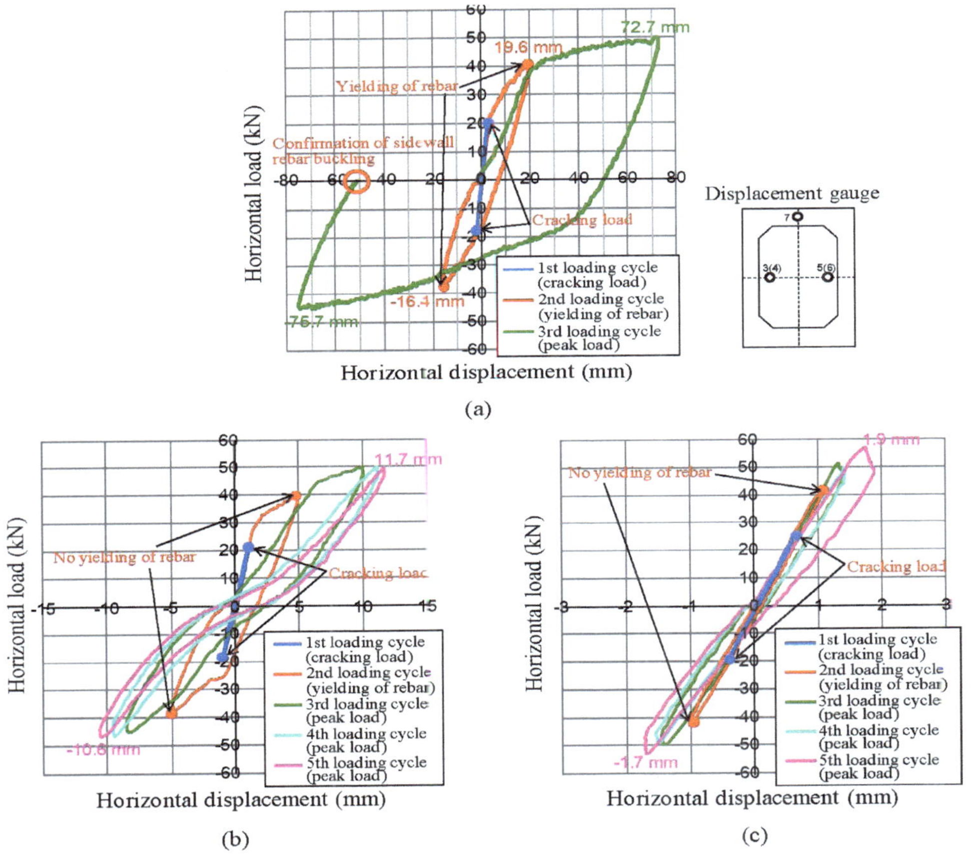

Figure 3.17: Load–displacement relationships (displacement gauge No. 7): (a) original pipe; (b) new pipe; (c) SPR renovated pipe

(6) Results of load-carrying capacity test

Table 3.15 shows the results of the load-carrying capacity tests. The load-carrying capacity of the original pipe that had undergone the reversed loading test was 183.8 kN, while that of a new pipe was 271.1 kN, higher than the former by a factor of 1.47. The load-carrying capacity of the pipe renovated by the SPR method was 298.5 kN, higher than that of the original pipe by a factor of 1.62. Because the maximum load of the renovated pipe is even higher than that of a new pipe, it can be stated that a well-renovated pipe is fully comparable to a new pipe in terms of strength when subjected to Level 2 strong earthquakes.

Table 3.15: Results of load-carrying capacity test

Test case	Type of pipe	Maximum load (kN)	Maximum load ratio	
			Ratio to original pipe	Ratio to new pipe
1	Original pipe	183.8		0.68
2	New pipe	271.1	1.47	
3	Renovated pipe	298.5	1.62	1.10

3.2 Structural Element Tests

3.2.1 Characteristics of SPR liner materials

The liner materials used in the SPR method consist of SPR fill mortar and the SPR profile. Both of them have been developed into a series of products with clear specifications on strength, pipe diameter, curvature of pipeline, etc., for renovation design of ageing sewers.

The SPR fill mortar, which is a fill material developed specifically for sewer renovation, is a resin-based polymer mortar composed of ordinary Portland cement, lightweight aggregate and admixtures such as acrylic emulsion (Photo 3.5, Table 3.16). The SPR profile, which is composed primarily of rigid polyvinyl chloride resin widely used as sewer pipe material, is designed to have appropriate flexibility, strength, chemical resistance, wear resistance and durability so as to meet the requirements for sewer renovation (Photo 3.6, Table 3.17).

(a) Conventional mortar (b) SPR mortar

Photo 3.5: Comparison of material characteristics by injecting conventional mortar and SPR mortar underwater

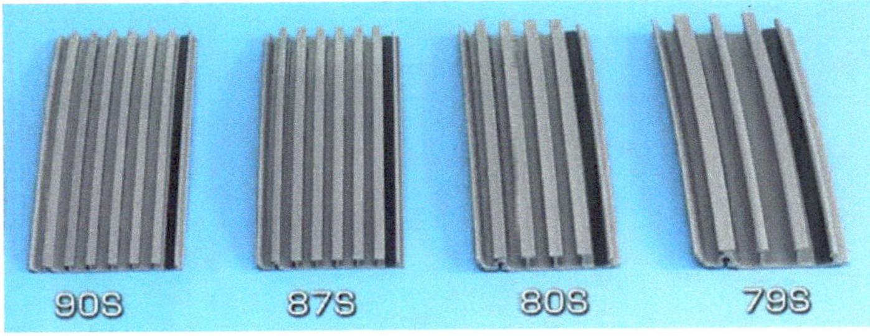

(a) Unreinforced profile for straight lining

(b) Steel-reinforced profile for large-diameter or non-circular pipe lining

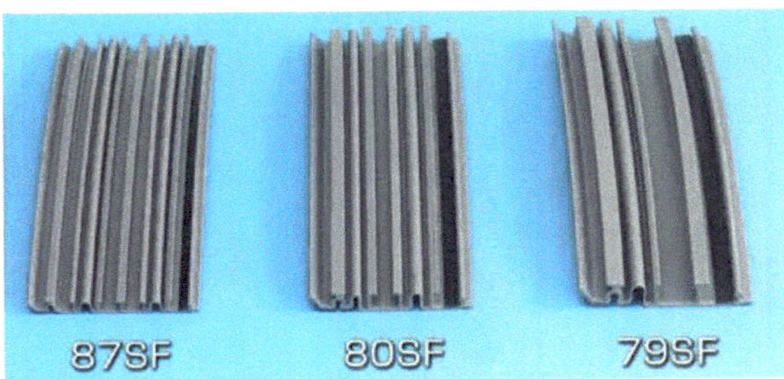

(c) Unreinforced profile for curved lining

(d) Steel-reinforced profile for curved lining

Photo 3.6: Various types of PVC profile

Table 3.16: Material properties of SPR fill mortar

Item	Mortar type		
	Mortar #2	Mortar #3	Mortar #4
Unit weight (t/m^3)	1.30 or higher	2.10 or higher	2.00 or higher
Compressive strength (N/mm^2)	$\sigma_7 = 12.0$ $\sigma_{28} = 21.0$	$\sigma_7 = 21.0$ $\sigma_{28} = 35.0$	$\sigma_7 = 30.0$ $\sigma_{28} = 55.0$
Flow value (mm)	290–370	250–380	280–370
Aggregate	Inorganic lightweight aggregate	Silica sand	Silica sand

Table 3.17: Material data of PVC profile

	Material data							
	Rigid polyvinyl chloride					Profile-reinforcing steel		
Serial number	Width (mm)	Thickness (mm)	Cross-sectional area (mm^2)	Tensile strength (N/mm^2)	Modulus of elasticity (N/mm^2)	Cross-sectional area (mm^2)	Yield strength (N/mm^2)	Modulus of elasticity (kN/mm^2)
#90S	90.0	9.0	291.28	37.2	2,350			
#90SW	90.0	9.0	291.28	37.2	2,350	38.2	205	165
#95S	95.0	8.0	263.56	37.2	2,350			
#87S	87.0	11.9	367.29	37.2	2,350			
#87SW	87.0	11.9	367.29	37.2	2,350	41	205	165
#80S	80.0	16.3	509.91	37.2	2,350			
#80SW	80.0	16.3	509.91	37.2	2,350	50	205	165
#79S	79.0	21.5	671.51	37.2	2,350			
#79SW	79.0	21.5	671.51	37.2	2,350	72.7	205	165
#792S	79.2	31.7	1033.66	37.2	2,350			
#792SU	79.2	31.7	1033.66	37.2	2,350	124.6	205	165

(1) The SPR fill mortar

The SPR fill mortar has the following characteristics:

1. Excellent post-hardening durability and stable strength
2. When injected, the mortar expands by 1.5 to 2.5% by volume and adheres to the inner surface of an existing sewer to be structurally integrated with it.

3. Adhesion to the original pipe is strong, and post-hardening water-tightness is excellent.
4. When injected underwater, the mortar pushes off the water and completely fills the space between the original pipe and the liner.

(2) The SPR Profile
The SPR profile is a strip of ribbed lining material that has a double-interlocking, compression-fit joint along both sides so that a liner pipe can be formed by spiral winding and interlocking the joints. The spiral-wound liner pipe achieves water-tightness by means of the interlocking between the main lock and sub lock joints and sealant connection, prevents slipping of interlocked joints and helps maintain pipe diameter.

3.2.2 Contents and purposes of structural element tests

(1) Identification of SPR mortar properties
In general, with the progress of cement hydration, concrete undergoes macroscopic volume reduction during and after the setting of the cementitious material. If the amount of autogenous shrinkage is large, cracking may occur and cause problems in not only strength but also durability.

By being injected into the space between the original pipe and the profile, the SPR mortar integrates with the existing pipe to form an integrated structure. Volume reduction due to autogenous shrinkage may give rise to voids and thus prevent structural integration. Therefore, an autogenous shrinkage test was conducted to make sure there would be no SPR mortar shrinkage having adverse effects on the performance of the composite pipe.

(2) Determination of SPR mortar strength
The SPR renovated pipe resists external forces as an integrated structure formed by integrating the original pipe and the liner. Although concrete surfaces are not roughened to increase adhesion with the original pipe, emulsion admixtures are used with mortar to increase adhesion. During mortar injection, measures to facilitate structural integration such as pressurisation are also taken. Thus, as mentioned in Section 3.1.1, the load-carrying mechanism of the SPR renovated pipe differs from that of a double-layered pipe. To check the strength of the interface between the original pipe and the liner, three tests were conducted:

1) Direct tension test
The direct tension test examines the tensile strength of the interface between the original pipe and the liner, namely, the bond strength of the SPR mortar.

2) Compressive shear test (adhesion test)
The compressive shear test examines the shear resistance of the interface between the original pipe and the liner.

3) Double shear test
The double shear test evaluates the shear resistance of the interfaces between concrete and mortar and between profile and mortar, which is an important shear transfer-related material property for an integral structure.
4) Verification of seismic performance of profile
Profile strip joints are required to withstand seismic ground motion without disconnecting and stay watertight even in the event of a strong earthquake. For the purpose of strength testing, seismic performance under the most adverse conditions, namely, a permanent strain of 1.5% in liquefied soil near a seawall under Level 2 earthquake loading, was verified experimentally.

(3) Test for evaluating the effectiveness of additional steel reinforcement in sewer renovation
In Japan, there are growing calls for government-subsidised projects for enhancing the earthquake resistance of existing sewers and for extending their service lives. Accordingly, it may be necessary to strengthen seismically vulnerable sewer pipes in sewer renovation, such as when the rebars in existing pipes are severely corroded or the sewer structure is made of plain concrete. In view of such possibilities, the effectiveness of embedding additional steel reinforcement in the renovation layer was evaluated experimentally.

3.2.3 Autogenous shrinkage test

(1) Test method
The test method is as follows:
1. Teflon sheet and polyester film are placed over the inner surface of the formwork to allow free deformation of mortar and to prevent the mortar and the formwork from coming into contact with each other.
2. An embedment-type strain gauge (KMC series strain gauge manufactured by Kyowa Electric Instruments) and a T-shaped thermocouple are installed in the middle region of the formwork (Fig. 3.18).
3. Upon completion of mortar placement, polyester film is placed over the surface of the mortar, and the formwork is completely covered with wrapping material and wet burlap to prevent moisture loss.
4. After two days of curing, the specimen is kept in a vinyl bag, without removing the formwork, for sealed curing.
5. Strain and temperature measurement is started immediately after mortar placement and continued until the 28th day (Photo 3.7). Note that an autogenous shrinkage test is performed in a thermostatic chamber under the condition of 20°C and 60%RH or higher.

6. By assuming the strain of a specimen as the sum of the thermal expansion strain and the autogenous shrinkage strain, the autogenous shrinkage strain can be obtained by Eq. (3.1), in which the self temperature compensation of the strain gauge is also taken into account:

$$\varepsilon_a = \varepsilon - (\alpha_x - \kappa)\Delta T \tag{3.1}$$

where ε_a = autogenous shrinkage strain; ε = measured strain; α_x = coefficient of linear expansion of specimen (strain/°C); k = self temperature compensation factor of strain gauge (strain/°C); ΔT = temperature change.

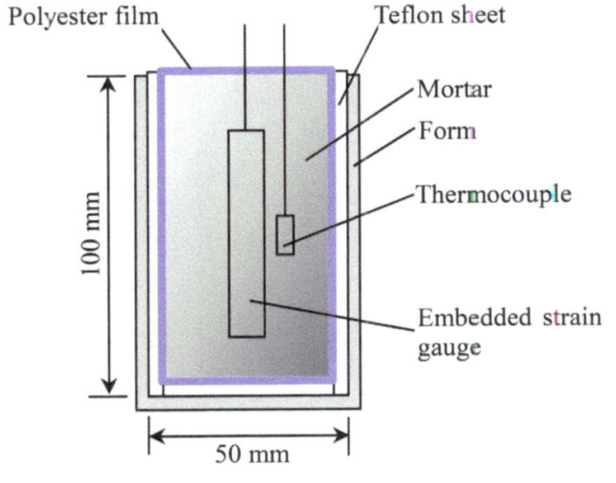

Figure 3.18: Autogenous shrinkage test

Photo 3.7: Mortar frameworks in autogenous shrinkage test

(2) Test results

Figure 3.19 shows the results of the autogenous shrinkage test for each of the three types of SPR mortar. For the calculation, a value of 15×10^{-6}/°C was assumed for the coefficient of linear expansion of mortar. The strain behaviours in the post-peak region after reaching the maximum expansion strain did not indicate any significant tendency of shrinkage that would likely have adverse effects on the original pipe, as the measured strains all remained in the region of thermal expansion.

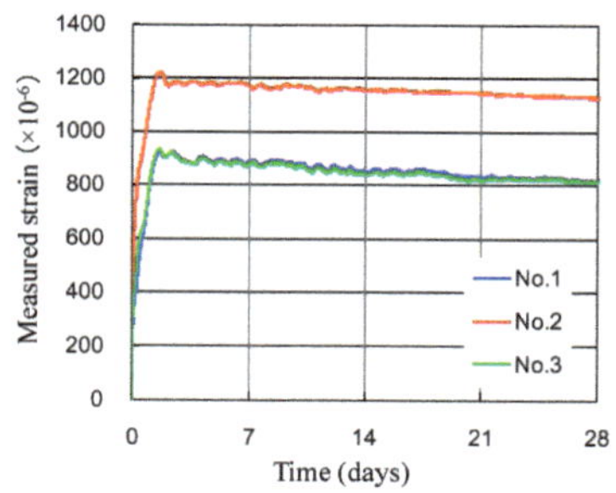

Specimen number		1	2	3	Average
Maximum expansion strain (μ) $(\alpha_x-\kappa)\,\Delta T$		924	1218	932	1025
Time to maximum strain (h:m) ΔT		39:03	42:18	38:48	—
Measured strain (μ) ε	7 days	889	1170	879	979
	14 days	848	1153	839	947
	21 days	828	1145	817	930
	28 days	820	1127	812	920
Autogenous shrinkage strain (μ) ε_a	7 days	-35	-48	-53	-45
	14 days	-76	-65	-93	-78
	21 days	-96	-73	-115	-95
	28 days	-104	-91	-120	-105

(a)

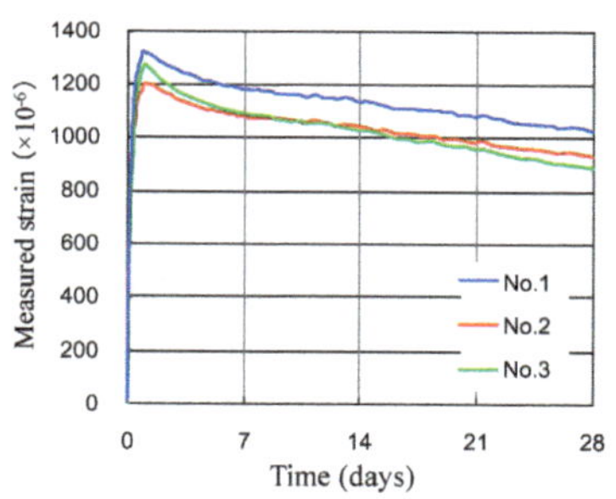

Specimen number		1	2	3	Average
Maximum expansion strain (μ) $(\alpha_x-\kappa)\,\Delta T$		1325	1203	1272	1267
Time to maximum strain (h:m) ΔT		21:30	13:30	16:30	—
Measured strain (μ) ε	7 days	1178	1077	1087	1114
	14 days	1137	1038	1027	1067
	21 days	1085	984	957	1009
	28 days	1025	928	890	948
Autogenous shrinkage strain (μ) ε_a	7 days	-147	-126	-185	-153
	14 days	-188	-165	-245	-199
	21 days	-240	-219	-315	-258
	28 days	-300	-275	-382	-319

(b)

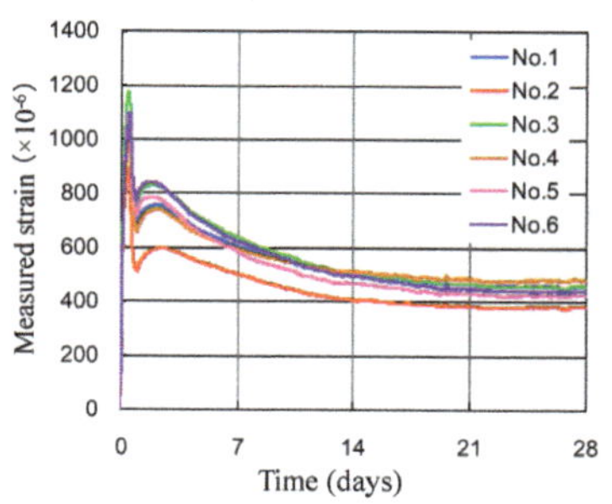

Specimen number		1	2	3	4	5	6	Average
Maximum expansion strain (μ) $(\alpha_x-\kappa)\,\Delta T$		1045	889	1173	1007	1089	1097	1050
Time to maximum strain (h:m) ΔT		11:59	10:26	10:29	10:23	10:31	12:15	—
Measured strain (μ) ε	7 days	604	500	637	594	584	621	590
	14 days	495	404	508	510	468	494	480
	21 days	467	383	465	485	431	450	447
	28 days	461	383	458	481	424	439	441
Autogenous shrinkage strain (μ) ε_a	7 days	-441	-389	-536	-413	-505	-476	-460
	14 days	-550	-485	-665	-497	-621	-603	-570
	21 days	-578	-506	-708	-522	-658	-647	-603
	28 days	-584	-506	-715	-526	-665	-658	-609

(c)

Figure 3.19: Results of autogenous shrinkage test: (a) mortar #2; (b) mortar #3; (c) mortar #4

3.2.4 Direct tension test

(1) Test method

In the direct tension test, the tensile strength of the base concrete–mortar interface is measured. The test method is as follows:

1. The base concrete mix is poured into a mould form 200 mm high and 100 mm in diameter. On the following day the form is removed, and the concrete is put into a vinyl bag and subjected to sealed curing until a material age of 28 days.
2. At the age of around 21 days, the specimen is cut at mid-height into two (upper and lower) equal halves.
3. At the age of 28 days, the lower half of the specimen is put into the mould form again. Then, the cut surface is cleaned with acetone and polished with sandpaper (150-grid polishing sandpaper), and mortar is poured. Note that the specimens used for the direct tension test were prepared in a thermostatic chamber under the condition of 20°C and 60%RH.
4. Upon completion of mortar placement, the upper surface is covered with a weighted glass plate and is cured at 20°C.
5. At the age of 28 days after pouring mortar (56 days after pouring concrete), jigs are attached to both ends of the specimen with epoxy resin-based adhesive as shown in Fig. 3.20, and the maximum load is determined through a tension test using an Instron Model 1125 testing instrument.
6. Based on the maximum load, the direct tensile strength of the interface between base concrete and mortar is obtained as

$$f_t = \frac{P_{\max}}{A} \tag{3.2}$$

where f_t = direct tensile strength; P_{max} = maximum load; A = fractured cross sectional area of the specimen.

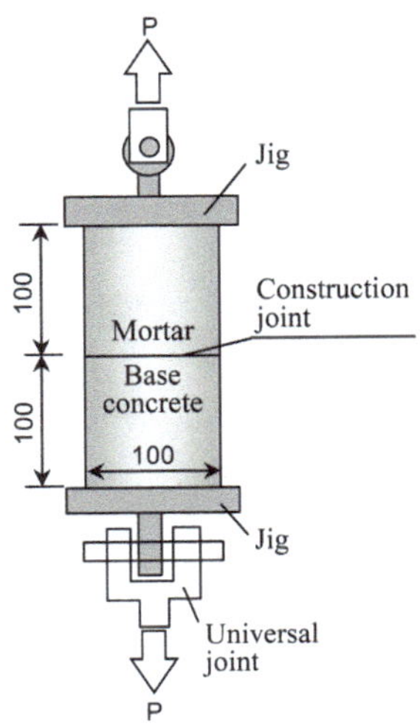

Figure 3.20: Tensile test to determine the direct tensile strength of the interface between base concrete and mortar

(2) Test results

Table 3.18 shows the results of the direct tension test.

Table 3.18: Results of direct tension test

	Mortar type						
	Specimen number	**Base concrete/ Mortar #2**		**Base concrete/ Mortar #3**		**Base concrete/ Mortar #4**	
Material property			**Average**		**Average**		**Average**
Direct tensile strength (N/mm^2)	1	0.74	0.72	1.77	1.70	1.17	1.05
	2	0.78		1.77		0.97	
	3	0.65		1.58		1.00	

3.2.5 Compressive shear test (adhesion test)

(1) Test method

A uniaxial compression test is conducted to investigate the shear resistance of the bonded joint between the base concrete and the injected mortar. The test method is as follows:

1. The base concrete mix is poured into a mould form 200 mm high and 100 mm in diameter. On the following day the form is removed, and the concrete is put into a vinyl bag and subjected to sealed curing until a material age of 28 days.
2. At the age of around 21 days, the specimen is cut into halves at mid-height at an angle of 45 degrees (Fig. 3.21).
3. At the age of 28 days, the lower half of the specimen is put into the mould form again. Then, the cut surface is cleaned with acetone and polished with sandpaper (150-grid polishing sandpaper), and mortar is poured. Note that the specimens used for the compressive shear test were prepared in a thermostatic chamber under the condition of 20°C and 60%RH.
4. Upon completion of mortar placement, the upper surface is covered with a weighted glass plate and is cured at 20°C.
5. At the age of 28 days after pouring mortar (56 days after pouring concrete), two strain gauges each are attached to the mid-height level of the mortar layer and the base concrete layer (Fig. 3.22).
6. The maximum compressive load is determined through a uniaxial compression test (Photo 3.8).
7. The shear strength of the construction joint (bond strength) is calculated from the maximum compressive load as

$$f_s = \frac{\dfrac{P_{max} \sin\theta}{A}}{\cos\theta} \tag{3.3}$$

where f_s = shear strength; P_{max} = maximum compressive load; A = cross sectional area of the specimen; θ = angle of base concrete and mortar interface (45 degrees).

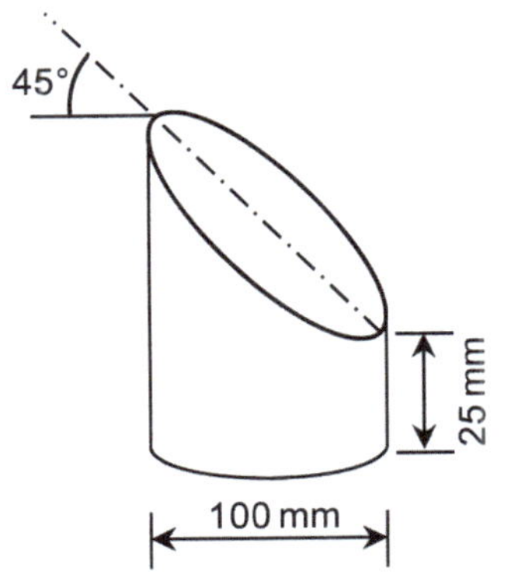

Figure 3.21: Shape of specimen cut at 45° for compressive shear test

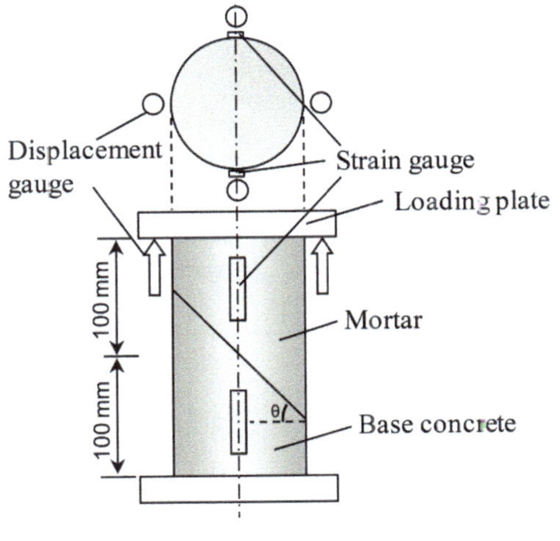

Figure 3.22: Setup for compressive shear test

8. The vertical displacement of the loading apparatus is measured during the test, and the shear displacement at the construction joint is calculated as

$$\Delta_s = (\Delta_1 - \Delta_2)\sin\theta \tag{3.4}$$

Δ_s = shear displacement of construction joint; Δ_1 = vertical displacement of loading apparatus; Δ_2 = deformation of specimen; θ = angle of base concrete and mortar interface (45 degrees).

Photo 3.8: Compressive shear test

(2) Test results

Table 3.19 shows the results of the compressive shear test.

Table 3.19: Results of compressive shear test

Material property	Specimen type: Specimen number	Base concrete/ Mortar #2	Average	Base concrete/ Mortar #3	Average	Base concrete/ Mortar #4	Average
Shear strength (N/mm^2)	1	13.1	11.9	19.6	20.0	32.0	30.9
	2	9.9		19.8		29.8	
	3	12.7		20.5		30.5	
	4					31.2	
Shear displacement of construction joint (mm)	1	0.12	0.09				0.158
	2	0.06				0.131	
	3	0.09				0.204	
	4					0.140	

3.2.6 Double shear test

(1) Test method

The shear strength test for the interface of concrete and mortar and the interface of mortar and profile was conducted in accordance with the *Test Method for Shear Strength of Steel Fibre Reinforced Concrete* (JSCE, 2002). Figure 3.23 illustrates the test setup, and Photo 3.9 shows the test jigs employed. By inducing shear failure along the dotted lines shown in Fig. 3.23, the shear strength of the concrete–mortar interface and the mortar–profile interface is calculated.

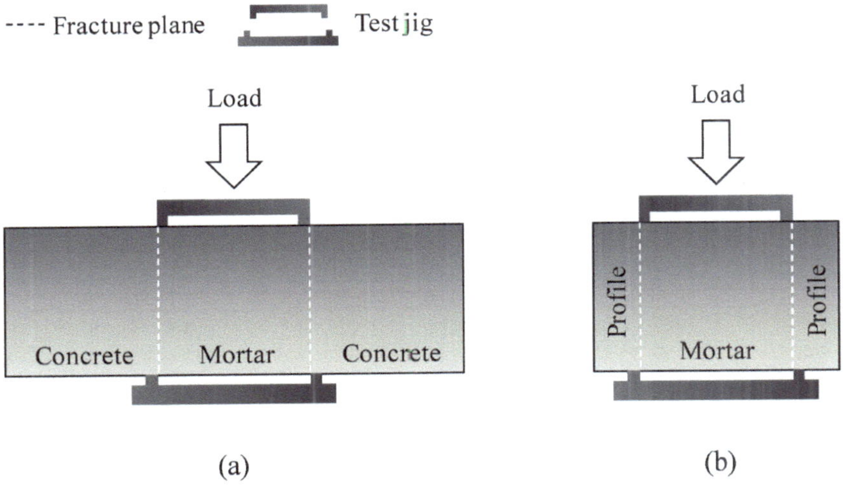

Figure 3.23: Test setup for double shear test: (a) shear strength test for the interface of mortar and concrete; (b) shear strength test for the interface of mortar and profile

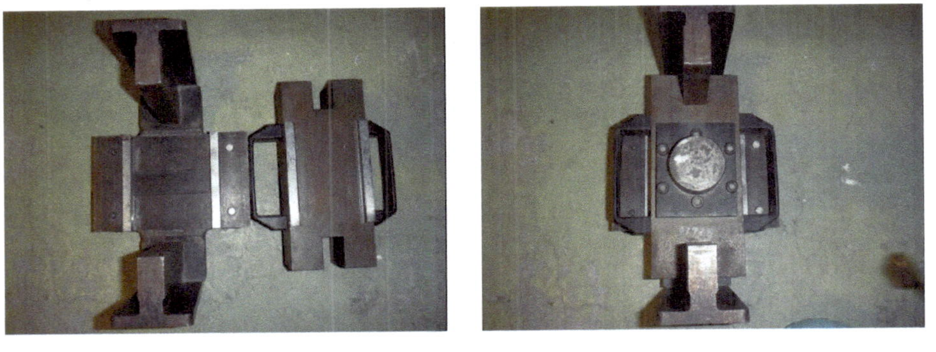

Photo 3.9: Test jigs for double shear test (edge width of loading plate ≒ 10 mm)

(2) Test cases and test specimens

Three cases were planned for the strength test of concrete and mortar (mortar #2 to #4), and nine cases for the strength test of mortar and profile (mortar #2 to #4 used with 80SW, 79SW and 792SU profiles). Table 3.20 presents details of the test specimens for the 12 test cases, and Fig. 3.24 shows the designs and geometric dimensions of the two types of test specimen.

Table 3.20: Test cases for double shear test

Case No.	SPR mortar	Profile	Profile thickness* (mm)	Mortar thickness (mm)	Concrete thickness* (mm)	Length of specimen (mm)	Quantity (pieces)
1	Mortar #2		0	100	100	300	3
2	Mortar #3		0	100	100	300	3
3	Mortar #4		0	100	100	300	3
4	Mortar #2	80SW	16.3	100		132.6	3
5		79SW	21.5	100		143	3
6		792SU	31.7	100		163.4	3
7	Mortar #3	80SW	16.3	100		132.6	3
8		79SW	21.5	100		143	3
9		792SU	31.7	100		163.4	3
10	Mortar #4	80SW	16.3	100		132.6	3
11		79SW	21.5	100		143	3
12		792SU	31.7	100		163.4	3

* 2 pieces (right and left) for each specimen

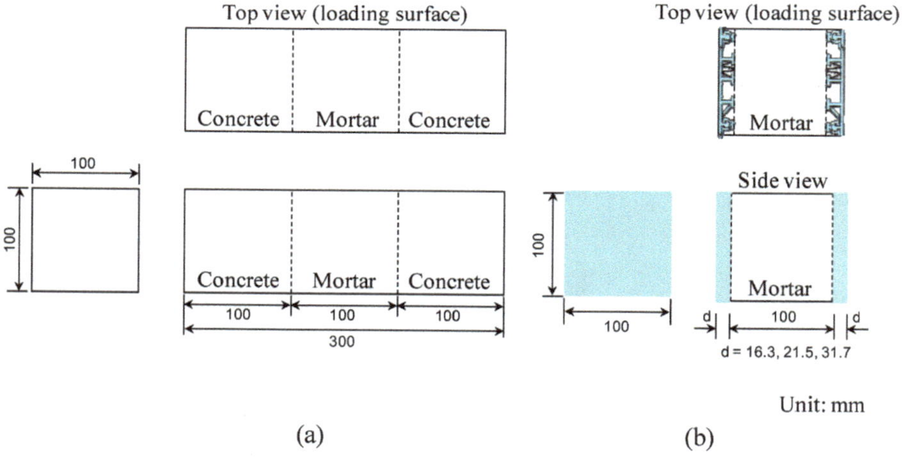

Figure 3.24: Test specimens for double shear test: (a) mortar and concrete; (b) mortar and profile

(3) Testing scenes

Photo 3.10 shows the test specimens under loading.

(a)

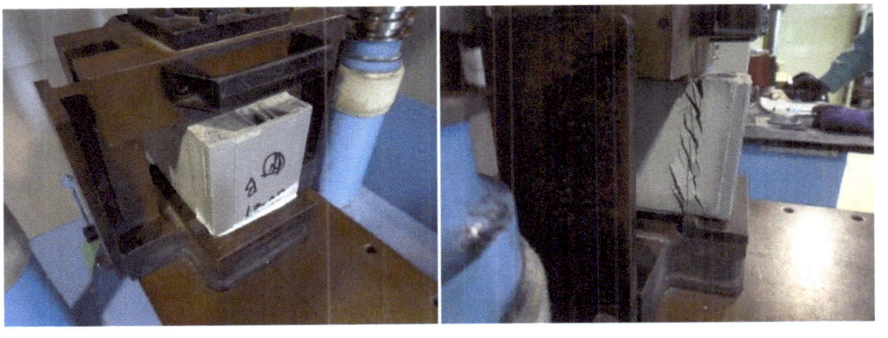

(b)

Photo 3.10: Double shear test in progress: (a) concrete–mortar specimen under loading; (b) mortar–profile specimen under loading

(4) Results of double shear test

Tables 3.21 and 3.22 show, respectively, the shear strengths obtained between mortar and concrete, and between mortar and profile. For comparison, the tensile strength of mortar and the bonding strength of mortar and concrete are presented in Table 3.23 for the three types of mortar.

As shown in Table 3.21, the shear strengths between mortar and concrete for mortar #2 to #4 are 2.35 N/mm^2, 3.98 N/mm^2 and 3.97 N/mm^2, respectively. Compared with the tensile strength of mortar and the bonding strength between mortar and concrete for mortar #2 in Table 3.23, the respective ratios of shear strength to tensile strength and to bonding strength are 1.3 and 3.3. Similarly, for mortar #3 and #4, these strength ratios are 1.4 and 2.3, and 0.7 and 3.8, respectively. Based on these values, it is clear that the shear strength of mortar and concrete is comparable to the tensile strength of mortar, and far larger than the bonding strength of mortar and concrete.

Table 3.21: Shear strength of mortar and concrete

	Specimen type						
	Specimen number	Base concrete/ Mortar #2		Base concrete/ Mortar #3		Base concrete/ Mortar #4	
Material property			Average		Average		Average
Shear strength (N/mm²)	1	2.62	2.35	3.90	3.98	3.67	3.97
	2	2.14		4.41		3.89	
	3	2.29		3.64		4.35	

Table 3.22: Shear strength of mortar and profile

	Specimen type							
	Profile	Specimen number	Mortar #2/ Profile		Mortar #3/ Profile		Mortar #4/ Profile	
Material property				Average		Average		Average
Shear strength (N/mm²)	#80SW	1	2.19	2.00	3.63	3.34	4.59	4.96
		2	1.71		2.93		4.60	
		3	2.09		3.47		5.70	
	#79SW	1	2.17	1.95	2.78	3.32	5.14	4.19
		2	1.87		2.75		3.77	
		3	1.80		4.42		3.65	
	#792SU	1	1.46	1.36	2.28	2.63	1.84	2.92
		2	1.71		3.25		3.77	
		3	0.91		2.36		3.14	

Table 3.23: Tensile strength of mortar and bonding strength of mortar and concrete

	Mortar type		
Material property	Mortar #2	Mortar #3	Mortar #4
Tensile strength of mortar (N/mm²)	1.83	2.92	5.46
Bonding strength of mortar and concrete (N/mm²)	0.72	1.70	1.05

Concerning the shear strength of mortar and profile, Table 3.22 shows that for the three types of profile, 80SW, 79SW and 792SU, the corresponding shear strength decreases in that order, which is due to the geometric details of these profiles. As shown, the shear strengths of mortar #3 and #4 are far larger than those of mortar #2, with mortar #4 being the largest among the three types of mortar.

3.2.7 Profile pullout test

(1) Test cases
Table 3.24 shows the test cases involving four types of unreinforced profile: #87S, #80S, #79S, and #90S; refer to Photo 3.6(a). Note that except for the #79S profile, the two inner diameters of the original pipe for each profile represent the minimum and maximum diameter applicable for renovation by using the designated profile. For the #79S profile, the applicable minimum diameter of 1000 mm was used.

Following the normal renovation process, each test specimen was produced by constructing a spiral-wound liner inside an original pipe and then filling the gap between the existing pipe and the profile liner with grout.

Table 3.24: Test cases for pullout test

Test case	Inner diameter of original pipe (mm)	Inner diameter of renovated pipe (mm)	Profile type	Pullout displacement corresponding to permanent ground strain of 1.5% (mm)	Number of specimens
1	450	410	#87S	36.5	1
2	600	550			1
3	700	640	#80S		1
4	900	820			1
5	1000	910	#79S		1
6	250	210	#90S	30.0	1
7	400	360		36.5	1

(2) Test method
In simulating a permanent ground strain of 1.5% due to a Level 2 earthquake, in the pullout test an axial displacement was enforced on the renovated pipe to observe the deformation behaviour of the profile liner, as shown in Fig. 3.25. As seen, two hydraulic jacks were employed in order to produce approximately equal amounts of displacement to the right and left sides of the pipe joint. The enforced displacement was measured with displacement gauges (4 points/cross section × 13 cross sections), as shown in Photo 3.11.

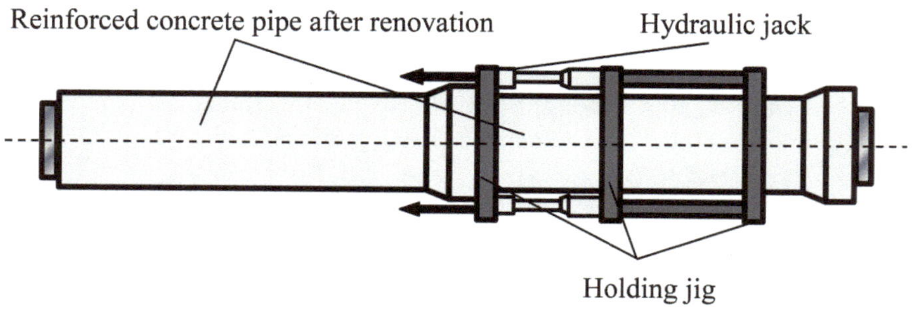

Figure 3.25: A test specimen and the test setup for pullout test

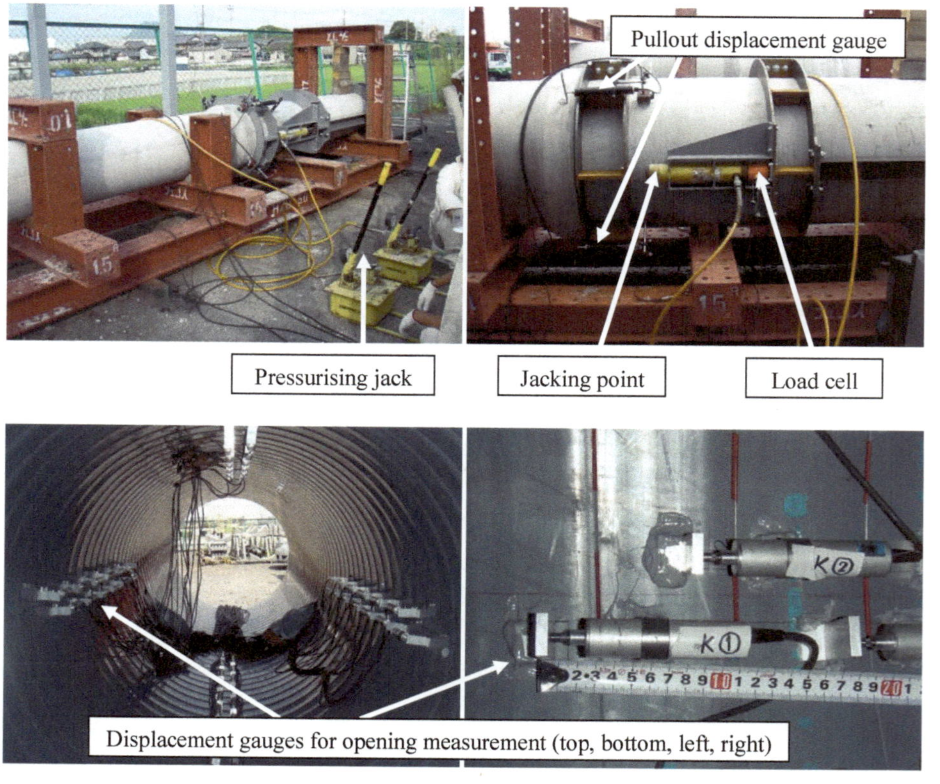

Photo 3.11: Pullout test in progress

(3) Test results

Table 3.25 shows the results of the pullout test. During the test, the following points were observed:

- In all the cases tested, the average of the pullout displacements measured from the four locations reached the permanent ground strain of 1.5% (equivalent to 36.5 mm for the diameters larger than 400 mm; and equivalent to 30.0 mm for the diameter of 250 mm). This fact shows that the renovated pipes possess the required displacement-following capability during strong earthquakes.
- In all the cases tested, the sliding displacement of the joints illustrated in Fig. 3.26 was not observed.
- Joint openings illustrated in Fig. 3.26 were observed for the three cases of ϕ250 mm (#90S), ϕ400 mm (#90S) and ϕ900 mm (#80S) at a few locations over the entire profile width including the middle and side regions. However, complete joint disconnection was not observed in any of the cases tested; refer to Photo 3.12.

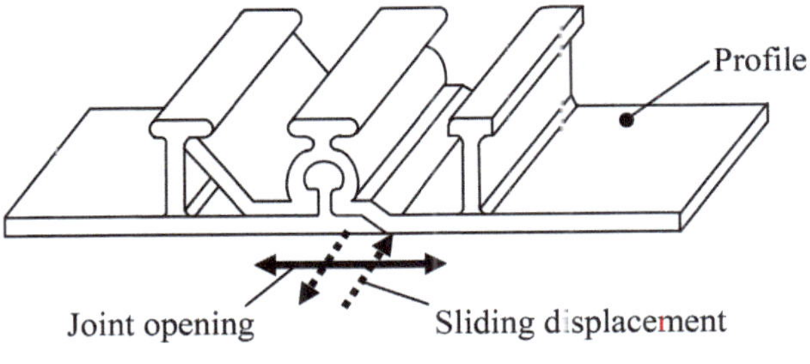

Figure 3.26: Deformation of compression-fit joint

Based on these results, it is concluded that the joint connections of SPR profiles basically satisfy the seismic resistance requirement for continuity during strong earthquakes.

Table 3.25: Results of pullout test

Test case	Inner diameter of original pipe (mm)	Profile type	Pullout displacement corresponding to permanent ground strain of 1.5% (mm)	Displacement at end of test (mm)	Joint continuity
1	450	#87S	36.5	38.02	Yes
2	600			37.50	Yes
3	700	#80S		37.02	Yes
4	900			38.46	Yes
5	1000	#79S		37.79	Yes
6	250	#90S	30.0	30.45	Yes
7	400		36.5	36.53	Yes

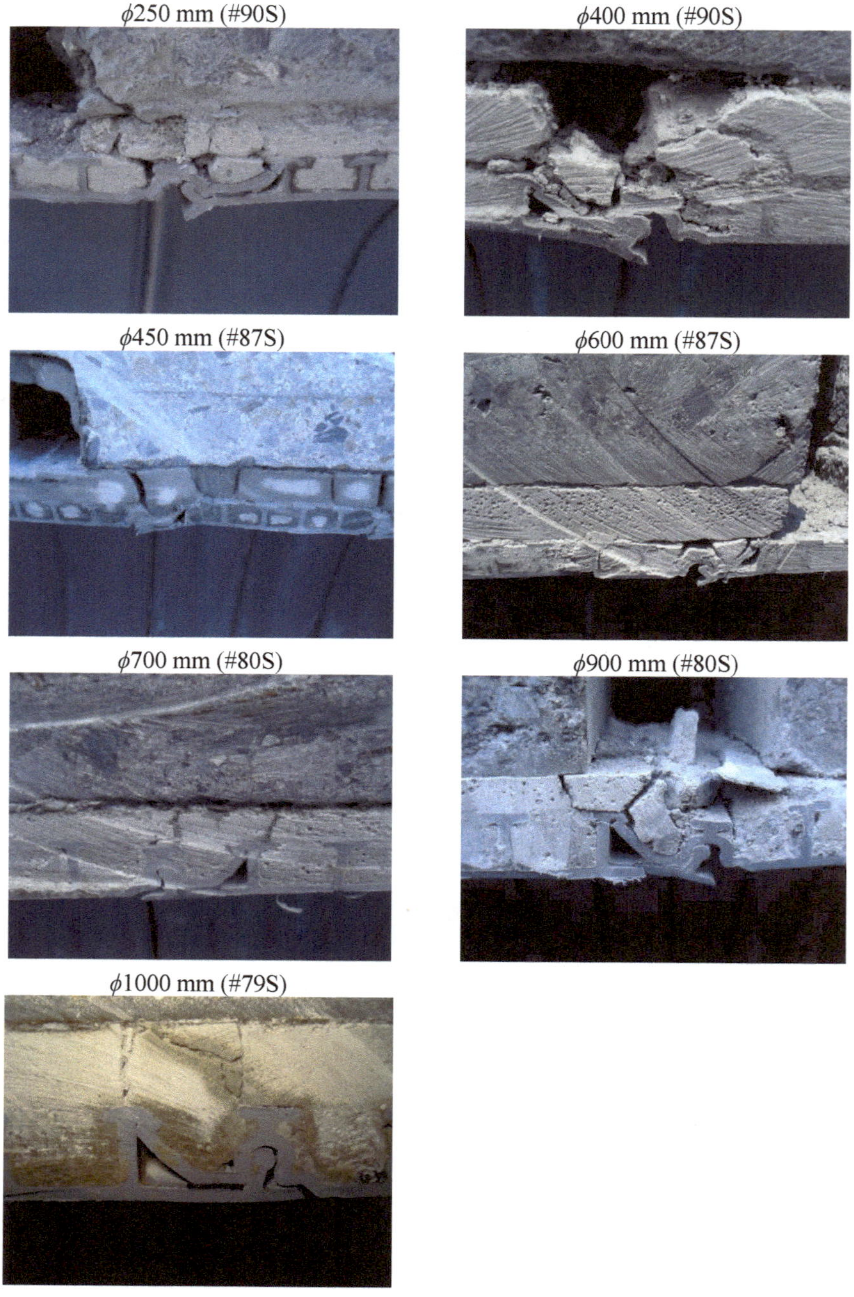

Photo 3.12: Joint conditions of all test cases observed after pullout test

3.2.8 Verification test on the effectiveness of additional rebar in SPR liner

(1) Test cases
In renovating ageing sewers, additional steel reinforcement is sometimes required for those sewers whose original rebars have been lost due to corrosion or which were originally made of plain concrete. To verify the effectiveness of additional rebars in the renovation layer, a series of flexural load-carrying capacity tests was conducted by using beam specimens designed to simulate an ageing pipe renovated by the SPR method.

Table 3.26 shows the test cases. For comparison, two types of beam specimen were prepared: one with additional rebars in the cement region of the renovation layer, and one without. To reflect the variations of thickness of the renovation layer in the top cover plate and the sidewalls, two types of thickness were assumed. A high-strength mortar, mortar #3, was used for the testing to comply with the general requirement of increasing the load-carrying capacity when additional steel reinforcement is needed in sewer renovation.

Table 3.26: Load-carrying capacity test of SPR renovated beam specimens

Test case	Specimen condition	Additional rebar	Mortar	Liner thickness (mm)	Number of specimens
1	SPR renovation	No	#3	55 (thin)	3
2				94 (thick)	3
3	SPR renovation + additional rebar	Yes		55 (thin)	3
4				94 (thick)	3

(2) Material properties of test specimens
Table 3.27 shows the material properties of the test specimens.

Table 3.27: Material properties of beam specimens for load-carrying capacity test

	Material type			
Material property	**Concrete**	**Mortar**	**Additional rebar**	**Reinforced profile (80SW)**
Compressive strength (N/mm^2)	36.4	46.7		
Tensile strength (N/mm^2)	3.50	2.48		
Poisson's ratio	0.20*	0.22*		
Unit weight (kN/m^3)	24.5*	21.0*		
Modulus of elasticity (kN/mm^2)	30.5	20.6	200	165
Yield strength (N/mm^2)			368	205

* Adopted design value

(3) Dimensions and structural details of test specimens

Figure 3.27 shows the dimensions and structural details of test specimens. Since a covered sewer made of plain concrete is a typical structure that requires additional reinforcement in the renovation layer, the original pipe portion of the beam specimen was made with plain concrete.

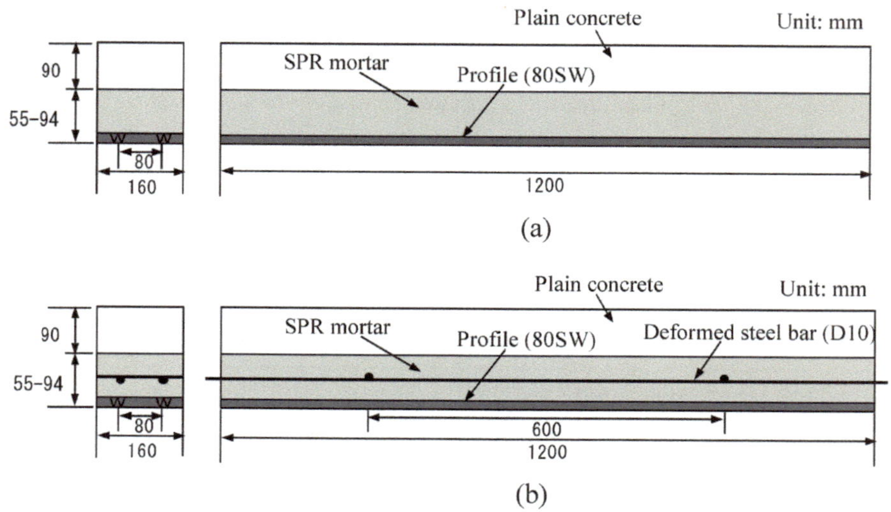

Figure 3.27: Specimens for testing the effect of additional steel reinforcement in the renovation layer: (a) SPR renovated specimen without additional rebar; (b) SPR renovated beam specimen with additional rebar

(4) Test method

A test specimen was subjected to a concentrated load at the midspan of the beam, and a fracture test was carried out to obtain the maximum load. Figure 3.28 illustrates the test condition, and Photo 3.13 shows the test setup.

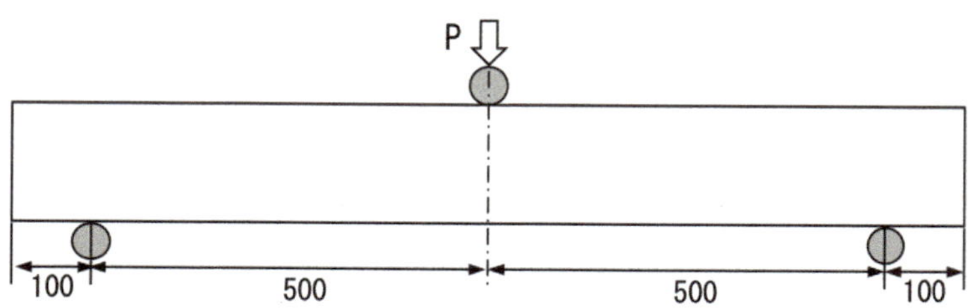

Figure 3.28: Test conditions of load-carrying capacity test

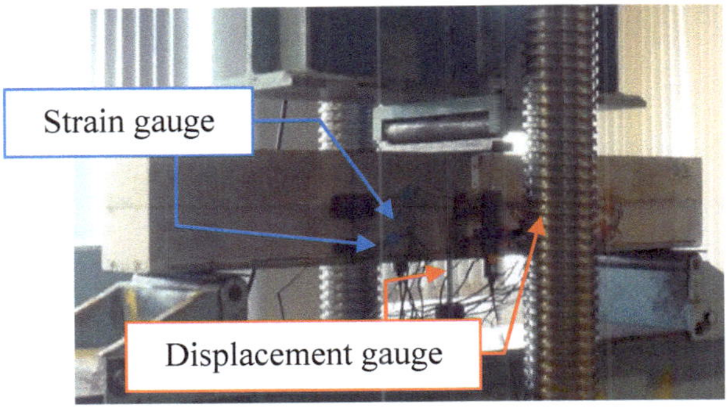

Photo 3.13: Setup of specimen for load-carrying capacity test

(5) Test results

Table 3.28 shows the test results of the load-carrying capacity test. By inserting additional rebars into the renovation layer, the load-carrying capacity of the SPR renovated beam specimens was increased to 1.8 times that of the specimens without additional rebars. This strengthening effect was consistent in both cases of a thin and a thick renovation layer. The load-carrying capacity in Cases 2 and 4 (equivalent to the normal thickness of a renovated top slab) was 30% larger than that in Cases 1 and 3 (equivalent to the normal thickness of a renovated sidewall). The effect of additional rebars was also evident in the fact that cracks were more distributed across the beam span. Yielding of the profile-reinforcing steel was also delayed by the added steel reinforcement in the renovation layer.

Table 3.28: Results of load-carrying capacity tests of SPR renovated beam specimens

Test case	Additional rebar	Liner thickness (mm)	Maximum load (kN)		Ratio of maximum loads	
			Measured value	Average	Additional rebar	Thickness of renovation layer
1	No	55 (thin)	30.2	27.9 (1)		
			26.0			
			27.4			
2		94 (thick)	35.8	35.8 (2)		$\frac{(2)}{(1)} = 1.3$
			40.7			
			30.9			
3	Yes	55 (thin)	50.5	50.0 (3)	$\frac{(3)}{(1)} = 1.8$	
			50.5			
			49.1			
4		94 (thick)	64.5	62.9 (4)	$\frac{(4)}{(2)} = 1.8$	$\frac{(4)}{(3)} = 1.3$
			63.1			
			61.0			

3.2.9 Review and certification

The certification programme of the Japan Institute of Wastewater Engineering and Technology certifies the performance of sewer pipes renovated by a renovation method. Table 3.29 summarises the items to be included in a certificate (JIWET, 2014).

Table 3.29: Review items and review methods employed by the Japan Institute of Wastewater Engineering and Technology

Review item	Review method
Integration of original pipe and liner	In an external pressure test in accordance with JISWAS A-1, external load is applied to the renovated pipe to verify that the strain behaviour at the interface between the original pipe and the liner is continuous with the strain behaviour of the original pipe. A fracture test is conducted on a renovated pipe specimen to verify that discontinuity of the interface between the original pipe and the liner does not occur at structural failure*.
Load-carrying capacity	For the purpose of evaluating load-carrying capacity, external pressure tests are conducted on renovated pipes to verify that they are as strong as or stronger than new pipes. The renovated pipes include severely fractured reinforced concrete pipes, intact precast box culvert pipes, thinned-wall precast box culvert pipes with and without internal steel reinforcement, and ceramic pipes.
Chemical resistance	A chemical resistance test is conducted in accordance with JISWAS K-1 to verify that chemical resistance requirements are met.
Wear resistance	A wear resistance test is conducted in accordance with JIS K 7204 to verify wear resistance by comparison with test results for rigid polyvinyl chloride pipes.
Watertightness	For the purpose of evaluating watertightness, tests are conducted under external and internal water pressure to verify that compression-fit joints of profile meet watertightness requirements.
Longitudinal earthquake resistance	Earthquake resistance is evaluated by inducing axial pullout displacement and flexural displacement at a reinforced concrete pipe joint of a renovated pipeline under the two assumed conditions specified below and checking watertightness. • Pullout due to a permanent ground strain of 1.5% • Bending angle corresponding to land subsidence (span length 30 m, subsidence 30 cm) due to liquefaction
Strength properties of fill material	Compressive strength is tested in accordance with JSCE G521, and Young's modulus is tested in accordance with JIS A1149.

*Regarding the continuity requirement, a no-tension interface model is employed by the SPR method in its structural analysis and renovation design of ageing sewers.

3.3 Basic Material Property Tests

To determine the basic material properties of the SPR fill mortar and the shear strength of the SPR renovation layer for design and renovation construction, the basic material tests listed in Table 3.30 are conducted.

Table 3.30: Basic material property tests of the SPR method

Test	Material	Applicable standard	Material property
Compressive strength test	Mortar	JIS A 1108 JIS A 1149	Compressive strength, modulus of elasticity, Poisson's ratio
Splitting tensile strength test	Mortar	JIS A 1113	Tensile strength
Fracture energy test	Mortar	JCI-S-001-2003	Fracture energy
Single shear test	SPR liner		Shear strength of liner

3.3.1 Test methods

The compressive strength test and the splitting tension test are conducted in accordance with relevant JIS standards. The test methods for the fracture energy of mortar and the shear strength of liner are explained below.

(1) Fracture energy test
In accordance with JCI-S-001-2003, a three-point bending test as shown in Figs. 3.29 and 3.30 was conducted on test specimens of mortar measuring 100×100×400 mm. The relation between the load and the crack mouth opening displacement (CMOD) was recorded and the fracture energy of mortar was calculated based on the load–CMOD curve (Fig. 3.31) as

$$G_F = \frac{0.75W_0 + W_1}{A_{lig}} \tag{3.5}$$

where W_0 = area under the load–CMOD curve; W_1 = work performed by the self weight of the test specimen and the loading instrument; and A_{lig} = fractured area.

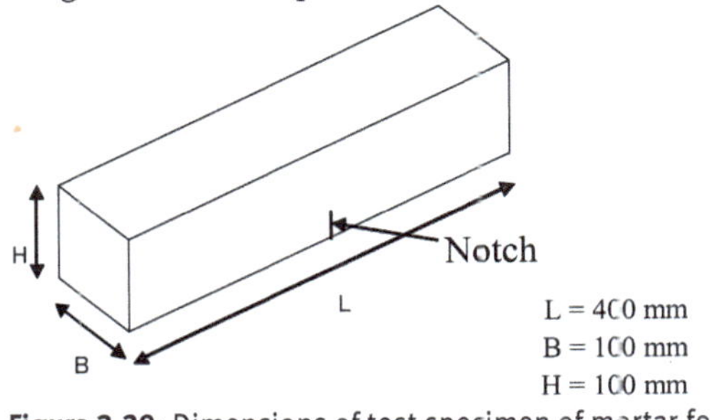

Figure 3.29: Dimensions of test specimen of mortar for fracture energy test

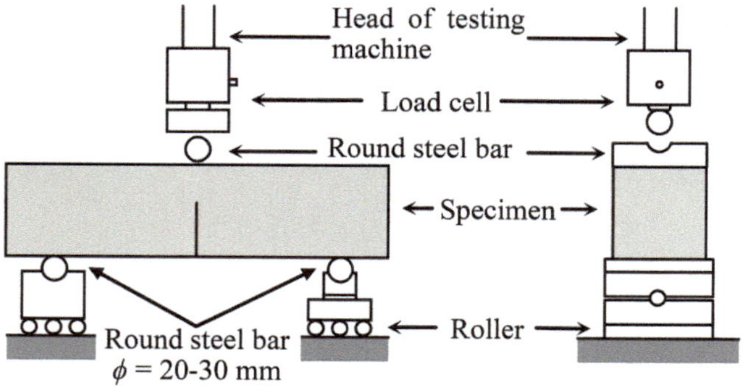

Figure 3.30: Test setup for fracture energy test

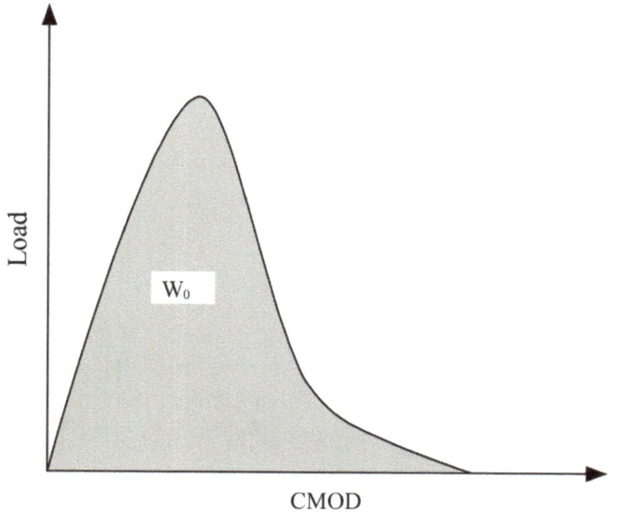

Figure 3.31: Load–CMOD curve

(2) Single shear test

To calculate the shear strength of the SPR renovation layer, composite test specimens of base concrete and renovation layer as shown in Fig. 3.32 were prepared, and single shear tests were conducted by using a large single shear testing machine designed for soil testing (load capacity 50 tons, Fig. 3.33). Table 3.31 lists the test cases, and Fig. 3.34 shows a schematic illustration of a shear failure of the specimen under the test condition. Based on the test results, the shear strength of the renovation layer was obtained by deducting the shear strength of the base concrete from that of the composite test specimen.

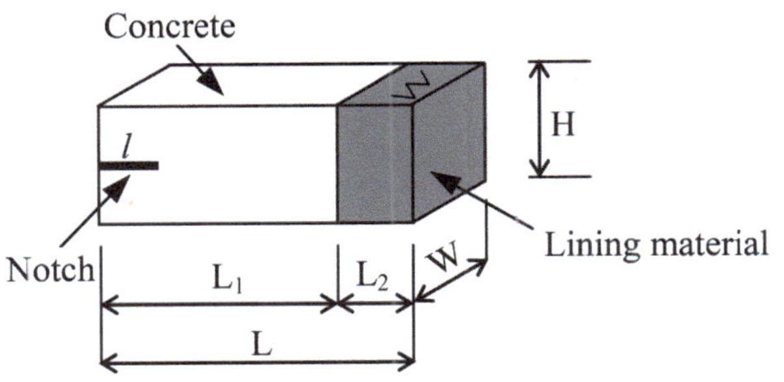

Figure 3.32: Specimen for shear test

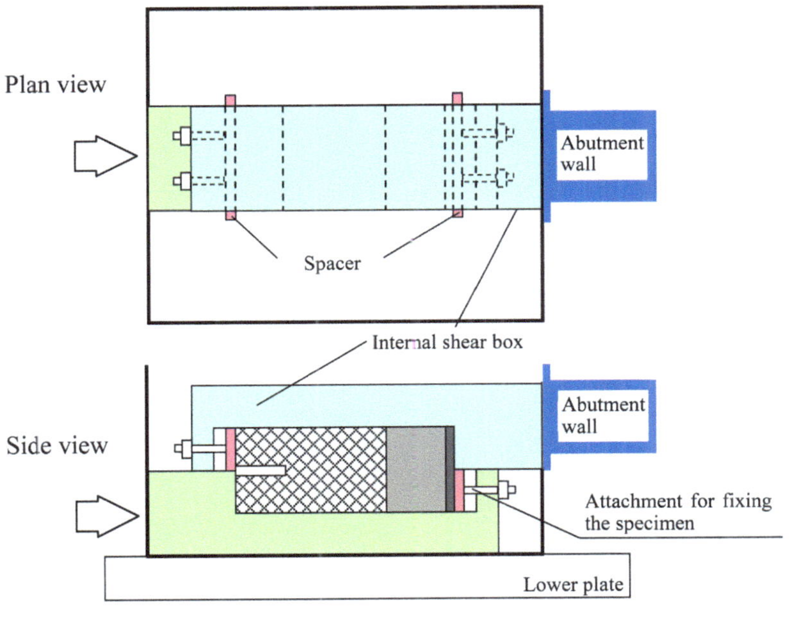

Figure 3.33: Configuration of shear testing machine and setup of specimen

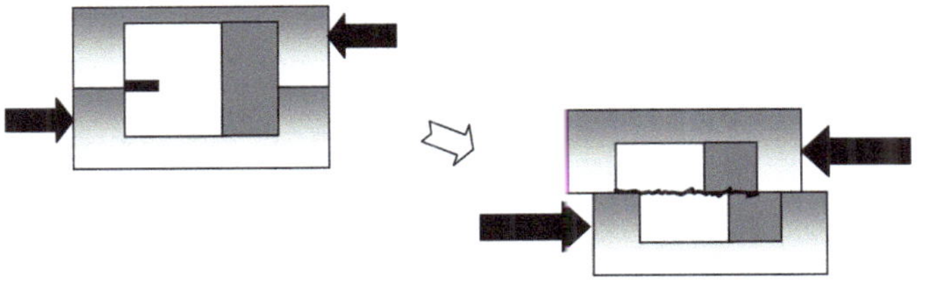

Figure 3.34: Schematic illustration of shear failure of a test specimen

Table 3.31: Test cases for single shear test

Test case	L (mm)	H (mm)	L_1 (mm)	L_1-l (mm)	L_2 (mm)	l (mm)	Number of specimens	Mortar type	Profile type
1*	110	140	110	95	0	15	4		
2	170	140	115	95	55	20	4	#2	80SW
3	130	140	114	95	16	19	4	#2	80SW
4*	130	140	130	112	0	18	4		
5	190	140	130	112	60	18	4	#2	79SW
6	150	140	128	112	22	16	4	#2	79SW
7*	190	140	190	175	0	15	4		
8	280	140	195	175	85	20	4	#2	792SU
9	230	140	198	175	32	23	4	#2	792SU
10	170	140	115	95	55	20	4	#2	80SW
11	131	140	115	95	16	20	4	#2	80SW
12	152	140	130	112	22	18	4	#2	79SW
13	167	140	135	112	32	23	4	#2	792SU
14*	100	70	100	85	0	15	4		
15	130	70	108.3	85	21.7	23.3	4	#3	79SW
16	130	70	98.5	85	31.5	13.5	4	#3	792SU
17	150	70	100	85	50	15	4	#3	79SW
18	150	70	100	85	50	15	4	#3	792SU
19	180	70	110	85	70	25	4	#3	79SW
20	180	70	110	85	70	25	4	#3	792SU
21	180	70	95	85	85	10	4	#3	79SW
22	180	70	95	85	85	10	4	#3	792SU
23*	130	70	130	110	0	20	4		
24	150	70	118.5	110	31.5	8.5	4	#3	792SU
25	150	70	128.3	110	21.7	18.3	4	#3	79SW
26	180	70	115	110	65	5	4	#3	79SW
27	180	70	115	110	65	5	4	#3	792SU
28	280	140	195	175	85	20	4	#4	792SU
29	230	140	198	175	32	23	4	#4	792SU

* Test specimen of base concrete for obtaining the shear strength of concrete

3.3.2 Results of basic material property tests

Tables 3.32 to 3.35 show the results of basic material property tests. It was confirmed that the target compressive strength set at the development stage was met. Table 3.36 presents the design material properties of the SPR mortar and profile, which are determined based on various material tests.

Table 3.32: Results of compressive strength test

	Mortar type					
	Mortar #2		Mortar #3		Mortar #4	
Material property	**Measured value**	**Average**	**Measured value**	**Average**	**Measured value**	**Average**
Compressive strength (N/mm²)	23.3	24.0 (21.0)	56.5	57.2 (35.0)	94.2	96.3 (55.0)
	23.3		56.0		96.8	
	23.7		58.6		92.2	
	25.3		57.6		102.0	
Static modulus of elasticity (kN/mm²)	6.69	7.43 (6.6)	22.5	22.4 (22.0)	28.6	28.9 (28.4)
	7.95		22.4		28.7	
	7.64		22.3		29.4	
Poisson's ratio	0.245	0.25	0.21	0.22	0.25	0.25
	0.260		0.23		0.25	
	0.245		0.23		0.24	

* A value in parentheses is a design value.

Table 3.33: Results of splitting tensile strength test

	Mortar type			
	Mortar #2		Mortar #3	Mortar #4
Material property	**ϕ50 × 100**	**ϕ100 × 150**	**ϕ50 × 100**	**ϕ50 × 100**
Tensile strength (N/mm²)	1.83*	2.45	2.92*	7.29
	2.36	2.45	3.12	6.30
	2.34	2.28	(1.70)	5.46*

* Adopted design value; the value in parentheses was invalid.

Table 3.34: Results of fracture energy test

	Mortar type					
	Mortar #2		Mortar #3		Mortar #4	
Material property	**Measured value**	**Average**	**Measured value**	**Average**	**Measured value**	**Average**
Fracture energy (N/m)	18.0	17.6	6.3	70.8	9.1	88.4
	16.6		6.3		8.7	
	19.1		7.6		8.1	
	18.3		8.4		10.2	
	15.9		6.8		8.1	

Table 3.35: Results of single shear test

	Profile	Material property		
		Shear strength (N/mm^2)		
Material type		**Mortar #2**	**Mortar #3**	**Mortar #4**
Mortar + Profile	#80SW	4.53	5.31	4.80
	#79SW			
	#792SU			
Profile alone	#80SW	1.13	1.29	2.31
	#79SW	1.71	1.90	2.73
	#792SU	2.84	3.15	

Table 3.36: Adopted values of material properties for design in the SPR method

Material property			**Mortar #2**	**Mortar #3**	**Mortar #4**
Compressive strength (N/mm^2)			21.0	35.0	55.0
Modulus of elasticity (kN/mm^2)			6.6	22.0	28.4
Unit weight (kN/m^3)			13.0	21.0	20.0
Poisson's ratio			0.25	0.22	0.25
Tensile strength (N/mm^2)			1.83	2.92	5.46
Fracture energy (N/m)			17.6	70.8	88.4
Shear strength (N/mm^2)	Mortar + Profile		4.53	5.31	4.80
	Profile alone	#80SW	1.13	1.29	2.31
		#79SW	1.71	1.90	2.73
		#792SU	2.84	3.15	

References

JIWET (2014). The Certification Report of Construction Technology (Sewer Engineering): The Renovation Method for Sewer Renewal (The SPR Method). Tokyo, Japan Institute of Wastewater Engineering and Technology.

JSCE (2002). Test Method for Shear Strength of Steel Fibre Reinforced Concrete. In: Standard Specifications for Concrete Structures: Test Methods and Specifications. Tokyo, Japan Society of Civil Engineers.

JSWA (2011). Design and Construction Guidelines for Sewer Pipe Rehabilitation. Tokyo, Japan Sewage Works Association.

Zihai Shi

4 Nonlinear Fracture Mechanics of Concrete

Ageing sewers may sustain varying degrees of material and structural damage, such as the general cracking of concrete due to the occurrence of numerous small cracks in the sewer structure, the localised fracturing of structural members due to a few large cracks, shear fractures at the joint interfaces of the top/bottom plates and the sidewalls, and the frequent loss of concrete cover (also as a result of cracking) and the corrosion of rebars. Clearly, the fracture mechanics of concrete plays an important role in the structural analysis and renovation design of ageing sewers, providing the theoretical basis for crack analysis in sewerage structures.

The theory of fracture mechanics of concrete, with its governing law for crack propagation based on the inelastic material behaviour exhibited in an extensive fracture process zone (FPZ) ahead of an open crack, is largely developed from the theory of linear elastic fracture mechanics (LEFM). This chapter introduces the fundamental concepts of LEFM and nonlinear fracture mechanics (NLFM) of concrete, and presents two major computational theories in crack analysis of concrete, namely, the discrete crack modelling approach and the smeared crack modelling approach.

Part I Fundamental Concepts of Linear Elastic Fracture Mechanics

4.1 Stress Intensity Factor and K-Controlled Crack-Tip Fields of LEFM

Consider a central crack of length $2a$ in an infinite plate subjected to tension, as shown in Fig. 4.1 where a combination of Cartesian coordinates and polar coordinates is used for denoting the near-tip fields. For a near-tip point $P(x, y)$, with $x = a + r\cos\theta$ and $y = r\sin\theta$, the well-known linear elastic solutions of its stress components are given by

$$\sigma_x = \sigma\frac{\sqrt{a}}{\sqrt{2r}}\cos\frac{\theta}{2}\left(1-\sin\frac{\theta}{2}\sin\frac{3\theta}{2}\right);\quad \sigma_y = \sigma\frac{\sqrt{a}}{\sqrt{2r}}\cos\frac{\theta}{2}\left(1+\sin\frac{\theta}{2}\sin\frac{3\theta}{2}\right)$$

$$\tau_{xy} = \sigma\frac{\sqrt{a}}{\sqrt{2r}}\sin\frac{\theta}{2}\cos\frac{\theta}{2}\cos\frac{3\theta}{2} \tag{4.1a-c}$$

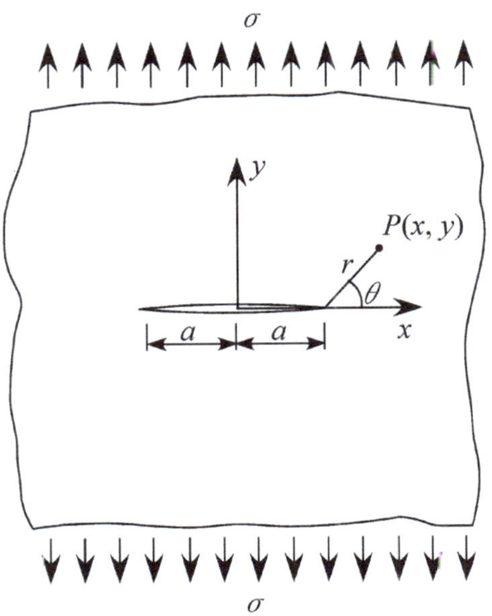

Figure: 4.1 An infinite plate with a central crack subjected to tension

Similarly, when the crack is subjected to shear as shown in Fig. 4.2, the stress components are obtained as

$$\sigma_x = -\tau\frac{\sqrt{a}}{\sqrt{2r}}\sin\frac{\theta}{2}\left(2+\cos\frac{\theta}{2}\cos\frac{3\theta}{2}\right);\quad \sigma_y = \tau\frac{\sqrt{a}}{\sqrt{2r}}\sin\frac{\theta}{2}\cos\frac{\theta}{2}\cos\frac{3\theta}{2}$$

$$\tau_{xy} = \tau\frac{\sqrt{a}}{\sqrt{2r}}\cos\frac{\theta}{2}\left(1-\sin\frac{\theta}{2}\sin\frac{3\theta}{2}\right) \qquad \text{(4.2a-c)}$$

As seen, under the assumption of elasticity, there exists an inverse-square-root singularity at the tip of the crack in the near-tip stress fields. It is known that this singularity-contained structure of stress distribution is invariant among the near-tip solutions of all the elastic crack problems, which differ from each other only by a constant. This single constant defines the amplitude of the stress in the vicinity of the crack tip, with which the crack-tip stress fields are uniquely determined. The significance of this constant in characterising the crack-tip fields in an elastic body was first recognised by Irwin (1957), who named it the stress intensity factor, K, which has become one of the most important fracture mechanics concepts in the theory of LEFM.

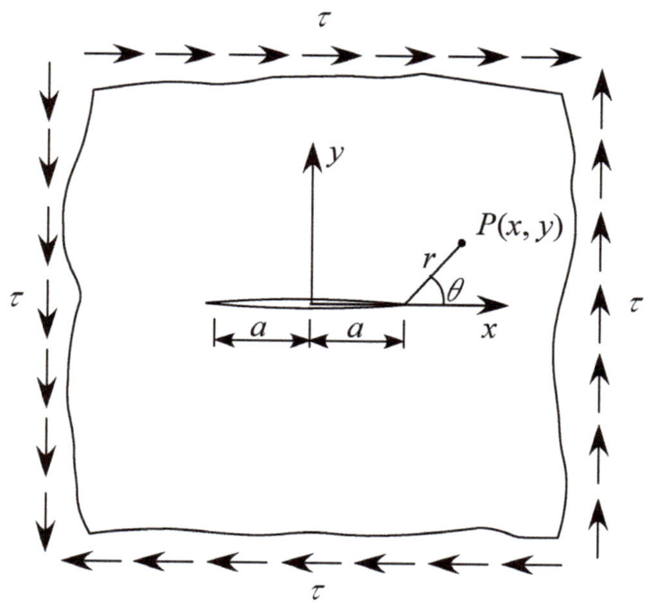

Figure 4.2: An infinite plate with a central crack subjected to shear

A crack may be subjected to three different types of loading that cause displacements of the crack surfaces, as shown in Fig. 4.3. Mode I loading, where the load is applied normal to the crack plane, tends to open the crack. Mode II loading refers to in-plane shear and causes the two crack surfaces to slide against each other. Mode III loading, where out-of-plane shear is applied, tends to tear the two crack surfaces apart. This last mode of deformation, called out-of-plane shear mode, does not occur in the plane elastic problem. The stress intensity factors for the three types of loading are denoted by K_I, K_{II}, and K_{III}, respectively. The crack-tip stresses for each mode of loading can now be rewritten in terms of the corresponding stress intensity factor as

Mode I:

$$\begin{aligned} \sigma_x &= \frac{K_I}{\sqrt{2\pi r}}\cos\frac{\theta}{2}\left(1-\sin\frac{\theta}{2}\sin\frac{3\theta}{2}\right); \\ \sigma_y &= \frac{K_I}{\sqrt{2\pi r}}\cos\frac{\theta}{2}\left(1+\sin\frac{\theta}{2}\sin\frac{3\theta}{2}\right); \\ \tau_{xy} &= \frac{K_I}{\sqrt{2\pi r}}\sin\frac{\theta}{2}\cos\frac{\theta}{2}\cos\frac{3\theta}{2} \end{aligned} \qquad (4.3a\text{-}c)$$

Mode II:

$$\sigma_x = -\frac{K_{II}}{\sqrt{2\pi r}}\sin\frac{\theta}{2}\left(2+\cos\frac{\theta}{2}\cos\frac{3\theta}{2}\right);$$

$$\sigma_y = \frac{K_{II}}{\sqrt{2\pi r}}\sin\frac{\theta}{2}\cos\frac{\theta}{2}\cos\frac{3\theta}{2};$$

$$\tau_{xy} = \frac{K_{II}}{\sqrt{2\pi r}}\cos\frac{\theta}{2}\left(1-\sin\frac{\theta}{2}\sin\frac{3\theta}{2}\right) \tag{4.4a-c}$$

where $\tau_{xz} = \tau_{yz} = 0$ for plane stress and plane strain; and $\sigma_z = 0$ for plane stress and $\sigma_z = \nu(\sigma_x + \sigma_y)$ for plane strain.

Mode III:

$$\tau_{xz} = -\frac{K_{III}}{\sqrt{2\pi r}}\sin\frac{\theta}{2};$$

$$\tau_{yz} = \frac{K_{III}}{\sqrt{2\pi r}}\cos\frac{\theta}{2} \tag{4.5a, b}$$

where $\sigma_x = \sigma_y = \sigma_z = \tau_{xy} = 0$.

Similarly, the crack-tip displacement fields are obtained as

Mode I:

$$u_x = \frac{K_I}{2G}\sqrt{\frac{r}{2\pi}}\cos\frac{\theta}{2}\left(\kappa-1+2\sin^2\frac{\theta}{2}\right);$$

$$u_y = \frac{K_I}{2G}\sqrt{\frac{r}{2\pi}}\sin\frac{\theta}{2}\left(\kappa+1-2\cos^2\frac{\theta}{2}\right) \tag{4.6a, b}$$

Mode II:

$$u_x = \frac{K_{II}}{2G}\sqrt{\frac{r}{2\pi}}\sin\frac{\theta}{2}\left(\kappa+1+2\cos^2\frac{\theta}{2}\right);$$

$$u_y = -\frac{K_{II}}{2G}\sqrt{\frac{r}{2\pi}}\cos\frac{\theta}{2}\left(\kappa-1-2\sin^2\frac{\theta}{2}\right) \tag{4.7a, b}$$

Mode III:

$$u_z = \frac{2K_{III}}{G}\sqrt{\frac{r}{2\pi}}\sin\frac{\theta}{2} \tag{4.8}$$

where G is the shear modulus; $k = 3 - 4\nu$ for plane strain, and $k = (3 - \nu)/(1 + \nu)$ for plane stress; and ν is Poisson's ratio. Note that the square-root of r characteristics of the near-tip displacement fields is the source of the inverse-square-root singularity of the near-tip stress fields.

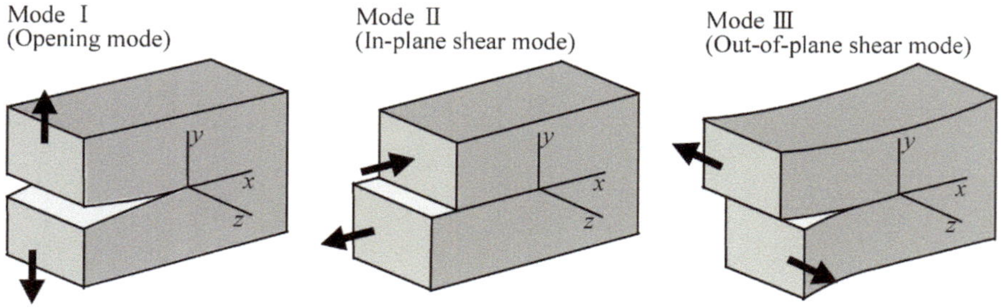

Figure 4.3: Three independent modes of deformation at the crack tip

By comparing Eqs. (4.3) and (4.4) with Eqs. (4.1) and (4.2), which are the exact solutions for an infinite plate with a central crack as $r \to 0$, the stress intensity factors for this particular problem are found to be

$$K_I = \sigma\sqrt{\pi a}\,; \quad K_{II} = \tau\sqrt{\pi a} \tag{4.9a, b}$$

As seen, the stress intensity factor is a function of the applied load and the size of the crack; when a finite body is involved, it is also a function of the geometric configuration of the problem. Obviously, the proportionality of K to the applied load reflects the linear nature of the theory of elasticity. In general, the stress intensity factors can be defined as

$$K_I = \lim_{r\to 0}\sqrt{2\pi r}\,\sigma_y(r,\theta = 0)\,;$$

$$K_{II} = \lim_{r\to 0}\sqrt{2\pi r}\,\tau_{xy}(r,\theta = 0)\,;$$

$$K_{III} = \lim_{r\to 0}\sqrt{2\pi r}\,\tau_{yz}(r,\theta = 0) \tag{4.10a-c}$$

According to Eqs. (4.10), the stress intensity factor governs the magnitude of the stress singularity at the crack tip. The larger the applied load and the size of the crack are (resulting in a larger K), the greater the rate of stress increase in the singularity-dominated zone becomes, provided that the crack remains stationary.

As seen in Eqs. (4.3)-(4.8), the stress intensity factor completely defines the crack-tip fields (or the stress redistribution due to the occurrence of the crack); if K is known, it is possible to obtain all the components of stress, strain, and displacement as functions of r and Θ. As a function of the applied load and the size of the crack as well as the geometry of the problem, the stress intensity factor represents both the strength of the crack-tip singularity and the mode of deformation, providing a link between the crack-contained local material behaviour and the global structural response. Based on the principle of linear superposition, the crack-tip stress fields of a mixed-mode crack can be written as

$$\sigma_{ij}(r,\theta)=\frac{1}{\sqrt{2\pi r}}\left\{K_I f_{ij}^{I}(\theta)+K_{II} f_{ij}^{II}(\theta)+K_{III} f_{ij}^{III}(\theta)\right\} \tag{4.11}$$

where the functions of $f_{ij}^{I}(\Theta)$, $f_{ij}^{II}(\Theta)$ and $f_{ij}^{III}(\Theta)$ are defined by Eqs. (4.3)-(4.5).

Although the K-controlled crack-tip fields ($r << a$) have been obtained using the theory of elasticity, these solutions do not reflect the physical reality of the material deformation at the tip of a crack in engineering materials, because no real material can sustain infinite stresses at the inverse-square-root singularity predicted by the stress solutions. It is known that the singularity in the elastic solutions is the result of two mathematical simplifications: the assumption that the configuration of a real crack (whose crack-tip radius is finite) is a sharp crack (whose crack-tip radius is zero), and the assumption that the inelastic material behaviour at the crack tip is elastic. To remove the crack-tip singularity, it is necessary to take into account inelastic material behaviour, such as plasticity in metals and fracture process zone in concrete, by introducing the inelastic-zone concept to the crack tip.

Despite the inaccuracy in describing the actual material behaviour in the close vicinity of the crack tip, the elastic solutions of Eqs. (4.3)-(4.8) are of great importance in LEFM. The singularity-containing solutions of the stress fields at the tip of a sharp crack in a linear elastic material reveal the nature of fracture in its ultimate form, i.e., causing unbounded stresses at the crack tip. With the help of these elastic solutions, Irwin envisioned the physical significance of the constants contained in these solutions and coined the term "stress intensity factor" to characterise the singularity at the crack tip, which has turned out to be one of the most important concepts in fracture mechanics. Furthermore, these elastic solutions may still be valid outside the inelastic zone at the crack tip, if the size of the zone is sufficiently small compared to the length dimensions of the cracked structure.

Outside the K-controlled crack-tip fields, the general solutions for stress or displacement components include higher order terms of r. For example, the solution of stress for any type of loading (K) is given by

$$\sigma_{ij}(r,\theta)=\frac{K}{\sqrt{2\pi r}}f_{ij}(\theta)+\sum_{m=0}^{\infty}A_m r^{\frac{m}{2}}h_{ij}^{(m)}(\theta) \tag{4.12}$$

where A_m is the amplitude for the m-th term and $h_{ij}^{(m)}$ is a dimensionless function of θ. Note that the higher order terms depend on the geometry of the problem. As $r\to 0$, the leading term approaches infinity while the other terms remain finite or approach zero.

Analytical and numerical methods for calculating stress intensity factors have been extensively discussed in many textbooks of fracture mechanics and there are also handbooks on K for different geometries and loading conditions. The present book focuses on the physical interpretation of the stress intensity factors and the K-controlled crack-tip fields because of their theoretical importance for the subsequent development of nonlinear fracture mechanics of concrete; for detailed coverage on stress intensity factors, refer to the textbooks (Sanford, 2003; Anderson, 2005).

4.2 Energy Principles

With the crack-tip fields now being completely defined in terms of the stress intensity factor, an energy description of the fracture process needs to be explained using Griffith's fracture theory (Griffith, 1921), whose innovative concept of introducing fracture energy into the study of cracked materials laid a solid foundation for the later development of fracture mechanics. It will be shown that the fracture energy required for creating a unit surface of an open crack, i.e., the energy release rate, has a close relationship with the stress intensity factor.

4.2.1 The Griffith fracture theory

Griffith derived his fracture criterion by considering the problem shown in Fig. 4.1, where an infinite plate of unit thickness that contains a crack of length $2a$ is subjected to uniform tensile stress σ. He first obtained the decrease of strain energy in the plate due to the introduction of the crack based on the stress analysis of Inglis (1913), as

$$U_a = -\frac{1}{E}\pi a^2\sigma^2 \quad \text{(plane stress)} \tag{4.13a}$$

$$U_a = -\frac{1-\nu^2}{E}\pi a^2\sigma^2 \quad \text{(plane strain)} \tag{4.13b}$$

where E = the modulus of elasticity. Griffith then introduced "the fundamental conception of the new theory", i.e., surface energy, which is the fracture energy required to form the open crack, as

$$U_s = 4a\gamma_s \tag{4.14}$$

where γ_s = the surface energy per unit area. Note that in Eq. (4.14) the crack surface is $4a$ because a crack has two surfaces. According to the Griffith energy criterion, crack growth is a natural course of energy transfer between the strain energy of an elastic body and the fracture energy required for creating a new crack surface. This requires the following condition to be satisfied:

$$\frac{d}{dA}(U_a + U_s) = \frac{1}{2}\frac{d}{da}(U_a + U_s) = 0 \tag{4.15}$$

Therefore, the critical load for crack propagation is obtained as

$$\sigma_c = \sqrt{\frac{2E\gamma_s}{\pi a}} \tag{4.16a}$$

in the case of plane stress, and

$$\sigma_c = \sqrt{\frac{2E\gamma_s}{(1-\nu^2)\pi a}} \tag{4.16b}$$

in the case of plane strain. Equation (4.16) is referred to as the Griffith fracture criterion, which is valid for ideally brittle materials.

4.2.2 The energy release rate *G*

The Griffith energy concept is of a global character: for a crack to propagate, the energy stored in the system must be sufficient to overcome the fracture energy of the material. Based on the Griffith fracture theory, Irwin generalised the concept by defining an energy release rate G, which is a measure of the energy available for a unit extension of the crack and represents "the force tending to cause crack extension" (Irwin, 1957). Figure 4.4 shows a cracked plate under a given load P, which causes the crack to extend by an amount dA. Since the load is fixed at P, the system is under load control. Obviously, the fracture energy dissipated in propagating the crack originates from the work done by the external force $Pd\Delta$ and the released elastic strain energy $-dU$ in the system, i.e.,

$$G = P\frac{d\Delta}{dA} - \frac{dU}{dA} \tag{4.17}$$

The strain energy of the elastic system is given by

$$U = \int_0^\Delta P d\Delta = \frac{P\Delta}{2} = \frac{CP^2}{2} \tag{4.18}$$

where the relation $\Delta = CP$ is employed; C is the compliance of the system, which is the inverse of the plate stiffness. Note that

$$\frac{d}{dA} = \frac{\partial}{\partial A} + \frac{dP}{dA}\frac{\partial}{\partial P} \tag{4.19}$$

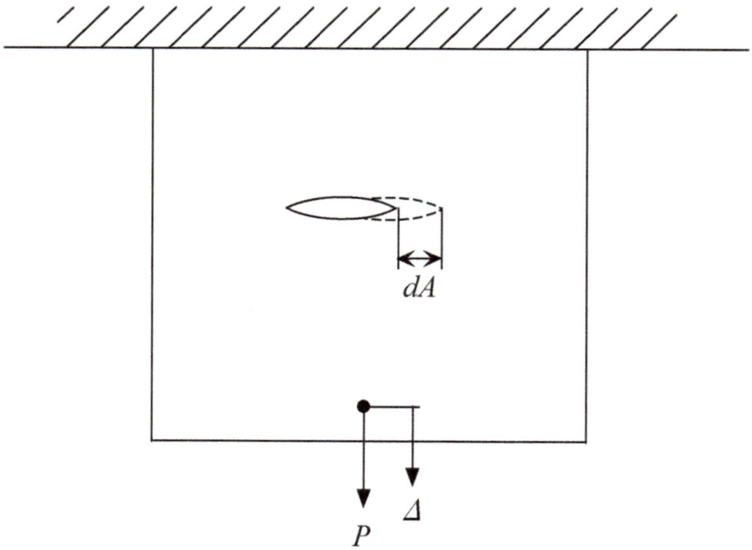

Figure 4.4: Cracked plate under a load *P*

After substituting Eq. (4.18) and $\Delta = CP$ into Eq. (4.17), the above derivative operations are carried out and the energy release rate is derived as

$$G = \frac{P^2}{2}\frac{\partial C}{\partial A} \tag{4.20}$$

It can be proved that Eq. (4.20) can be obtained under the condition of displacement control as well. Theoretically, if the derivative of the compliance of a test specimen with respect to crack length can be derived experimentally or numerically by a FE solution, then the energy release rate can be calculated from Eq. (4.20).

4.2.3 Relationship between *K* and *G*

Two fracture parameters have been introduced so far: the stress intensity factor K and the energy release rate G. While the first parameter uniquely defines the near-tip stress and displacement fields, the second one represents the crack driving force to open that crack. Obviously, a relationship exists between the two. The following deduction is based on Irwin's crack closure analysis (Irwin, 1957).

Figure 4.5 shows an infinite plate of unit thickness with fixed ends that contains a crack of size a. Since the problem is a mode-I fracture, the terms K and G are denoted by K_I and G_I, respectively. Suppose that compressive forces are applied to the crack edge over an infinitesimal distance Δa to close the crack. Since the work required to close the crack is equal to the fracture energy needed to extend the crack, the energy release rate can be calculated from

$$G_I = \lim_{\Delta a \to 0} \frac{\Delta W}{\Delta a} = \lim_{\Delta a \to 0} \frac{2}{\Delta a} \int_0^{\Delta a} \frac{1}{2} \sigma_y u_y dr \tag{4.21}$$

where ΔW is the required work, and u_y is the displacement in the y direction. Note that the factor 2 arises because the crack has two surfaces, and that the incremental work at r is calculated based on the fact that the stresses increase from zero.

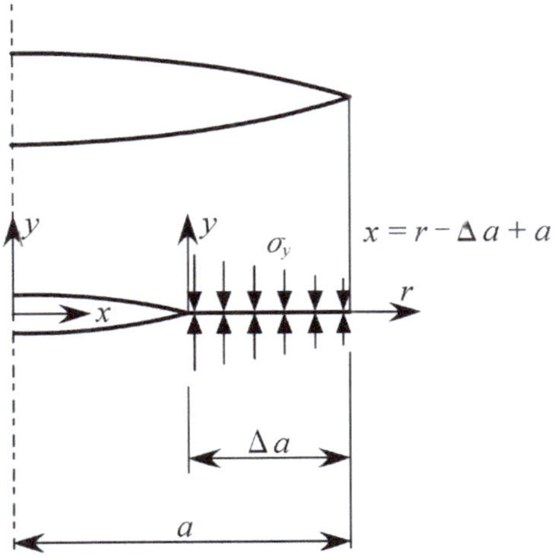

Figure 4.5: Crack closure analysis

For the infinite plate under tension, the value of u_y is given by

$$u_y = \frac{2\sigma}{E}\sqrt{a^2 - x^2} = \frac{2K_I}{E}\sqrt{\frac{a^2 - x^2}{\pi a}} \tag{4.22}$$

Substituting $x = r - \Delta a + a$ into Eq. (4.22) and neglecting second-order terms lead to

$$u_y = \frac{2K_I}{E\sqrt{\pi}}\sqrt{2\Delta a - 2r + \frac{2r\Delta a}{a} - \frac{r^2}{a}} \cong \frac{2K_I}{E\sqrt{\pi}}\sqrt{2(\Delta a - r)} \tag{4.23}$$

The closing force is given by

$$\sigma_y = \frac{K_I}{\sqrt{2\pi r}} \tag{4.24}$$

By using Eqs. (4.23) and (4.24), Eq. (4.21) becomes

$$G_I = \lim_{\Delta a \to 0} \frac{2K_I^2}{E\pi\Delta a} \int_0^{\Delta a} \sqrt{\frac{\Delta a - r}{r}} dr \tag{4.25}$$

which leads to

$$G_I = \frac{K_I^2}{E} \quad \text{(plane stress)} \tag{4.26a}$$

and to

$$G_I = (1-\nu^2)\frac{K_I^2}{E} \quad \text{(plane strain)} \tag{4.26b}$$

The obtained relation between G and K clearly reveals that the stress intensity factor has an energy basis. Similarly, it can be shown that for mode II:

$$G_{II} = \frac{K_{II}^2}{E} \quad \text{(plane stress)} \tag{4.27a}$$

$$G_{II} = (1-\nu^2)\frac{K_{II}^2}{E} \quad \text{(plane strain)} \tag{4.27b}$$

and for mode III:

$$G_{III} = \frac{1}{1-\nu}\frac{K_{III}^2}{E} \quad \text{(plane stress)} \tag{4.28a}$$

$$G_{III} = (1+\nu)\frac{K_{III}^2}{E} \quad \text{(plane strain)} \tag{4.28b}$$

The total energy release rate in combined crack modes can be obtained by superposition:

$$G = G_I + G_{II} + G_{III} = \frac{1}{E}(K_I^2 + K_{II}^2 + \frac{1}{1-\nu}K_{III}^2) \quad \text{(plane stress)} \tag{4.29a}$$

$$G = G_I + G_{II} + G_{III} = \frac{1-\nu^2}{E}(K_I^2 + K_{II}^2 + \frac{1}{1-\nu}K_{III}^2) \quad \text{(plane strain)} \tag{4.29b}$$

Since the K-controlled near-tip fields are uniquely defined for linear elastic materials, it can be proved that Eqs. (4.26)-(4.28) are valid for all geometries of specimens and types of cracks. In fact, G may even be considered as a physical parameter defined at the tip of a crack using the near-tip stress and displacement fields, as shown in Eq. (4.21). Hence, its relations with K must be invariant as the near-tip fields are uniquely established, regardless of any specific changes that may occur to the problem. For plates of finite size the influence of the geometric shape of the problem is reflected in the stress intensity factor K. Therefore, it is possible to obtain G for a problem of finite size by using the relationships between G and K.

4.2.4 The criterion for crack propagation

Since the stress at the tip of a crack in a K-controlled stress field is unbounded, a criterion for crack propagation cannot be stipulated in terms of the tip stress, but has to be described by using either of the two fracture parameters introduced so far, i.e., the energy release rate or the stress intensity factor, which are equivalent based on Eqs. (4.26)-(4.28). For ideally brittle materials, the Griffith criterion in Eq. (4.16a) can be rewritten in terms of the critical energy release rate, G_{Ic}, as

$$\sigma_c = \sqrt{\frac{EG_{Ic}}{\pi a}} \tag{4.30}$$

where

$$G_{Ic} = \frac{dU_s}{dA} = 2\gamma_s \tag{4.31}$$

Substituting Eq. (4.26a) into Eq. (4.30), the critical load is then given in terms of the critical stress intensity factor, K_{Ic}, as

$$\sigma_c = \frac{K_{Ic}}{\sqrt{\pi a}} \tag{4.32}$$

or

$$K_I = K_{Ic} \tag{4.33}$$

Equation (4.33) states that a crack propagates whenever K_I is equal to the threshold value, K_{Ic}, which is a measure of fracture toughness and is regarded as a material fracture property based on LEFM. In principle, the criterion for mode-II or mode-III shear-type fracture can also be expressed as

$$K_{II} = K_{IIc} \tag{4.34}$$

$$K_{III} = K_{IIIc} \tag{4.35}$$

For real engineering materials, stresses cannot increase unboundedly and yielding takes place in metallic materials to form a plastic zone ahead of an open crack. Therefore, the K-controlled stress states in Eqs. (4.3)-(4.8) do not describe the stress distribution inside the plastic zone. This raises a serious question: Is the K criterion still a valid criterion for crack growth?

It turns out that, if the plastic zone is sufficiently small compared to all the length dimensions in the structure and test specimen, the stress intensity factor is still a valid fracture parameter that can uniquely characterise the condition for crack propagation, even if a fracture actually nucleates inside the plastic zone. This requirement for the smallness of the plastic zone restricts its existence well within the singularity-controlled region where Eqs. (4.3)-(4.8) are valid except for the plastic zone, as shown in Fig. 4.6. Obviously, in a K-controlled stress field it is always possible to associate the condition for crack propagation with a threshold value of K, whether the crack originates from the plastic zone or from the singularity-dominated region, implying the validity of Eqs. (4.33)-(4.35).

If the plastic zone is not small compared with the characteristic length of the problem (such as the crack size), there is no longer a singularity-dominated region where the stress varies as $1/\sqrt{r}$. Consequently, the stress intensity factor loses its physical significance as a fracture parameter, and the stresses outside the plastic zone are described by Eq. (4.12) in which higher-order terms become significant. This implies that LEFM is no longer valid and the problem has to be analysed based on NLFM.

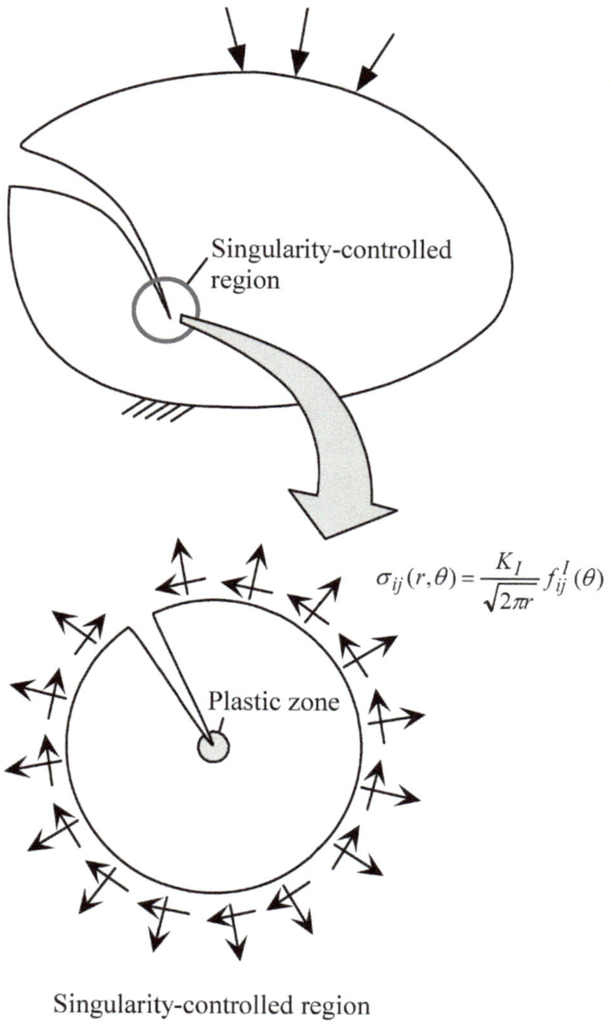

Figure 4.6: Schematic illustration of K-controlled near-tip stresses with the presence of a plastic zone

Part II Fundamental Concepts of Nonlinear Fracture Mechanics of Concrete

4.3 Fracture Process Zone and Tension-Softening Phenomenon

Concrete is a heterogeneous material consisting of aggregates and cement pastes bonded together at the interface, and the material is inherently weak in tension due to the limited bonding strength and various pre-existing microcracks and flaws formed during hardening of the matrix. The tensile strength of concrete approximately ranges from 8 to 15% of its compressive strength. Under external loading a tension zone forms near the crack tip, in which complicated micro-failure mechanisms take place. These fracture processes include microcracking, crack deflection, crack branching, crack coalescence and debonding of the aggregate from the matrix, which are examples of inelastic toughening mechanisms that coexist with a crack when it propagates. In concrete the inelastic zone at the crack tip is extensively developed and therefore, in principle, LEFM cannot be used to study the fracture of concrete.

Figure 4.7 schematically illustrates the formation of an inelastic zone in concrete, which is known as a fracture process zone (FPZ) that can be roughly divided into a bridging zone and a microcracking zone, along with two idealizations of the FPZ. It is known that bridging is a result of the weak interface between the aggregates and the cement pastes, and is an important toughening mechanism in concrete. Within the damage zone the effective modulus of elasticity is reduced from that of the undamaged material E to E^*, if the process zone is modelled as a region of strain softening, as shown in Fig. 4.7(b). Inspired by the concept of the Dugdale and Barenblatt cohesive zone models for studying small-scale plasticity at the crack tip (Dugdale, 1960; Barenblatt, 1962), Hillerborg and his colleagues envisioned a fictitious crack in place of the physical FPZ and subjected it to closure tractions, as shown in Fig. 4.7(c) (Hillerborg, Modeer and Peterson, 1976). As is illustrated in the figure, the closure stress associated with the bridging grains and microcracks is a maximum f_t at the tip of the FPZ and decreases to zero at the continuous crack tip where the crack opening displacement (COD) reaches its critical value W_c, beyond which an open crack forms. Known as the tension-softening phenomenon, the relation between the closure stress and the COD, which completely defines the fracture energy of concrete, describes the local material behaviour inside the FPZ when fracture takes place in concrete.

Figure 4.8 presents a numerically obtained load-deformation relation of a laboratory size, notched plain concrete beam under bending, and the growth of a FPZ at the notch tip based on Hillerborg's fictitious crack model (FCM). The correspondence between the various points of the figure is meant to convey a clear picture of the FPZ and how it develops during beam failure. As seen, in the pre-peak region the tip stress at the notch reaches the tensile strength of concrete at point A, signalling the initiation of a FPZ. Upon reaching the peak load at point B, the FPZ has grown to a length of 50 mm and the tensile stress at the notch tip decreases to $0.55f_t$. The tip stress of the notch drops to zero at point *C* in the post-peak region, as the FPZ stretches to its full length of 150 mm, reaching approximately three fourths of the beam height. Beyond point *C*, a fully-open crack appears ahead of the notch and continues to extend until the beam breaks into two at point *D*. The dashed line drawn across the load-deformation curve divides the structural response into two regions. As seen, the pre-peak nonlinearity and the tension softening in the pre- and the early post-peak regions (somewhere above point *C*) are mainly the work of microcracking. The remaining structural response before final failure at point *D* is the result of the fracture processes that involve mainly aggregate interlocking and other frictional effects.

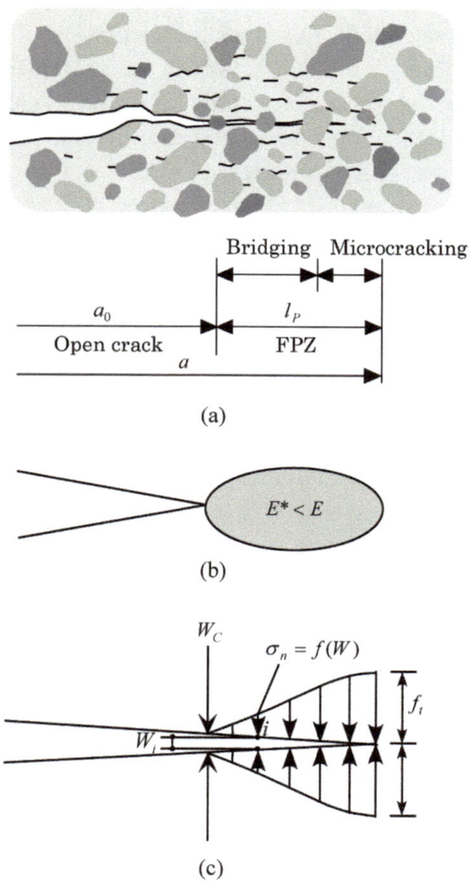

Figure 4.7: Concept of the FPZ and tension softening in concrete: (a) the FPZ in front of an open crack, (b) a region of strain softening with reduced effective modulus of elasticity inside the FPZ, and (c) Hillerborg and colleagues' fictitious crack model with tension softening inside the FPZ

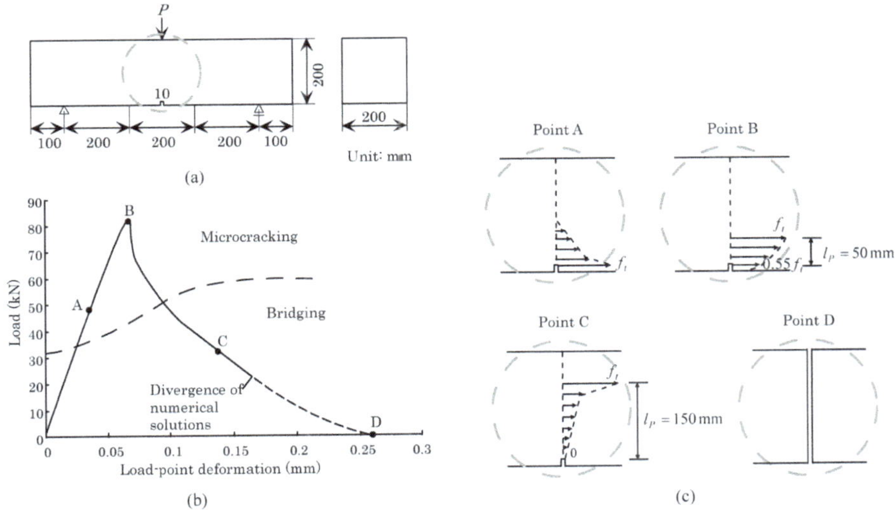

Figure 4.8: Typical load–deformation relation of a notched concrete beam under bending and development of the FPZ in front of the notch: (a) a notched beam; (b) load–displacement relation; and (c) development of the FPZ at points A, B, C and D

As clearly shown by this example, with the presence of such a large FPZ at the notch tip the fracture of concrete cannot be described by a single fracture parameter, such as K_{Ic} or G_{Ic}, as in the case of brittle materials. The fracture behaviour and structural response of the beam can only be described when the FPZ and the tension-softening characteristics are taken into account based on the principles of NLFM of concrete, which will be discussed below.

4.4 Fracture Energy G_F and Tension-Softening Law

As discussed above, a concrete fracture initiates in the FPZ ahead of an open crack through complicated micro-failure processes, and the fracture energy is consumed in overcoming the resistance of various toughening mechanisms to form an open crack at the end of the FPZ. The amount of fracture energy required to break a unit area of concrete is generally regarded as a material property (although it varies slightly with size) that determines the fracture behaviour of the material through the fundamental relationship between the cohesive stress and the COD in the FPZ, which is known as the tension-softening law of concrete. Just like the constitutive relationship of a continuous material that governs the fundamental material behaviour (whether it is elastic or inelastic), the tension-softening law with the fracture energy as its defining characteristic is the constitutive relationship for the material in the FPZ that describes the transitional material behaviour from the continuous state to the discontinuous state. In the following, the fracture energy is discussed with reference to the numerical example of Fig. 4.8 and the standard RILEM test method for its measurement (RILEM, 1985). Tension-softening relations are then introduced using experimental and numerical methods.

4.4.1 Fracture energy G_F

In the load-displacement relations as shown in Fig. 4.8, the area enclosed by the response curve and the horizontal axis represents the work done by the external load to fracture the beam. Suppose that the crack growth is stable and the work done by the external load is spent solely in crack propagation. Based on the Griffith energy criterion, crack growth in an elastic body in the equilibrium state is a natural process of energy transfer between the strain energy of the body and the fracture energy required for creating a new crack surface so that a state of minimum potential energy is achieved for the system at a given load level. In the present case, the work is consumed in breaking the unnotched part of the beam cross-section, i.e., the ligament ahead of the notch. Denoting the work of the external load by W_F and the ligament area by A_{lig}, the energy needed to create a crack of unit area, G_F, is obtained as

$$G_F = \frac{W_F}{A_{lig}} \tag{4.36}$$

In fracture mechanics of concrete, G_F represents the mode-I fracture energy, which is often simplified as fracture energy.

As mentioned before, at point C of the response curve in Fig. 4.8 the FPZ has grown to its full length with the COD of the fictitious crack at the end of the FPZ reaching its critical value W_c, with which the tensile stress decreases to zero and an open crack forms. Along this FPZ a full relationship between the cohesive stress and the COD is obtained and plotted in Fig. 4.9, which is known as the tension-softening relation of concrete. Obviously, this relationship can also be obtained from the stress-COD relations at the notch tip, where an open crack has just been created by fracturing the intact material there. Note that the area enclosed by this tension-softening curve with the horizontal axis in Fig. 4.9 is exactly the fracture energy defined by Eq. (4.36), i.e.,

$$G_F = \int_0^{W_c} \sigma_n(W)dW \tag{4.37}$$

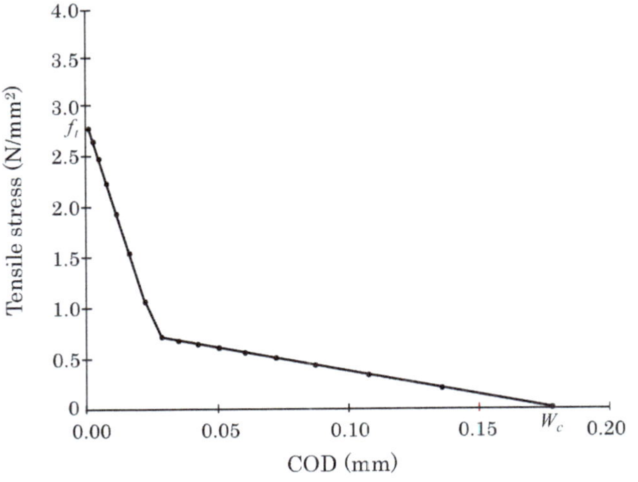

Figure 4.9: Relationships between the cohesive stress and the COD along the FPZ at point C of the response curve in Fig. 4.8

In 1985, the RILEM Technical Committee 50-FMC (Fracture Mechanics of Concrete) proposed a test method for determining the fracture energy G_F of mortar and concrete by means of stable three-point bend tests on notched beams (RILEM, 1985), as shown in Fig. 4.10. The size of the beam depends on the maximum size of aggregate, d_a, according to Table 4.1. As seen from the table, the depth of the beam is at least six times the aggregate size, and the span to depth ratio can vary between 4 and 8. To ensure that the work performed by the external load will be spent entirely in fracturing the beam at the notch, a notch size as large as half the beam depth is used. With such a large notch at the midspan, the maximum bending moment is expected to be kept reasonably low to ensure that all deformations outside the fracture zone are purely elastic due to the low stresses there.

To ensure stable crack propagation during the test, the testing machine has to be stiff enough or furnished with a closed-loop servo control. Recommendations with regard to the stiffness of the machine are given in RILEM (1985). Note that a test is regarded as stable if the load and the deformation change slowly during the whole test, i.e., without any sudden jump. To ensure pure bending the supports and loading arrangements should be arranged in such a manner that the forces acting on the beam are statically determined, as shown in Fig. 4.10(a). During the test the vertical displacement at midspan is measured and continuously plotted along with the gradually applied load P until complete fracture of the specimen. As shown in Fig. 4.10(b), the area under the load-deformation curve, W_0, represents the work done by the external load to fracture the beam. Besides this major work of fracture, there are other types of loads which also perform work of fracture that is secondary but may not be negligible. These include the self-weight of the specimen and the loading equipment carried by it. Therefore, the fracture energy should be calculated as

$$G_F = \frac{W_F}{A_{lig}} = \frac{W_0 + mg\delta_0}{(H - a_0)B} \tag{4.38}$$

where W_F is the total work done by all the loads involved, mg is the weighted sum of the loads that perform the secondary work, δ_0 is the displacement at failure, and H, B and a_0 are respectively the specimen height, width and notch size.

From a fracture mechanics point of view the RILEM method is conceptually clear, simple and accurate as a test method for measuring the fracture energy of mortar and concrete. From an experimental point of view, the fracture test on a three-point-bend beam is both practical and reliable. For these reasons the principles embodied in the RILEM method still serve as the guidelines for designing test methods for the measurement of this fracture-related material constant for various types of structural concrete.

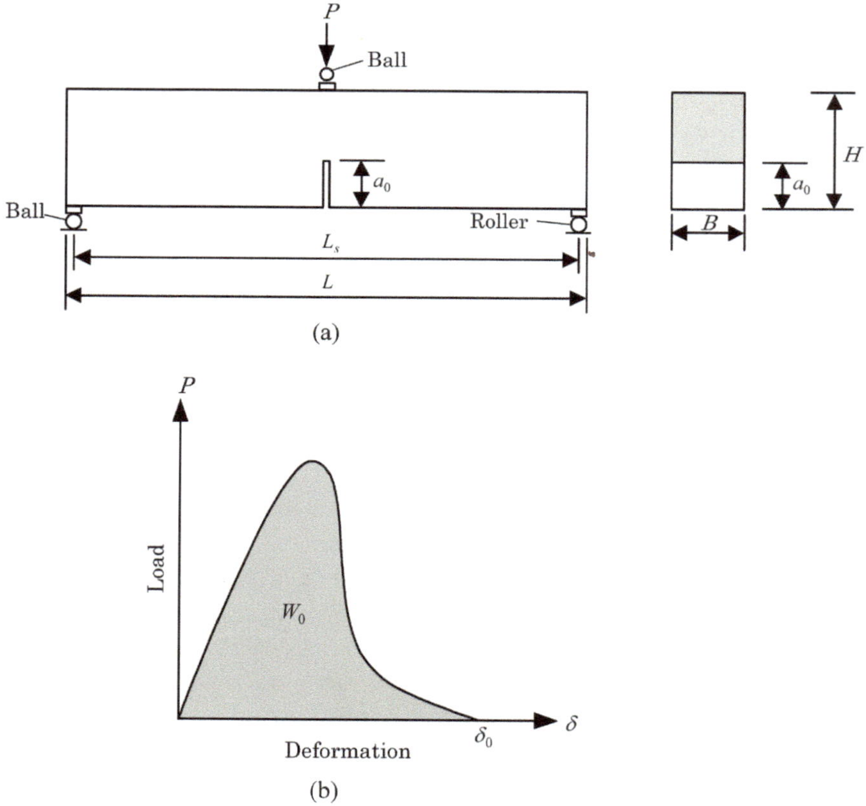

Figure 4.10: Determination of fracture energy G_F based on the RILEM method: (a) notched beam under three-point bending, and (b) load–deformation relation

4.4.2 Tension-softening law

Through the great efforts of many researchers the concept of the tension-softening law of concrete has been firmly established as the basis for the fracture mechanics of concrete, and the general shape of this constitutive relation has been obtained by both experimental and numerical methods.

Table 4.1: Recommended sizes of beams for measuring G_F based on the RILEM method (RILEM, 1985)

d_a* (mm)	Depth, H (mm)	Width, B (mm)	Length, L (mm)	Span, L_s (mm)
1.0-1.6	100 ± 5	100 ± 5	840 ± 10	800 ± 5
16.1-32	200 ± 5	100 ± 5	1190 ± 10	1130 ± 5
32.1-48	300 ± 5	150 ± 5	1450 ± 10	1385 ± 5
48.1-64	400 ± 5	200 ± 5	1640 ± 10	1600 ± 5

*d_a: Maximum size of aggregate

Figure 4.11 presents the experimental results obtained from a series of direct tension tests on normal-weight concrete ($\gamma = 2370$ kg/m³, $f_c = 47$ MPa, $E = 39$ GPa, and $f_t = 3.2$ MPa) by a group of scientists from the Delft University of Technology in the 1980s, where the data points of individual experiments represent the relation between the normalised stress, σ/f_t, and the COD, W (Hordijk, 1991). A regression analysis on these test data was carried out, and the optimum relation (standard deviation = 0.052) was given by

$$\frac{\sigma}{f_t} = \left[1 + \left(c_1 \frac{W}{W_c}\right)^3\right] \exp\left(-c_2 \frac{W}{W_c}\right) - \frac{W}{W_c}\left(1 + c_1^3\right) \exp\left(-c_2\right) \tag{4.39}$$

The best-fit curve was obtained for $c_1 = 3$, $c_2 = 6.93$ and $W_c = 160$ mm, and the corresponding fracture energy was found to be $G_F = 0.1$ N/mm based on Eq. (4.37), resulting in a relation of $W_c = 5.14G_F/f_t$.

Because of the difficulty of carrying out a stable uniaxial tension test to obtain the tension-softening relation, Roelfstra and Wittmann (1986) proposed an inverse modelling method for determining this relationship using the fictitious crack approach and data fitting. In this method a conventional fracture test (such as a bending test) is simulated numerically by assuming and modifying a bilinear tension-softening relation that contains four parameters as shown in Fig. 4.12 through iterative computations to reduce the differences between the experimental and numerical results. They found that the best-fitting response curve (such as a load-deflection curve) could be obtained when a good approximation to the actual tension-softening characteristics was reached. The original formulation has been modified by other researchers to improve the convergence of solutions, and the revised formulation by Nomura et al. is outlined below (Nomura, Mihashi, Suzuki, et al., 1990).

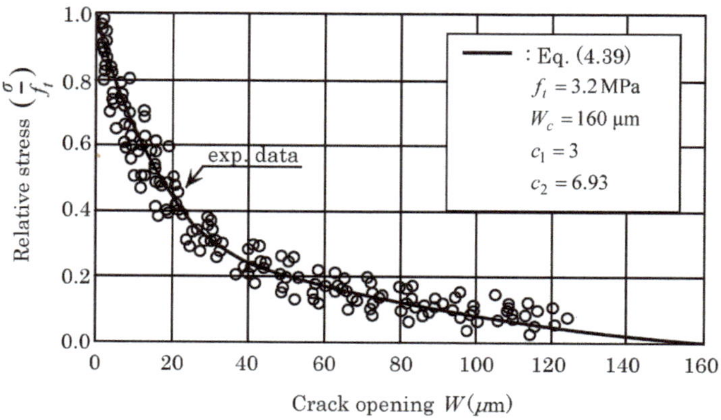

Figure 4.11: Regression analysis for the σ-w relation; data points and fitted relation (Hordijk, 1991)

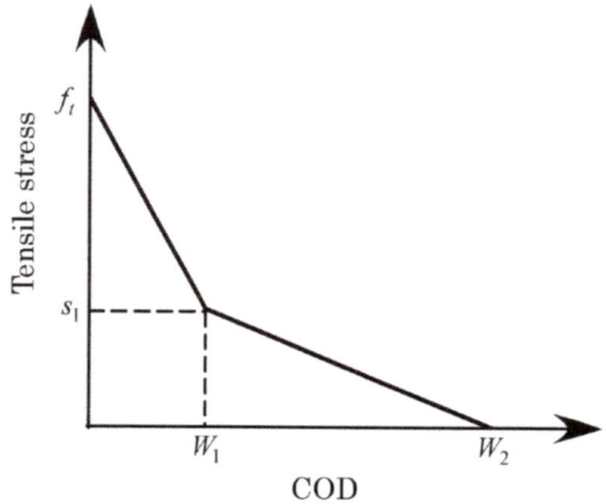

Figure 4.12: Definition of f_t, s_1, W_1, and W_2 in the bilinear softening diagram

Let $F_E(i)$ represent the experimentally obtained load-deformation curve recorded in a time sequence (i = 1, 2, ..., r) at a given location. The numerically-calculated load-deformation curve at the same spot is denoted by $F_M(i; \{x\})$, which is obtained by assuming a bilinear softening relation with the four parameters $\{x\} = f_t, s_1, W_1, W_2$. By minimising the following sum of the square of differences of the two curves, the parameters for an optimum softening relation can be determined:

$$S(\{x\}) = \sum_{i=1}^{r} \left[F_E(i) - F_M(i; \{x\})\right]^2 \rightarrow \min \tag{4.40}$$

Let $\{x^{(k)}\}$ and $\{\Delta x^{(k)}\}$ represent, respectively, the kth estimation of the softening parameters and the corresponding incremental modification. The next estimation is then given by $\{x^{(k+1)}\} = \{x^{(k)}\} + \{\Delta x^{(k)}\}$. The incremental modification $\{\Delta x^{(k)}\}$ is estimated from the following iterative equation:

$$\left\{\Delta y^{(k)}\right\} \cong \left[J^{(k)}\right]\left\{\Delta x^{(k)}\right\} \tag{4.41a}$$

where

$$\Delta y_i^{(k)} = F_E(i) - F_M\left(i;\left\{x^{(k)}\right\}\right) \tag{4.41b}$$

$$J_{ij}^{(k)} = \left[\frac{\partial F_M(i;\{x\})}{\partial x_j}\right]_{x_j = x_j^{(k)}} \tag{4.41c}$$

The flow-chart for calculating the parameters of an optimum softening relation is illustrated in Fig. 4.13. For details of the solution procedure, refer to Nomura et al. (1990).

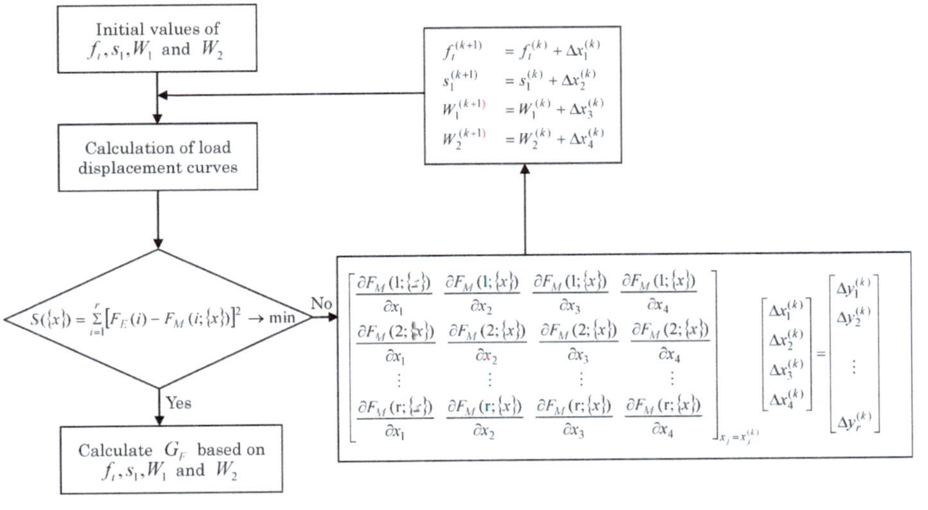

Figure 4.13: Flowchart of the numerical procedure for inverse modelling

Hordijk compared the experimental results presented in Fig. 4.11 with the optimum bilinear relations obtained by Wittmann et al. for a number of concrete mixes and loading rates, in which the stress at the break-point was fixed at $f_t/4$, and found agreement between them (Wittmann, Rokugo, Bruhwiler, et al. 1988; Hordijk, 1991). He observed that these optimum relations with only limited differences from each other appear to be a good approximation of the directly determined softening curve given by Eq. (4.39), as shown in Fig. 4.14 where the averaged bilinear relation is presented. As seen, both the initial slope and the slope of the second part of the descending branch coincide very well, while the parameter W_c is found to be coincidentally the same.

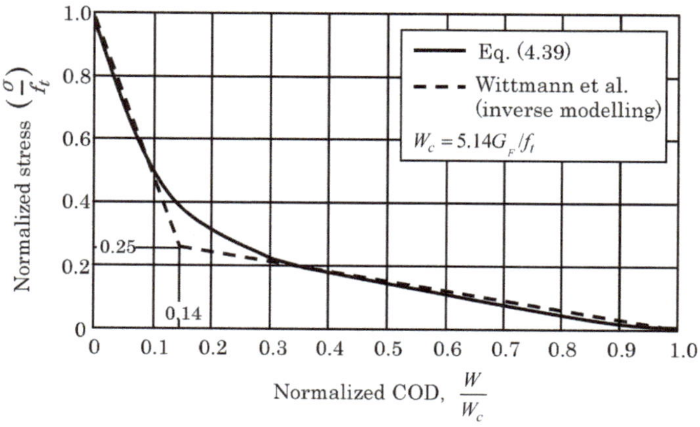

Figure 4.14: Bilinear softening relation by inverse modelling as compared with the relation obtained from uniaxial tensile tests (Hordijk, 1991)

The success of the inverse modelling method discussed above highlights the relevance of adopting a bilinear relation to represent the real tension-softening characteristics of concrete, which, according to Cotterell and Mai (1996), "has the advantage of simplicity and has enough parameters to enable the load-deflection curve to be accurately predicted by numerical simulation." Figure 4.15 presents two types of bilinear relation that have been proposed for a wide range of concrete. Realistic predictions of fracture processes and structural responses in various concrete structures have been obtained by using these models. When these proposed relations fail to represent the real tension-softening characteristics for a particular type of concrete, the inverse modelling method can be employed to obtain an optimum softening relation for that particular material. With a bilinear softening relation the mode-I fracture energy is given by

$$G_F = \frac{1}{2}\left(f_t W_1 + s_1 W_c\right) \tag{4.42}$$

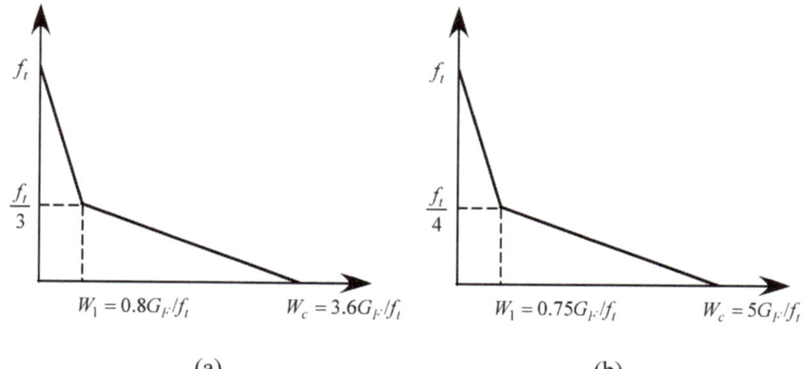

Figure 4.15: Proposed bilinear tension-softening models (a) by Petersson (1981) and (b) by Rokugo et al. (1989)

Part III Two Numerical Modelling Theories for Crack Analysis of Concrete

4.5 The Discrete Crack Modelling Approach

In this approach, a crack is treated as a geometric entity and the FPZ forms a part of the boundary condition, including the geometric shapes and cohesive forces acting on the crack surface. From an engineering point of view, this is the most accurate physical model to study cracks because the approach reflects most closely the physical reality of the cracked concrete and its numerical formulation represents an accurate mathematical description of the problem. Crack propagation results in a continuous change of the FE topology, which affects the element connectivity and the stiffness matrix. Applications of the discrete modelling approach for crack analysis of concrete have greatly increased in recent years, and some discrete crack models have been incorporated into some commercial FE codes that have crack analysis functions.

4.5.1 Fictitious crack model by Hillerborg and colleagues

As shown in Fig. 4.7(c), the fictitious crack model (FCM) proposed by Hillerborg, Modeer and Petersson (1976) for studying crack formation and growth in concrete brought two new concepts to the study of concrete fracture. The first concept is the physical presence of a FPZ ("microcracked zone" in their original words) in front of an open crack as a result of strain localization. The second concept is the existence of a tension-softening law inside the FPZ, stipulating the relationship between the cohesive force and the crack-opening displacement (COD). These concepts have provided the theoretical basis of the fracture mechanics of concrete.

Hillerborg and his co-workers developed these concepts based on experimental evidence of uniaxial tension tests on concrete bars that were performed by various researchers in the late 1960s. These experimental studies showed that crack formation in concrete is an evolutionary process of strain build-up in the pre-peak region, and strain localization and formation of a visible crack in the post-peak region, as illustrated in Fig. 4.16. Therefore, in their modelling of concrete fracture they imagined the zone of strain localization as a fictitious crack that "corresponds to a microcracked zone with some remaining ligaments for stress transfer." They further suggested that this fictitious crack "may be looked upon as a reality. Stresses may be present in a microcracked zone as long as the corresponding displacement is small." As shown in Fig. 4.16, for a bar of length L the total deformation in the post-peak region is given by

$$\Delta L = \delta_e + \delta_{ine} + W \tag{4.43}$$

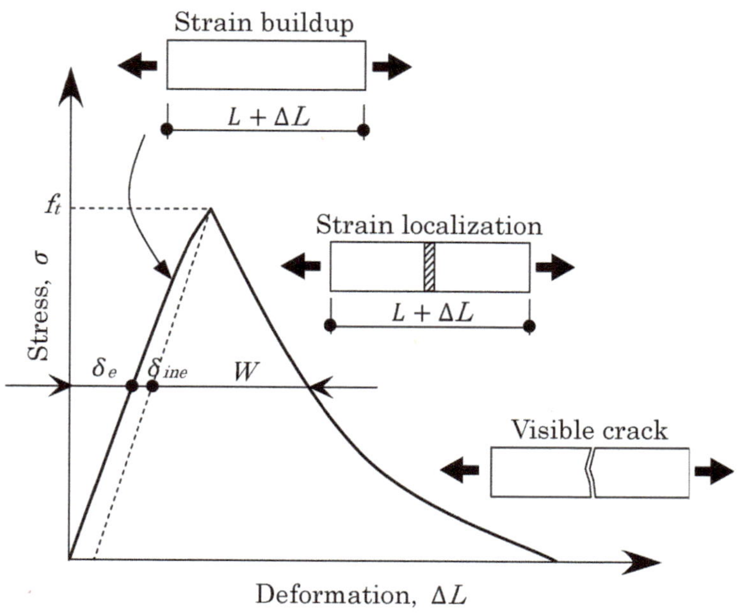

Figure 4.16: The evolution of crack formation

where d_e is the elastic deformation, d_{ine} is the inelastic deformation, and W is the COD of the fictitious crack. For the crack to propagate, the stress at the crack tip was assumed to reach the tensile strength of the material. They further assumed that the stress transferred through the fictitious crack was a function of the COD:

$$\sigma = f(W) \tag{4.44}$$

where $f(W)$ is a tension-softening curve unique to the material that must be determined from experiments, as discussed in the preceding section. As W reaches a critical value, W_c, the transferred stress drops to zero and the fictitious crack becomes a stress-free open crack.

For simplicity, Hillerborg and his colleagues further assumed that the inelastic deformation in the test specimen was negligible, i.e., the behaviour of the bulk material was linear elastic. Based on this assumption, Eq. (4.43) can be rewritten as

$$\Delta L = \delta_e + W = L\frac{\sigma}{E} + W = L\frac{f(W)}{E} + W \tag{4.45}$$

where E is the elastic modulus. Hence, if E and $f(W)$ are given, the load-deformation curve of the tensile bar is completely determined. As simple as Eq. (4.45) is, it underlines the general principles for crack analysis in concrete. First, it illustrates the fundamental importance of the tension-softening law: without it, crack analysis in its strict sense cannot be carried out. Second, it shows that after cracking, the general deformation of the cracked body is composed of two parts, i.e., the deformation of the elastic body and the crack-opening width of the crack. Third, the elastic body outside the cracking zone is subjected to the stress transferred through the fictitious crack, and therefore, its deformation is also a function of the COD.

As such, the essence of crack analysis is to determine the general deformation of the body after the occurrence of cracks. If the total deformation of the cracked body is determined, strain and stress analyses can then be carried out without any particular difficulty. It should be noted that, though the assumption of linear elasticity for the bulk material makes the problem easier to solve, it is not an essential requirement and the FCM can be applied to more general situations including nonlinear behaviour of the bulk material. In general, however, the following hypotheses are used when applying the FCM in crack analysis of concrete:

1. The bulk material behaviour is isotropic linear elastic, and hence is defined by elastic modulus E and Poisson's ratio ν.
2. A crack initiates at a point where the maximum principal stress at that point reaches the tensile strength f_t, and the crack forms normal to the direction of the principal tensile stress.
3. As a crack forms and its crack opening increases, stress transfer takes place across the crack surfaces until a critical COD is reached. For pure mode-I opening, the stress transferred between the surfaces of the crack is a function of the COD, as defined by Eq. (4.44).

It is obvious that Eqs. (4.43) and (4.45) are derived based on the known cracking behaviour of a tensile bar, which is idealized as the crack forms perpendicular to the bar axis and its crack width is uniform. In most crack analysis problems, however, this is not the case because the growth of a fictitious crack is a highly nonlinear process which usually requires numerical procedures to find the solution.

4.5.2 Numerical formulation of a single-crack problem

The analytical concept of the crack-tip-controlled modelling of a single crack, formulated and referred to as the crack equations by Ohtsu (1990), is discussed below. The method offers an easy way of regulating cracking behaviour, which will be very useful when modelling multiple cracks. Figure 4.17 shows a single crack of the mode-I type, propagating in the direction normal to the maximum principal stress at the tip of the crack. The basic equations governing the next-step crack growth (by moving the present tip of the crack to the next node along the presumed future crack path) are obtained using the principle of superposition in elasticity. Assuming the solid body to be linear elastic, the overall equations are derived by linear combinations of individual solutions that make use of the influence coefficients due to unit external loads (Fig.

4.17(a)) and unit cohesive forces at the fictitious crack (Fig. 4.17(b)), respectively. Note that the external loads and the cohesive forces are unknowns and will be determined by solving the following crack equations.

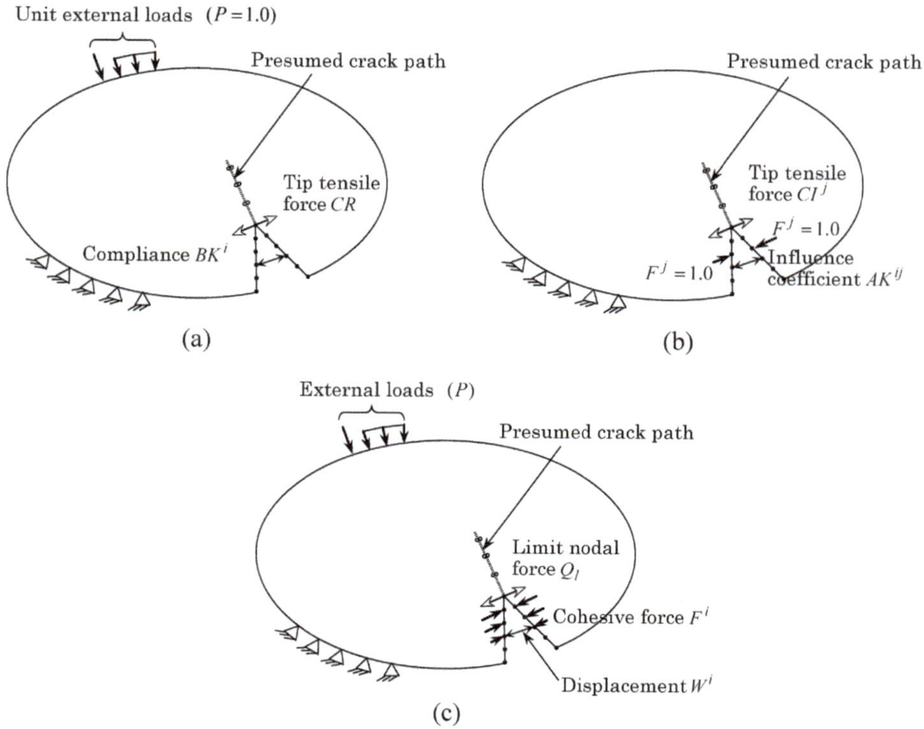

Figure 4.17: Concept of the crack-tip-controlled modelling of a single crack: (a) forces and displacements at the crack due to unit external loads, (b) forces and displacements at the crack due to a pair of unit cohesive forces, and (c) load condition for crack propagation

As shown in Fig. 4.17, the crack and its fictitious prolongation are represented by separating pairs of dual nodes, a special kind of node consisting of two nodal points of the same coordinates that are rigidly connected to each other before separation: refer to Fig. 4.18 for crack path modelling using dual nodes. In formulating crack equations, subscript l represents the limit value of a nodal force, and superscripts i and j denote the i-th and j-th nodes. Let CR represent the reaction at the tip of the crack under a unit external load, and CI^i denote the tip force due to a pair of unit cohesive forces at the i-th node of the fictitious crack. For the crack to propagate, the tip tensile force must reach the tensile strength of concrete, which is given by

$$Q_l = CR \cdot P + \sum_{i=1}^{M} CI^i F^i \tag{4.46}$$

where Q_l is the limit nodal force (the tensile strength of concrete times the surface area apportioned to a nodal point); P is the external load; F^i represents the pair of cohesive forces at the i-th node; and M is the number of nodes inside the fictitious crack. Equation (4.46) is the condition for stable crack propagation.

The COD at the i-th node is given by

$$W^i = BK^i \cdot P + \sum_{j=1}^{M} AK^{ij} F^j \tag{4.47}$$

where $i = 1, ..., M$. Here, BK^i is the compliance at the i-th node due to the external load P. The influence coefficient AK^{ij} is the displacement at the i-th node due to a pair of unit cohesive forces at the j-th node. By virtue of the reciprocity theorem, $AK^{ij} = AK^{ji}$. The geometric shape of the fictitious crack is determined by Eq. (4.47). Note that CR, CI^i, BK^i and AK^{ij} are all determined by linear elastic FE computations based on the FE models shown in Fig. 4.17(a) and (b).

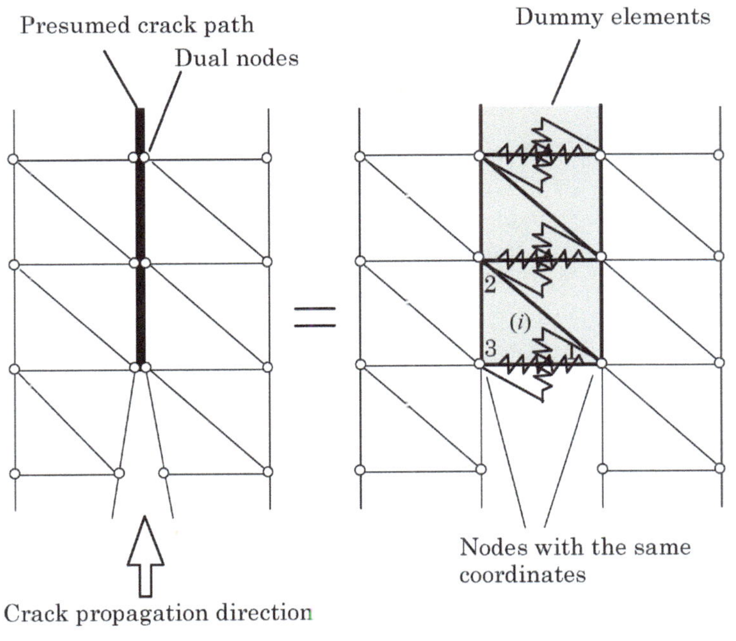

Figure 4.18: Crack-path modelling with dummy elements and dual nodes: Dummy element (*i*) is composed of nodes 1 to 3, and node 1 and node 3 share the same coordinates.

Along the fictitious crack, the cohesive forces F^i and the crack-opening displacements W^i must obey the tension-softening law of concrete

$$F^i = f(W^i) \tag{4.48}$$

where i = 1, ..., M. Equations (4.46)-(4.48) form the so-called crack equations, which are composed of the crack growth condition, the crack shape function, and the cohesive force-COD relation of the fictitious crack. These equations represent a mathematical formulation of the crack propagation in a single-crack problem. With the number of equations ($2M$ + 1) matching the number of unknowns ($2M$ + 1), the problem can be uniquely solved to obtain the external load P and the cohesive forces F, since these equations are linearly independent. Then, the stress and displacement fields due to the next-step crack propagation can be calculated under the newly-obtained boundary condition, as shown in Fig. 4.17(c). This computational process is continued until structural failure. As such, the crack equations are the constitutive equations for crack growth in concrete, and the crack analysis is a continuing process of finding the loading history and the equilibrium states of a propagating crack.

As cracks in concrete structures are generally curvilinear due to the existence of shear force on the crack surface and their exact locations are unknown prior to numerical analysis, remeshing is necessary for modelling discrete crack propagation. A simple mesh modification method is introduced, which employs triangular elements that are connected in a regular manner (Shi, 2009). As shown in Fig. 4.19(a)-(c), after obtaining a new stress field the next-step crack propagation is set in the direction normal to the tip tensile force Q, and the presumed future crack path marked by a row of dual nodes ahead of the crack is then shifted parallel by moving the nearest nodes of corresponding elements to the new positions indicated in each case. Except a few interchanges between the normal nodes and the dual nodes in the two cases shown in Fig. 4.19(b) and (c), no additional nodes and meshes need to be generated in this process. Though Fig. 4.19(a)-(c) only illustrate the cases when the crack curves to the left, the same rules apply when the crack curves to the right. As shown in Fig. 4.19(d), in the equivalent cases of Fig. 4.19(b) and (c) reconstructions of the two meshes on the right side of the crack are required prior to constructing the new crack path.

Note that this simple scheme should be used with caution because it may result in ill-shaped elements. This problem can be avoided by readjusting the shapes of the modified elements with poor aspect ratio after remeshing, while keeping the newly-revised future crack path unchanged. Apparently, some additional computations are required to recalculate the stress and displacement fields before moving on to the next step of the crack analysis.

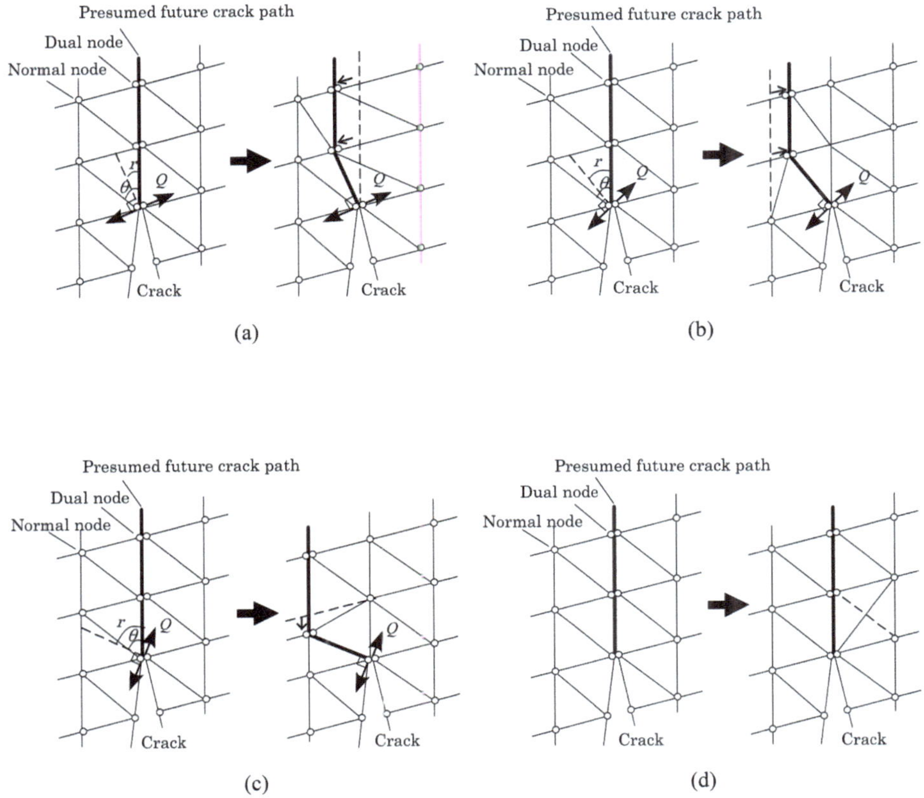

Figure 4.19: Remeshing scheme: (a) for left-curving without interchanges of normal and dual nodes $(r \leq \theta/2)$; (b) for left-curving with interchanges of normal and dual nodes $(\theta/2 < r \leq \theta)$; (c) for left-curving with interchanges of normal and dual nodes $(r > \theta)$; and (d) for right-curving: changing element compositions and forming new meshes

4.5.3 Numerical formulation of a multiple-crack problem

The fundamental difference between single-crack and multiple-crack problems is whether the next-step cracking behaviour is predictable or not. The mathematical formulation for single-crack problems discussed in the previous section is based on the mode of crack propagation, which is the only valid mode for a single crack under tension. When multiple cracks are involved, the next-step cracking behaviour cannot be uniquely determined because there are multiple potential cracking modes for the given problem, including crack propagation, crack arrest and crack closure. Therefore, crack equations for a multiple-crack problem must be established by taking into account these potential cracking modes.

The following formulations are based on the single-active-crack modes in which only one crack is assumed to be propagating. Figure 4.20 illustrates two cracks of the mode-I type, crack A and crack B, where crack propagation is set in the direction normal to the tensile force at the tip of each fictitious crack. In formulating crack equations, subscripts a and b represent, respectively, crack A and crack B, and l stands for the limit value of a nodal force. Superscripts i, j and k denote the corresponding nodes at designated cracks. For clarity, the cohesive forces and the CODs of the inactive crack are marked by asterisks. To begin with, crack A is assumed to be the sole propagating crack. Hence, the tensile force at its tip must reach the nodal force limit Q_{la}, given by

$$Q_{la} = CR_a \cdot P_a + \sum_{i=1}^{N} CI_{aa}^{i} F_a^{i} + \sum_{j=1}^{M} CI_{ab}^{j} F_b^{*j} \tag{4.49}$$

where N and M are the number of nodes inside each fictitious crack, respectively. Note that the tensile forces at the tip of crack A, CR_a, CI^i_{aa} and CI^j_{ab}, are due to a unit external load, a pair of unit cohesive forces at the i-th node of crack A, and a pair of unit cohesive forces at the j-th node of crack B, respectively. The external load P_a is the required load for propagating crack A, while crack B remains inactive. It should be noted that the tip force components due to the cohesive forces of crack B in Eq. (4.49) represent crack interaction.

The CODs along the two fictitious cracks are given by

$$W_a^i = BK_a^i \cdot P_a + \sum_{k=1}^{N} AK_{aa}^{ik} F_a^k + \sum_{j=1}^{M} AK_{ab}^{ij} F_b^{*j} \tag{4.50}$$

$$W_b^{*j} = BK_b^j \cdot P_a + \sum_{i=1}^{N} AK_{ba}^{ji} F_a^i + \sum_{k=1}^{M} AK_{bb}^{jk} F_b^{*k} \tag{4.51}$$

where $i = 1, \ldots, N$; $j = 1, \ldots, M$. Here, the compliances BK^i_a at crack A and BK^j_b at crack B are due to the external load. The influence coefficients AK^{ik}_{aa} and AK^{ij}_{ab} are the CODs at the i-th node of crack A due to a pair of unit cohesive forces at the k-th node of crack A, and a pair of unit cohesive forces at the j-th node of crack B, respectively. Similarly, the influence coefficients AK^{ji}_{ba} and AK^{jk}_{bb} represent the CODs at the j-th node of crack B due to a pair of unit cohesive forces at the i-th node of crack A, and a pair of unit cohesive forces at the k-th node of crack B, respectively. According to the reciprocity theorem, $AK^{ik}_{aa} = AK^{ki}_{aa}$, $AK^{jk}_{bb} = AK^{kj}_{bb}$, and $AK^{ij}_{ab} = AK^{ji}_{ba}$.

Finally, imposing the tension-softening law of concrete along each fictitious crack leads to

$$F_a^i = f(W_a^i) \tag{4.52}$$

$$F_b^{*j} = f(W_b^{*j}) \tag{4.53}$$

where $i = 1, \ldots, N$; $j = 1, \ldots, M$. Equations (4.49)-(4.53) form the crack equations, stipulating the conditions for crack A to propagate. With the number of equations

$(2N + 2M + 1)$ matching the number of unknowns $(2N + 2M + 1)$, the problem is solved uniquely to obtain the external load P_a, the cohesive forces and CODs at the two cracks.

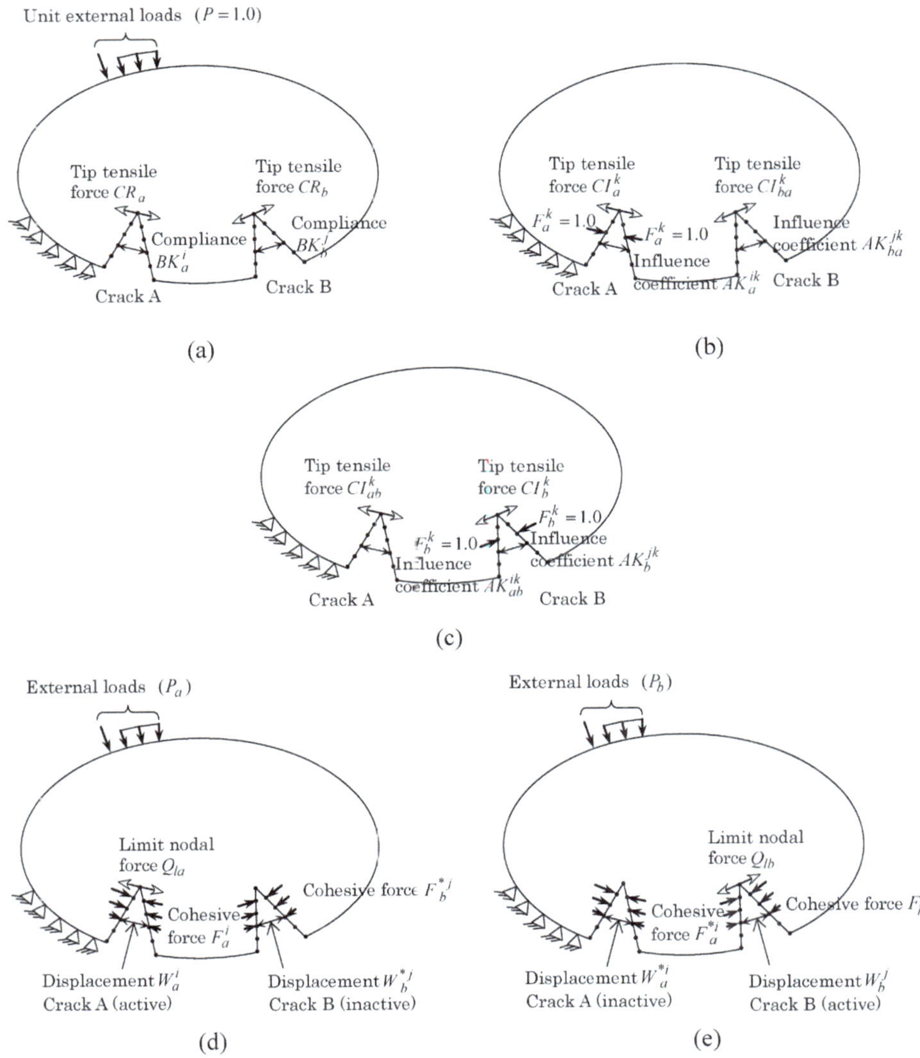

Figure 4.20: Crack-tip-controlled modelling of multiple cracks: (a) forces and displacements at the cracks due to unit external loads, (b) forces and displacements at the cracks due to a pair of unit cohesive forces at crack A, (c) forces and displacements at the cracks due to a pair of unit cohesive forces at crack B, (d) load condition for the growth of crack A, and (e) load condition for the growth of crack B

Alternatively, when crack B is assumed to be the only active crack, the crack equations are derived as

$$Q_{lb} = CR_b \cdot P_b + \sum_{i=1}^{N} CI_{ba}^{i} F_a^{*i} + \sum_{j=1}^{M} CI_{bb}^{j} F_b^{j} \tag{4.54}$$

$$W_a^{*i} = BK_a^{i} \cdot P_b + \sum_{k=1}^{N} AK_{aa}^{ik} F_a^{*k} + \sum_{j=1}^{M} AK_{ab}^{ij} F_b^{j} \tag{4.55}$$

$$W_b^{j} = BK_b^{j} \cdot P_b + \sum_{i=1}^{N} AK_{ba}^{ji} F_a^{*i} + \sum_{k=1}^{M} AK_{bb}^{jk} F_b^{k} \tag{4.56}$$

$$F_a^{*i} = f(W_a^{*i}) \tag{4.57}$$

$$F_b^{j} = f(W_b^{j}) \tag{4.58}$$

where the external load P_b is the required load for propagating crack B, while crack A is assumed to be inactive. Equations (4.54)-(4.58) are the crack equations that set the conditions for crack B to propagate.

Equations (4.49)-(4.58) are the two sets of crack equations required for modelling two discrete cracks. The influence coefficients employed in the crack equations are determined by linear elastic FE computations based on the FE models shown in Fig. 4.20(a-c). Upon solving the two sets of crack equations, the true cracking mode can be identified based on the minimum load criterion, which predicts the onset of crack propagation at the minimum load, i.e.,

$$P = \min(P_a, P_b) \tag{4.59}$$

After setting up the true crack paths for the next-step crack propagation, the stress and displacement fields are calculated under the condition of the obtained load and cohesive forces, as shown in Fig. 4.20(d, e). This process can be repeated until structural failure.

The theoretical basis for the minimum load criterion can be found in the Griffith energy principle, which states that crack extension occurs when the energy available for crack growth is sufficient to overcome the resistance of the material. Obviously, the Griffith energy principle is satisfied at the minimum load for crack propagation. Hence, the minimum load criterion is equivalent to an energy criterion for the growths of multiple cracks.

Obviously, the above solution procedure can be readily extended to problems with an arbitrary number of cracks. The flowchart of the computational procedure is shown in Fig. 4.21. As seen, numerical results are checked to eliminate invalid solutions upon solving the crack equations and obtaining the stress fields. These invalid solutions are encountered when an assumed cracking mode is irrelevant to the problem, and are manifested either by the tip tensile force exceeding the tensile strength at an inactive crack, or by the overlapping of the crack surfaces with negative CODs obtained at an inactive crack. In a situation like this, the crack tip is readjusted by releasing or closing the tip nodes and the problem is recalculated, as illustrated in Fig. 4.21. To close the tip of a crack, the two disconnected nodes next to the tip of the crack are reconnected, while the previous normal traction acting at these nodes is referred to as the transient material strength of the cracked material (based on the fact that the material damage due to cracking is irreversible). By readjusting the crack-tip position, other cracking modes with geometrically admissible strain fields will emerge, which include simultaneous propagations of several cracks, and crack growth accompanied by crack closure. For further details of the solution procedures and how to extend them to solve a mixed-mode fracture problem, refer to Shi (2009).

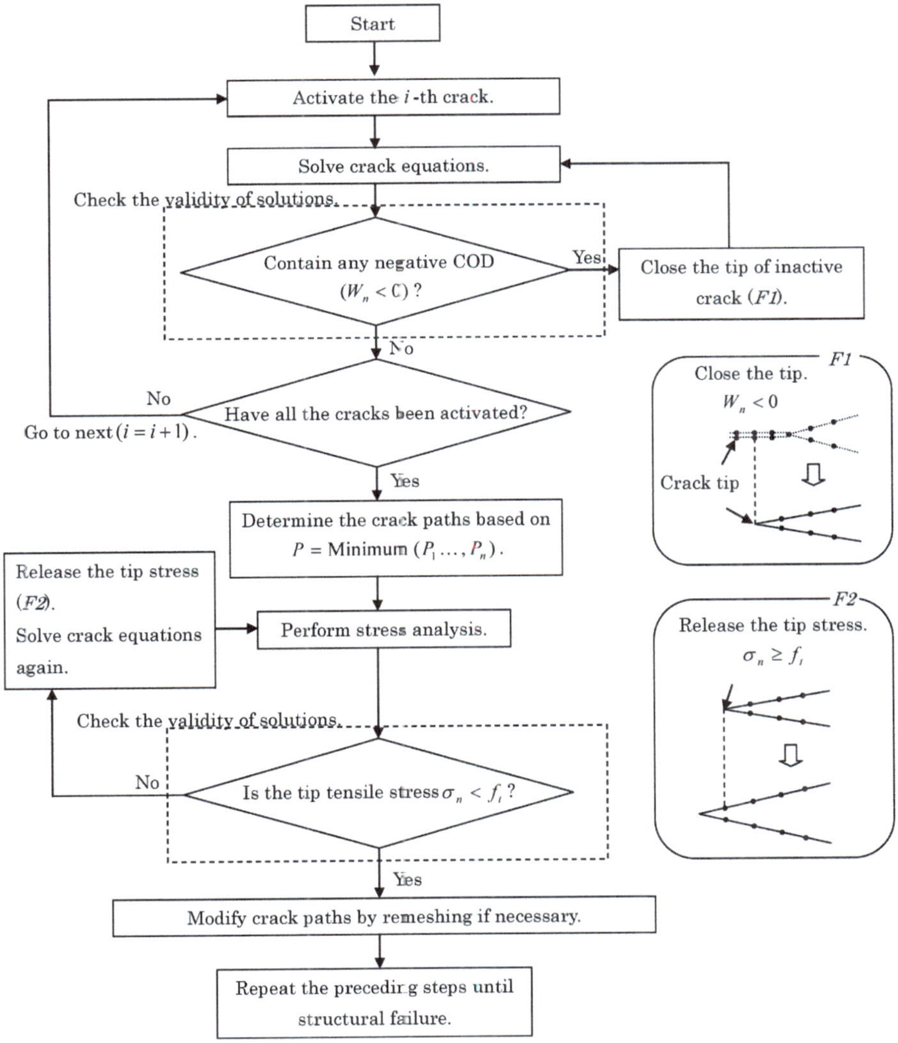

Figure 4.21: Solution procedure for the crack-tip-controlled modelling of multiple cracks based on the single-active-crack modes

4.6 The Smeared Crack Modelling Approach

In this approach, after cracking of concrete the stiffness of a cracked element and its material strength are reduced to simulate the damaging effect of cracking on the element, which is assumed to be caused by numerous microcracks continuously distributed or smeared over the region. Obviously, this smeared-crack concept is a computational convenience and the method is best suited to studying the effect of fracture on general structural behaviours, such as load-carrying capacity and structural stability. The advantage of the method lies in its simplicity and cost-effectiveness since the topology of the FE mesh remains unchanged due to the continuity assumption for the cracked concrete. This modelling approach does have some weaknesses, such as its tendency to cause diffused crack patterns and the directional bias. Smeared crack models are widely used in engineering practice and have been incorporated in commercial FE codes that have crack analysis functions.

4.6.1 Crack band model

Based on the continuity assumption for cracked concrete in the smeared crack approach, Bazant and Oh (1983) modelled the fracture process zone (FPZ) by a band of uniformly distributed imaginary cracks with a fixed width of w_c, as shown in Fig. 4.22. Stable crack propagation is then simulated by progressively reducing the stiffness and strength of the cracked material within this band, based on a strain-softening relationship. For concrete, the band width is normally assumed to be three times the aggregate size. In numerical simulation the width of the FPZ is assumed to be constant to avoid spurious mesh sensitivity. This ensures that the energy dissipation due to fracture per unit length of crack is a constant, which is equal to the fracture energy of the material, G_F.

For simplicity, the cracking in Fig. 4.22 is assumed to be the mode-I type of fracture, i.e., no shear stress on the crack surface. If concrete is idealised as a homogeneous material, then the effect of cracking is simulated by changing the isotropic elastic moduli matrix to an orthotropic one, thus reducing the stiffness in the direction normal to the crack plane. The softening behaviour of concrete is modelled by superimposing the fracturing strain ε_f on the elastic strain as

$$\begin{Bmatrix} \varepsilon_x \\ \varepsilon_y \\ \varepsilon_z \end{Bmatrix} = \frac{1}{E}\begin{bmatrix} 1 & -\nu & -\nu \\ -\nu & 1 & -\nu \\ -\nu & -\nu & 1 \end{bmatrix}\begin{Bmatrix} \sigma_x \\ \sigma_y \\ \sigma_z \end{Bmatrix} + \begin{Bmatrix} 0 \\ 0 \\ \varepsilon_f \end{Bmatrix} \tag{4.60}$$

where σ_x, σ_y, σ_z = the principal stresses; ε_x, ε_y, ε_z = the principal strains; ε_f= the fracture strain by the opening of microcracks; E = the modulus of elasticity; and ν = Poisson's ratio. Note that $\varepsilon_f = \delta_f/w_c$, where $\delta_f = \Sigma\delta^i_f$ is the sum of all the openings of individual microcracks intersecting the z-axis over the crack band width w_c.

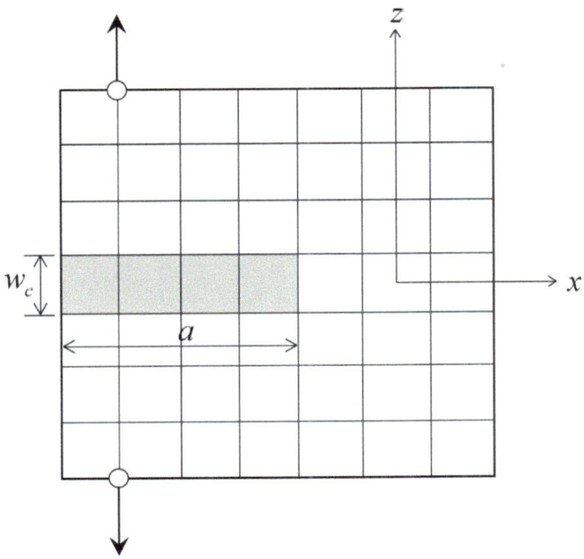

Figure 4.22: Crack band model for simulating crack propagation in a tension plate with a crack of length *a*

The fracture starts when the maximum principal stress σ_z at the crack tip reaches the tensile strength of concrete, f_t. As the microcracks in the crack band start to open, a strain-softened FPZ gradually develops with increasing ε_f and decreasing σ_z. Although a realistic strain-softening relationship between σ_z and ε_f is nonlinear, the following linear approximation is often adopted in practice:

$$\varepsilon_f = f(\sigma_z) = \frac{1}{C_f}(f_t - \sigma_z) \tag{4.61}$$

where $C_f = f_t/\varepsilon_0$; ε_0 = fracture strain at the end of the strain-softening relation; refer to Fig. 4.23(a). Theoretically, the microcracks coalesce into a continuous open crack when σ_z vanishes with the total strain reaching ε^f_z. Substituting Eq. (4.61) into Eq. (4.60) leads to

$$\begin{Bmatrix} \varepsilon_x \\ \varepsilon_y \\ \varepsilon_z \end{Bmatrix} = \frac{1}{E}\begin{bmatrix} 1 & -\nu & -\nu \\ -\nu & 1 & -\nu \\ -\nu & -\nu & \dfrac{E}{E_t} \end{bmatrix}\begin{Bmatrix} \sigma_x \\ \sigma_y \\ \sigma_z \end{Bmatrix} + \begin{Bmatrix} 0 \\ 0 \\ \varepsilon_0 \end{Bmatrix} \tag{4.62}$$

where E_t is the tangent softening modulus of the postpeak σ_z-ε_z relation, as shown in Fig. 4.23(b). Note that E_t is defined as

$$\frac{1}{E_t} = \frac{1}{E} - \frac{1}{C_f} \leq 0 \tag{4.63}$$

By definition, the fracture energy G_F is obtained by the following integral:

$$G_F = w_c \int_0^{\varepsilon_0} \sigma_z(\varepsilon) d\varepsilon \qquad (4.64)$$

Based on the linear strain-softening relationship of Fig. 4.23, G_F is expressed as

$$G_F = \frac{1}{2} f_t \varepsilon_0 w_c \text{ or } G_F = \frac{f_t^2}{2C_f} w_c \text{ or } G_F = \frac{1}{2}\left(\frac{1}{E} - \frac{1}{E_t}\right) f_t^2 w_c \qquad (4.65\text{a-c})$$

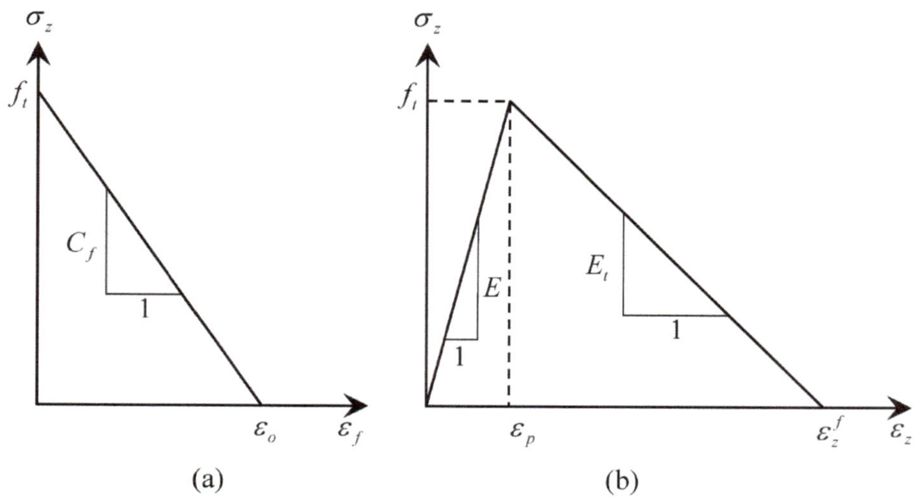

Figure 4.23: The stress–strain relationship in the crack band model: (a) linear strain-softening relation; and (b) stress–total strain relation

With the known material properties f_t, G_F and w_c, the basic parameters of the strain-softening relationship in Fig. 4.23 are obtained as

$$C_f = \frac{f_t^2}{2G_F} w_c;\ \varepsilon_0 = \frac{f_t}{C_f} = \frac{2G_F}{f_t w_c};\ E_t = \frac{f_t^2 E}{f_t^2 - 2EG_F / w_c} \qquad (4.66\text{a-c})$$

As shown from the above formulations, application of the crack band model requires the size of the FE mesh to be limited to w_c. In later development of the smeared modelling approach, this requirement has been eased to allow the effect of cracking to be smeared over a zone of the finite element with the average strain-softening relationship being adjusted to conserve the fracture energy G_F, as shown in Fig. 4.24 (a). The length of the smeared crack zone is referred to as the crack characteristic length, h_c. For parallel cracks in a rectangular element, the characteristic length is usually taken as the height of the element; for inclined cracks it is often defined as the square root of the element area or as the maximum projected length of the element in the direction parallel to the crack; referred to Fig. 4.24(b) and (c). The empirical nature of these assumptions stems from the fact that the smeared crack approach is an approximate modelling method, and therefore, the adequacy of model parameters such as the crack characteristic length should be carefully verified before solving practical problems. Note that in the case of $h_c = w_c$, a smeared crack model becomes the crack band model.

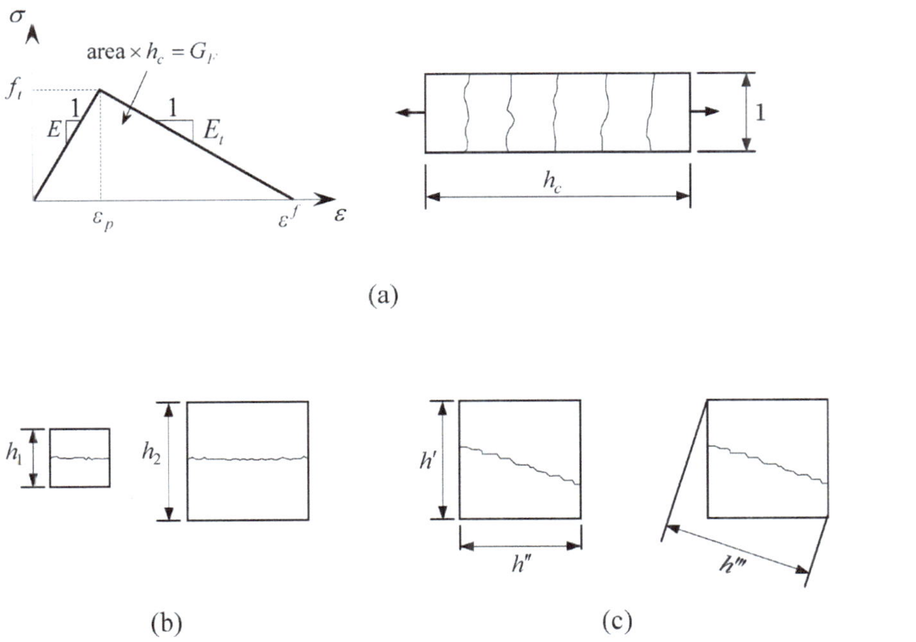

Figure 4.24: The average stress–strain relation in a smeared crack model and the crack characteristic length: (a) stress–strain relation in a cracked element; (b) crack characteristic length for parallel cracks, $h_c = h_1, h_2$ etc.; (c) crack characteristic length for inclined cracks, $h_c = \sqrt{h'h''}$, h''', etc.

4.6.2 Non-orthogonal crack model

Due to aggregate interlocking, shear tractions develop on the crack surfaces, causing the principal stress direction to rotate. Hence, it is possible that under continued loading the current principal stress at a stress integration point or Gauss point reaches the tensile strength of concrete again in a direction deviating from the normal of the originally formed crack plane. In the non-orthogonal smeared crack model proposed by De Borst and Nauta (1985), a second crack is allowed to form after the change in principal stress directions has exceeded a threshold value, such as 30 or 45 degrees. If on subsequent loading a further rotation in principal stress directions would occur, even a third crack is allowed to develop. This modelling concept is different from the crack band model and most of the other computational models in the smeared crack approach in which a subsequent crack is allowed to occur only in the direction perpendicular to the original crack plane.

The non-orthogonal crack model is formulated below to obtain an incremental stress-strain relationship. As with the crack band model, the total crack strain increment is decomposed into a concrete strain increment $\Delta\varepsilon^{co}$ and a crack strain increment $\Delta\varepsilon^{cr}$:

$$\Delta\varepsilon = \Delta\varepsilon^{cr} + \Delta\varepsilon^{co} \tag{4.67}$$

The crack strain increment $\Delta\varepsilon^{cr}$ is contributed by all the individual cracks at a particular integration point as

$$\Delta\varepsilon^{cr} = \Delta\varepsilon_1^{\ cr} + \Delta\varepsilon_2^{\ cr} + \ldots \tag{4.68}$$

where $\Delta\varepsilon_1^{cr}$ is the strain increment of the first crack; $\Delta\varepsilon_2^{cr}$ is the strain increment of the second crack, and so on, with each strain increment being determined in the global Cartesian coordinates.

Following algebraic manipulation, the overall relation between global stress and strain is obtained as

$$\Delta\sigma = \left\{D^{co} - D^{co}N\left[D^{cr} + N^T D^{co} N\right]^{-1} N^T D^{co}\right\}\Delta\varepsilon \tag{4.69}$$

where D^{co} is the constitutive matrix of uncracked concrete given by

$$D^{co} = \frac{E}{1-\nu^2}\begin{bmatrix} 1 & \nu & 0 \\ \nu & 1 & 0 \\ 0 & 0 & \frac{1}{2}(1-\nu) \end{bmatrix} \tag{4.70}$$

and D^{cr} is the constitutive matrix of local cracks given by

$$D^{cr} = \begin{bmatrix} D_1^{cr} & 0 & \cdots \\ 0 & D_2^{cr} & \cdots \\ \cdots & \cdots & \cdots \end{bmatrix} \tag{4.71}$$

where the number of columns and rows of D^{cr} depends on the number of cracks at the Gauss point. As shown by the zero off-diagonal terms, coupling effects between different cracks are ignored. For the i-th crack, D_i^{cr} is given by

$$D_i^{cr} = \begin{bmatrix} D_i^I & 0 \\ 0 & D_i^{II} \end{bmatrix} \tag{4.72}$$

where D_i^I is a mode-I stiffness modulus, and D_i^{II} is a mode-II shear stiffness modulus. Again, no coupling is considered between shear and normal strain on the crack plane. Details of this stiffness will be explained later.

The transformation matrix $N = [N_1\ N_2\ ...]$ combines all the individual crack transformation matrices, in which

$$N_i = \begin{bmatrix} \cos^2\theta_i & -\sin\theta_i\cos\theta_i \\ \sin^2\theta_i & \sin\theta_i\cos\theta_i \\ 2\sin\theta_i\cos\theta_i & \cos^2\theta_i - \sin^2\theta_i \end{bmatrix} \tag{4.73}$$

where Θ_i is the inclination angle of the normal of the *i*-th crack with the *x*-axis, as shown in Fig. 4.25. Again, the size of *N* depends on the number of cracks at the Gauss point.

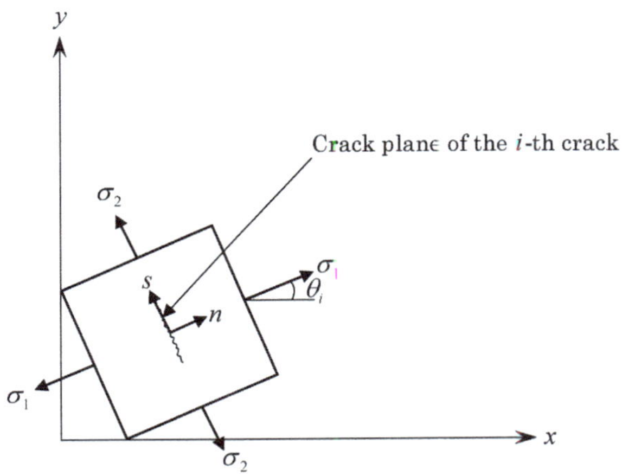

Figure 4.25: Local and global coordinates at the *i*-th crack

For the linear strain-softening relationship shown in Fig. 4.26, the mode-I stiffness modulus $D^I_{i,l}$ is obtained as

$$D^I_{i,l} = \frac{EE_t}{E - E_t} = -\frac{f_t^2 h_c}{2G_F} \tag{4.74}$$

Note that $D^I_{i,l}$ is negative in the present formulation, while its counterpart C_f in the crack band model is defined as positive. For the bilinear strain-softening relationship of Fig. 4.26, which was proposed by Cai et al. (2006), the fracture energy G_F is obtained as

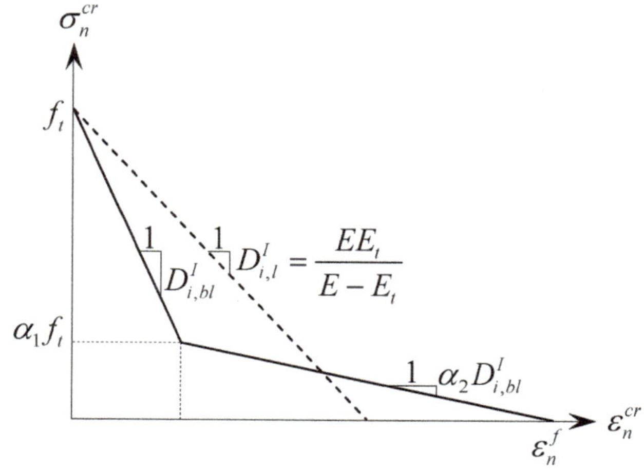

Figure 4.26: Linear and bilinear strain-softening relationships in local coordinates of the *i*-th crack

$$G_F = -\frac{h_c}{2}\left[\frac{(\alpha_1 f_t + f_t)(1-\alpha_1)f_t}{D^I_{i,bl}} + \frac{\alpha_1^2 f_t^2}{\alpha_2 D^I_{i,bl}}\right] = -\frac{f_t^2 h_c}{2}\left[\frac{\alpha_2 + (1-\alpha_2)\alpha_1^2}{\alpha_2 D^I_{i,bl}}\right] \tag{4.75}$$

from which the mode-I stiffness modulus $D^I_{i,bl}$ is obtained as

$$D^I_{i,bl} = \frac{\alpha_2 + (1-\alpha_2)\alpha_1^2}{\alpha_2}\left(-\frac{f_t^2 h_c}{2G_F}\right) = \frac{\alpha_2 + (1-\alpha_2)\alpha_1^2}{\alpha_2} D^I_{i,l} \tag{4.76}$$

where α_1 and α_2 are bilinear softening shape parameters defined in Fig. 4.26: α_1 = ratio of the tensile strength at the concave point, and α_2 = ratio of the mode-I stiffness modulus of the first slope to that of the second slope. Note that for $\alpha_1 = 0$ or $\alpha_2 = 1$, $D^I_{i,bl} = D^I_{i,l}$, the strain softening becomes linear. Figure 4.27 shows how the shapes of the bilinear diagram change and their relationship with the linear mode-I softening modulus $D^I_{i,l}$ as α_1 is fixed at 1/3 while α_2 is assigned values of 0.1, 0.2 and 0.3, respectively. Although in engineering practice linear softening relations are frequently employed, bilinear softening relations are sometimes desirable because they lead to better results, especially when conducting laboratory tests.

A simple shear retention factor β for reducing the tangential shear modulus of the fracture plane is often used in modelling concrete cracking, with which the mode-II shear stiffness modulus can be expressed as

$$D_i^{II} = \beta G = \beta\frac{E}{2(1+\nu)} \tag{4.77}$$

where G is the elastic shear modulus. For more realistic modelling, Cai proposed defining β as a decreasing function of the crack normal strain as

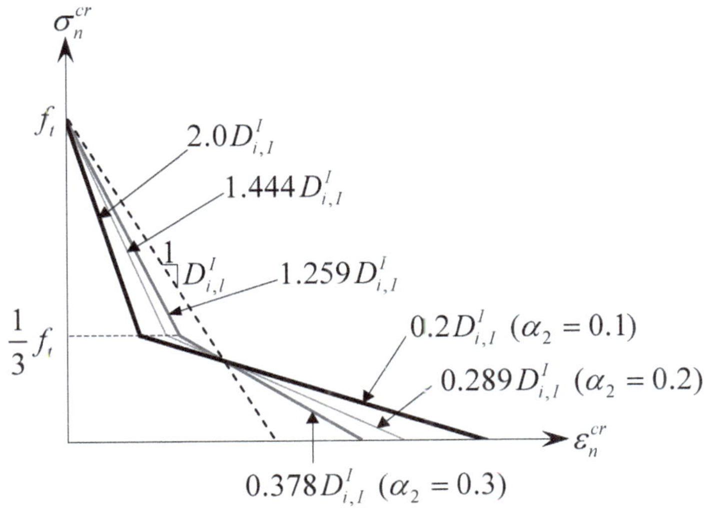

Figure 4.27: Bilinear mode-I strain-softening relations for $\alpha_1 = 1/3$; $\alpha_2 = 0.1$, 0.2 and in local coordinates of the *i*-th crack (Cai, et al., 2006)

$$\beta = \beta_{\max}\left(1 - \frac{\varepsilon_n^{cr}}{\varepsilon_n^{f}}\right)^p \tag{4.78}$$

where p is a constant defining the shear softening shape (Cai, Robberts and van Rensburg, 2006). As shown in Fig. 4.28, if $p = 0$, $\beta = \beta_{max}$ (constant); if $p = 1$, shear softening is linear; if $p = 2$, shear softening is nonlinear. Normally, β_{max} of less than 0.5 is applied to concrete. Figure 4.29 shows the stress-strain relationship of a cracked element with unloading, reloading, closing and reopening strategies. Note that there are two approaches for a partially opened crack to close, i.e., the secant approach returning back to the origin and the elastic unloading approach with immediate closing of the crack. A computational flowchart for crack analysis using a smeared crack model is illustrated in Fig. 4.30 for reference.

It should be pointed out that a close interrelationship exists between the non-orthogonal crack model and the other types of smeared crack model (De Borst and Nauta1985). For instance, by restricting the number of cracks allowed to occur at each Gauss point to one crack in any principal stress direction with the condition of $h_c = w_c$, a crack band model is obtained. By restricting the second and the third crack to occur normal to the existing crack plane, the orthogonal smeared crack models can be derived. Obviously, a rotational crack model can be obtained by constantly renewing the principal stress directions at Gauss points and by performing crack analysis accordingly.

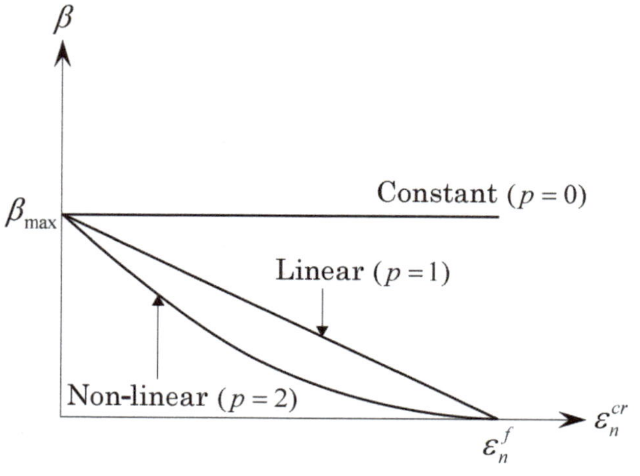

Figure 4.28: Relationship between shear retention factor and crack normal strain (Cai, et al., 2006)

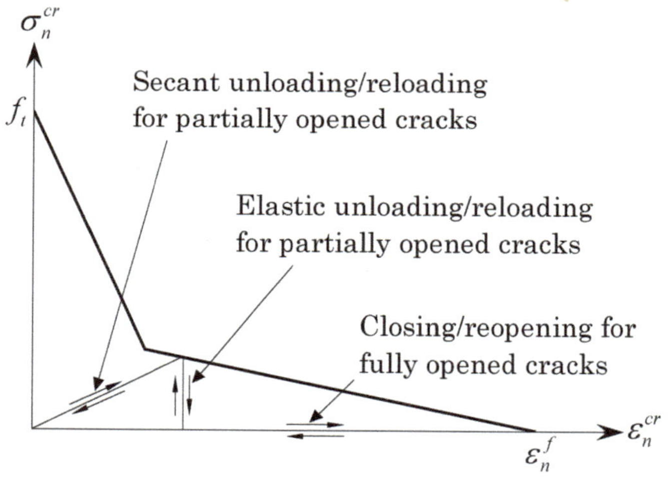

Figure 4.29: Stress–strain relationship of a cracked element with unloading, reloading, closing and reopening crack response

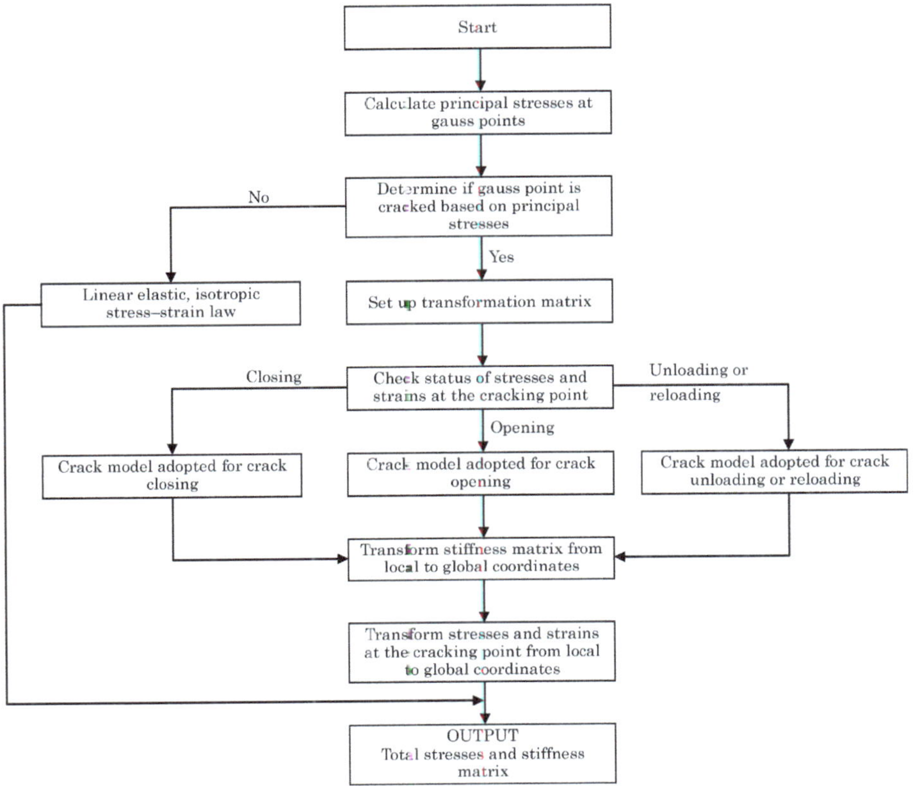

Figure 4.30: A general computational flowchart for crack analysis using the smeared crack modelling approach

4.6.3 Localised smeared crack model using the secant modulus of elasticity for strain softening

In the crack band model and the non-orthogonal crack model, the stress-strain relations of cracked elements are formulated in Eqs. (4.62) and (4.69) using the tangent softening modulus E_t of the postpeak stress-strain relation. Although theoretically sound, this type of formulation is known to have a potential problem of non-convergence in its numerical solutions, especially when a crack problem becomes highly nonlinear with extensive propagation of cracks. To overcome this problem, Bhattacharjee and Leger (1994) proposed formulating the cracked stress-strain relation using a secant modulus of elasticity E_n for the postpeak region, as shown in Fig. 4.31. By continuously reducing E_n and removing the excessive stress during loading, the condition of strain softening can be strictly imposed with less likelihood of stress fluctuation along the softening slope of E_t near a convergence point; refer to Fig. 4.31(b).

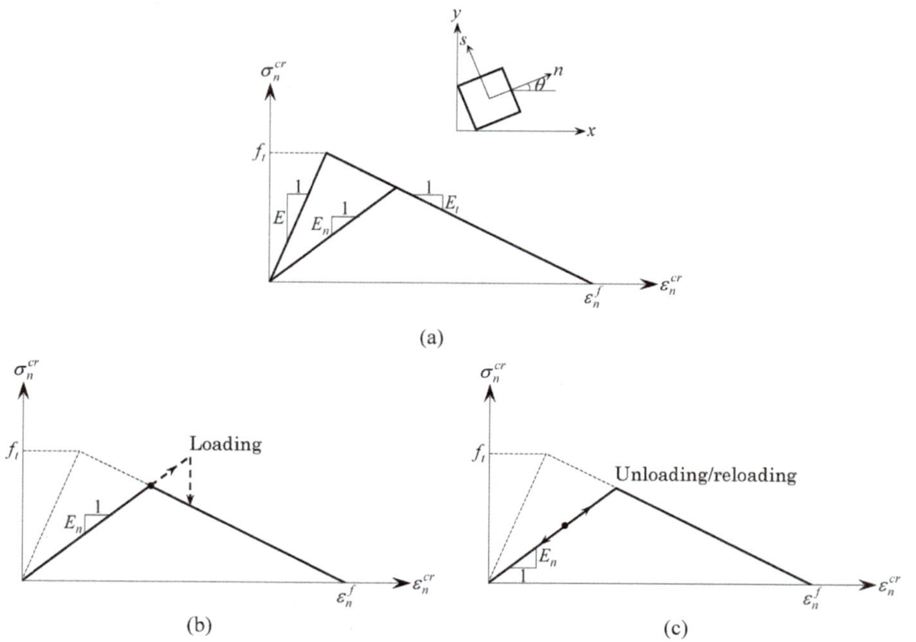

Figure 4.31: Stress–strain relationship in local coordinates using the secant stiffness E_n for strain softening: (a) stress–strain relationship with E, E_t and E_n; (b) loading with E_n, and (c) unloading/reloading with E_n

For a two-dimensional plane stress condition, the constitutive matrix relating local stresses and strains is defined as

$$D_{ns} = \frac{E}{1-\eta\nu^2}\begin{bmatrix} \eta & \eta\nu & 0 \\ \eta\nu & 1 & 0 \\ 0 & 0 & \mu\dfrac{1-\eta\nu^2}{2(1+\nu)} \end{bmatrix} \tag{4.79}$$

where

$$\eta = \frac{E_n}{E};\mu = \frac{1+\nu}{1-\eta\nu^2}\left(\frac{\eta\varepsilon_n - \varepsilon_s}{\varepsilon_n - \varepsilon_s} - \eta\nu\right);(0 \le \mu \le 1) \tag{4.80a, b}$$

where ε_n, ε_s = normal strain components in the local axis normal to and parallel with the crack plane, respectively. Note that when $\eta = 1$ and $\mu = 0$, Eq. (4.79) maintains the pre-softening isotropic elastic stress-strain relationship. The local constitutive relationship matrix, D_{ns}, can be transformed into the global coordinates as

$$[D]_{xy} = [T]^T [D]_{ns} [T] \tag{4.81}$$

where

$$[T] = \begin{bmatrix} \cos^2\theta & \sin^2\theta & \sin\theta\cos\theta \\ \sin^2\theta & \cos^2\theta & -\sin\theta\cos\theta \\ -2\sin\theta\cos\theta & 2\sin\theta\cos\theta & \cos^2\theta - \sin^2\theta \end{bmatrix} \tag{4.82}$$

With increasing strain softening, the damaged modulus of elasticity, E_n, and hence the parameters η and μ decrease gradually, and may eventually reach zero after complete fracture ($\varepsilon_n > \varepsilon^f_n$). The constitutive matrix in Eq. (4.79) is updated as the values of parameters η and μ change. During unloading and reloading, the secant modulus E_n remains unchanged, as shown in Fig. 4.31(c); the parameter μ, however, may vary during that process based on the actual strain values.

References

Anderson, T. L. (2005). Fracture Mechanics - Fundamentals and Applications, 3rd ed., Taylor and Francis.

Barenblatt, G. I. (1962). The mathematical theory of equilibrium cracks in brittle fracture. Advances in Appl. Mech., 7, 55-129.

Bazant, Z. P. and Oh, B. H. (1983). Crack band theory for fracture of concrete. RILEM Mat. and Struct., 16, 155-177.

Bhattacharjee, S. S. and Leger, P. (1994). Application of NLFM models to predict cracking in concrete gravity dams. Structural Engineering, 120(4), 1255-1271.

Cai, Q., Robberts, J. M. and van Rensburg, B. W. J. (2006). Cracking in concrete using smeared cracking finite element modelling. South African Journal of Science, 102, 548-556.

Cotterell, B. and Mai, Y. W. (1996). Fracture Mechanics of Cementitious Materials, 1st ed., Blackie Academic & Professional.

De Borst, R. and Nauta, P. (1985). Non-orthogonal cracks in a smeared finite element model. Engineering Computations, 2, 36-46.

Dugdale, D. S. (1960). Yielding of steel sheets containing slits. J. Mech. Phys., Solids, 8, 100-104.

Griffith, A. A. (1921). The phenomena of rupture and flow in solids. Phil. Trans. Royal Society, Series A, 221, 163-198.

Hillerborg, A., Modeer, M. and Peterson, P. E. (1976). Analysis of crack formation and crack growth in concrete by means of fracture mechanics and finite elements. Cement and Concrete Research, 6(6), 773-782.

Hordijk, D. A. (1991). Local Approach to Fatigue of Concrete. Ph. D. Thesis, Delft University of Technology.

Inglis, C. E. (1913). Stresses in a plate due to the presence of cracks and sharp corners. Trans. Inst. Naval Architects, 55, 219-230.

Irwin, G. R. (1957). Analysis of stresses and strains near the end of a crack traversing a plate. J. Applied Mechanics, 24, 361-364.

Nomura, N., Mihashi, H., Suzuki, A., et al. (1990). Mechanism of brittleness in high strength concrete based on nonlinear fracture mechanics. J. Struct. Constr. Engng, AIJ, 416, 9-14.

Petersson, P. E. (1981). Crack Growth and Development of Fracture Zones in Plain Concrete and Similar Materials. Report TVBM-1006. Division of Building Materials, Lund Institute of Technology, Sweden.

Ohtsu, M. (1990). Tension softening properties in numerical analysis. Colloquium on fracture mechanics of concrete structures. JCI Committee Report, 55-65.

RILEM (1985). Determination of the fracture energy of mortar and concrete by means of three-point bend tests on notched beams. Materials and Structures, 18, 285-290.

Roelfstra, P. E. and Wittmann, F. H. (1986). Numerical method to link strain softening with failure of concrete. In: Fracture Toughness and Fracture Energy of Concrete. F. H. Wittmann ed., Elsevier Science Publishers B. V., Amsterdam, 163-175.

Rokugo, K., Iwasa, M., Suzuki, T., et al. (1989). Testing methods to determine tensile strain softening curve and fracture energy of concrete. In: Fracture Toughness and Fracture Energy-Test Methods for Concrete and Rock. H. Mihashi, H. Takahashi, and F. H. Wittmann eds., Balkema Publishers, 153-163.

Sanford, R. J. (2003). Principles of Fracture Mechanics, Pearson Education.

Shi, Z. (2009). Crack Analysis in Structural Concrete—Theory and Applications, Elsevier.

Wittmann, F. H., Rokugo, K., Bruhwiler, E., et al. (1988). Fracture energy and strain softening of concrete as determined by means of compact tension specimens. Materials and Structures, 21, 21-32.

Zihai Shi, Jianhong Wang

5 Structural Analysis Theories of Composite Pipes as Semi-Composite Structure in Sewer Renovation

5.1 Review of Code Requirements

5.1.1 Outline of Guidelines for sewer renovation by the composite pipe method

The general practice of sewer renovation in Japan, from site survey and structural design to renovation, is subject to the Japan Sewage Works Association's *Design and Construction Guidelines for Sewer Pipe Rehabilitation* (Guidelines) (JSWA, 2011). According to the Guidelines, *a renovated sewer pipe must be structurally comparable with or superior to a newly constructed sewer pipe in terms of its load-carrying capacity, material and structural endurance, and hydraulic capacity*. In the following, this requirement is simply referred to as the structural comparability requirement. To avoid possible confusion between the composite pipe method as defined in the Guidelines and the semi-composite pipe structure as actually employed in the structural analysis of the method, an outline of the Guidelines on related subjects is summarised below.

(1) The composite pipe method
The scope of application of the Guidelines includes two types of renovation methods, namely, the independent (or stand-alone) pipe method and the composite pipe method. The definition of an independent pipe is a pipe which is constructed inside the existing one and is designed to resist all the design loads independently, and the renovated pipe must satisfy the structural comparability requirement stated above. On the other hand, a composite pipe is defined as a composite structure constructed by rigidly attaching a renovation layer to the existing pipe, and the renovated sewer satisfying the structural comparability requirement is designed to bear the external loads by the combined resistance of the two structural components. Figure 5.1 shows schematic illustrations of the two renovation methods.

Based on the extensive track record of sewer renovation constructions in Japan, the Guidelines state that the two renovation methods should be applied to rigid pipes, of which the majority are reinforced concrete structures and the rest are masonry and vitrified clay pipes.

The Guidelines define a renovated pipe as a composite pipe when it is constructed by employing the pipe-reforming method, as discussed in Section 1.3 of Chapter 1. In this type of rehabilitation approach, a liner pipe is constructed inside an existing sewer by interlocking strips of surface materials that are made of polyvinyl chloride or polythene resin materials, and the annular space behind the liner is filled with cementitious grout under pressure to form a highly integrated structure, as shown in Fig. 5.2.

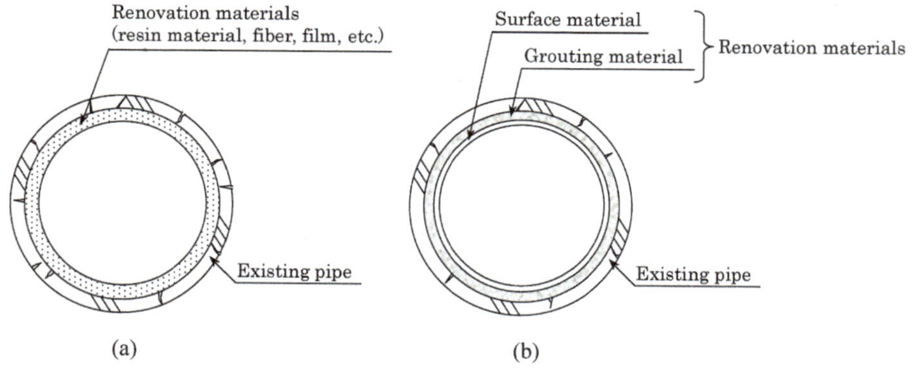

Figure 5.1: Two types of sewer renovation method defined by the Guideline: (a) the independent pipe method; (b) the composite pipe method

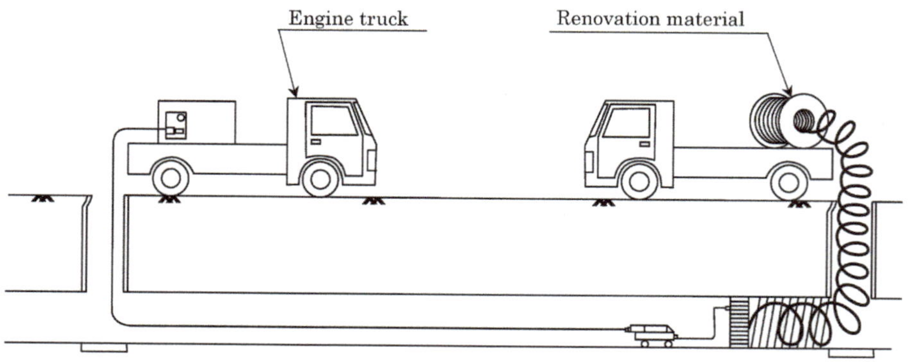

Figure 5.2: Schematic illustration of the pipe-reforming method

There are several construction techniques in the pipe-reforming method for liner fabrication, which include the spirally-wound lining, the assembled panel lining, and the assembled segment lining. Depending on site conditions, some of these techniques can be used without stopping the flow of wastewater in the sewer during construction.

The Guidelines also define a third type of renovation method, the double-layered pipe method, in which the existing pipe and the liner pipe made of resin materials work together to resist design loads. Unlike the composite pipe method, there is no bonding strength between the two pipes. Due to the lack of a generally-accepted structural analysis theory and design theory for this type of structure, application of the double-layered pipe method is not covered by the Guidelines.

(2) Requirements on strength and complete integration of structural components
The basic performance requirements for sewer pipes can be summarised as sufficient strength against various loads, sufficient watertightness against wastewater leakage, and sufficient cross-sectional area for the required drainage capacity. Based on these fundamental requirements for sewers in general, the Guidelines present specific performance requirements for the composite pipe method. Among them, the requirements on load-carrying capacity and the complete integration of structural components are explained below, because they influence the structural modelling of a composite pipe and the development of the corresponding structural analysis theories.

1) Strength requirement
The Guidelines specify two methods for strength verification and they can be used independently:

(a) Analytical approach
The strength requirement of a composite pipe structure can be evaluated using the limit state design theory. The ultimate load-carrying capacity of the pipe should be evaluated by analytical or numerical methods based on proper modelling of the existing pipe with its structural and material damage reflected.

(b) Experimental approach
External pressure tests should be performed on a composite pipe according to the JSWA standard: *JSWAS A-1, Reinforced Concrete Sewerage Pipes* (JSWA, 2003), and the obtained failure load must exceed the nominal value of failure load for a new pipe specified in the standard (or constructed according to relevant building codes). The composite pipe should be prepared by first loading a new pipe up to the failure load, then renovating the fractured pipe following standard renovation procedures. Grouting materials used for filling the annular gap behind the liner must possess good fluidity for sufficient fill-in, and a high bonding strength with the existing pipe with little hardening shrinkage. The target compressive strength of grout materials should be verified by the JSCE-G521 standard: *The Compressive Strength Test of Prepacked Concrete.*

2) Complete integration of structural components
The Guidelines require complete integration between the renovation layer and the existing pipe, and the mandatory confirmation method is by performing the direct tension test according to JIS A 1171, which was introduced in Chapter 3, Section 3.2.4, on a test specimen made of base concrete and mortar, as shown in Fig. 3.20. The test should show that failure of the test specimen is not caused by interfacial debonding at the concrete-mortar interface, but by fracture of the base concrete.

(3) Assessment of structural integrity of existing pipes
Based on a site survey and detailed investigations on an ageing sewer, a structural assessment of the pipe should be performed, focusing on the drainage capacity and structural performance of the existing pipe. This is especially true for the composite

pipe method because in general, the residual strength of the ageing pipe contributes greatly to the structural strength of the composite pipe. Hence, the structural assessment of an existing sewer can strongly influence the composite pipe design and the structural strength of the renovated sewerage structure. If local damage or fracture has occurred to the existing pipe, the Guidelines require that the fractured pipe structure be adequately modelled in the structural analysis and design of the composite pipe.

5.1.2 Basic code requirements for composite structural members

In the design of composite structures, the limit state design theory has been adopted in the current codes of practice as the basic design method. Most composite structural members used in modern bridges, buildings and various utility structures are steel-concrete members, which are cost-effective structural systems. As shown in Fig. 5.3, these structural members include composite columns, composite beams and composite slabs that are formed by connecting a steel component to a concrete component either by encasing one component into the other, or by rigid mechanical connectors. As steel and concrete are, respectively, the most effective engineering materials in carrying tension and in resisting compression, these composite members make the best use of the effective material properties of both steel and concrete.

Obviously, what makes a composite member work, i.e., its individual components function together as one structural member to resist design actions, lies in an efficient connection of the steel to the concrete. It is this connection that allows the transfer of forces and characterises the composite member. As shown in Fig. 5.3(a), the common types of composite column include concrete encased composite columns, and rectangular and circular concrete-filled steel tubular columns, which have been increasingly used in various modern structures for their high structural performance. Due to the large contact area between the structural steel and the concrete, a large bonding strength between the two components can develop, which in many cases is sufficient to ensure the composite action. If necessary, mechanical shear connectors can also be welded to the steel section to increase force transfer. In order to qualify as composite columns, there are detailed code requirements on the material strength of each structural component, the minimum cross-sectional area of the structural steel, and other structural details.

As shown in Fig. 5.3(b), a composite beam can be a fully encased one or can be formed by attaching a concrete slab to the top flange of a steel beam using mechanical shear connectors. In simply-supported bridges incorporating composite beams, the concrete slab is subjected to compressive forces, and this slab is supported typically by steel I-section components; the high strength of steel in tension complements the comparatively high strength of concrete in compression. The connection between the steel and concrete is achieved through mechanical shear connectors, which allow the shear transfer of the forces in the concrete to the steel and vice versa, and which also prevent vertical separation of the concrete and steel components. A similar composite structural type can be found in composite slabs with profiled steel sheeting, as shown in Fig. 5.3(c). There are many types of mechanical shear connectors as illustrated in Fig. 5.4. Among them the stud shear connectors of Fig. 5.4(a) and 5.4(b), which possess a head and a plain shank connected to the steel component by welds or bolts, are frequently employed. In general, mechanical shear connectors have to be designed according to building codes to satisfy the strength, serviceability and construction criteria.

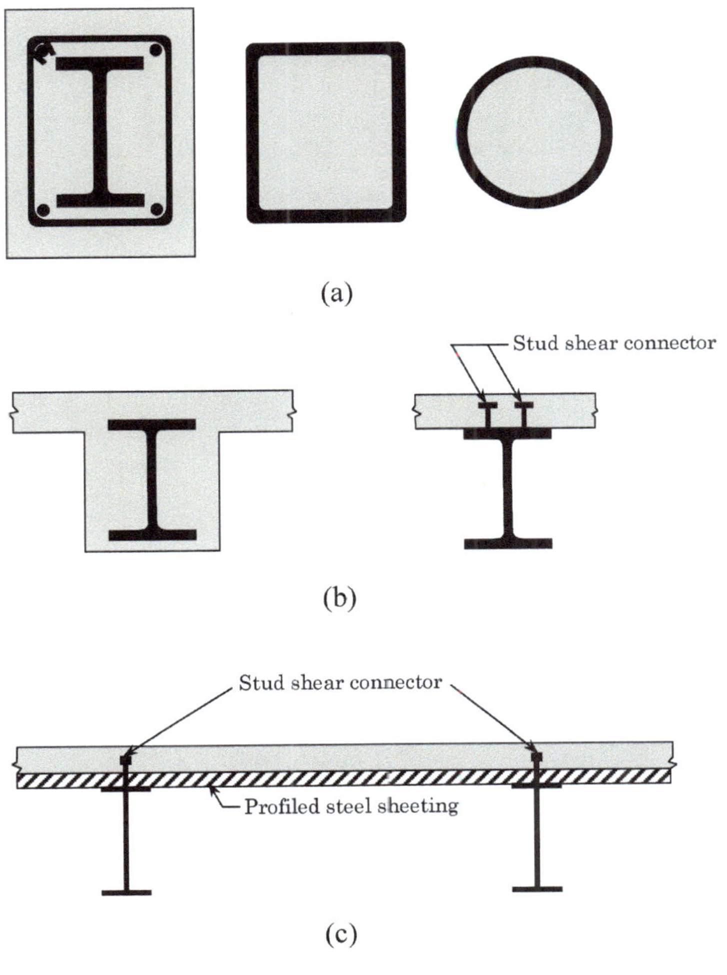

Figure 5.3: Composite structural members: (a) composite columns; (b) composite beams; (c) composite slabs

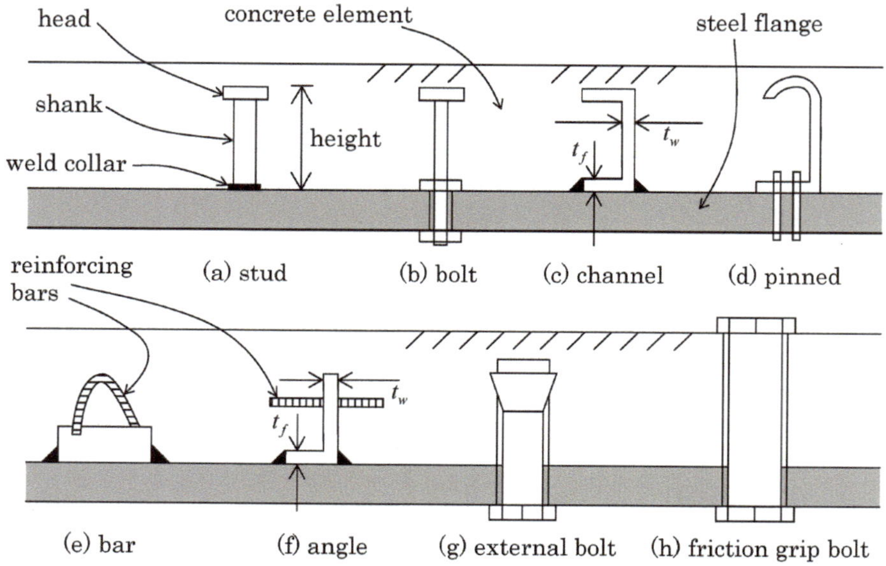

Figure 5.4: Mechanical shear connectors

In summary, there are two main methods for shear connection in composite structural members: the bond or adhesion at the steel-concrete interface in fully-encased columns and beams, and the mechanical shear connectors in uncased composite beams and slabs. Bond stress in steel-concrete composite structures has been well studied and the following comments are drawn from *Composite Structures of Steel and Concrete* (Johnson, 2004).

1. In the design, it is necessary to restrict bond stress to a low value, to provide a margin for the incalculable effects of shrinkage of concrete, poor adhesion to the underside of steel surfaces, and stresses due to variations of temperature.
2. Research on the ultimate strength of encased beams has shown that, at high loads, calculated bond stresses have little meaning, due to the development of cracking and local bond failures.
3. Codes of practice do not allow ultimate-strength design methods to be used for composite beams without shear connectors.
4. Tests on uncased composite beams show that, at low loads, most of the longitudinal shear is transferred by the bond at the interface, that bond breaks down at higher loads, and that once broken it cannot be restored.
5. For uncased beams, the most practical form of shear connection is some form of dowel welded into the top flange of the steel member and subsequently surrounded by in situ concrete when the floor or deck slab is cast.

5.2 No-Tension Interface Modelling and Fracture-Mechanics Based Numerical Analysis Theories

5.2.1 No-tension interface modelling and the semi-composite pipe structure

Although the Guidelines require a direct tension test to be performed on concrete-mortar test specimens to ensure no occurrence of interfacial debonding at the ultimate failure of the test specimen, cases of debonding during fracture tests at the interface of pipe specimens renovated by various pipe-reforming methods have been reported. Photo 5.1 shows two typical examples of fractured circular and box-culvert pipe specimens renovated using different pipe-reforming methods. In each case localised debonding of the interface between the original pipe and the renovation layer had occurred at high loads. This type of interfacial debonding in renovated pipe specimens shows that the requirement for complete integration in the composite pipe method may not always be satisfied.

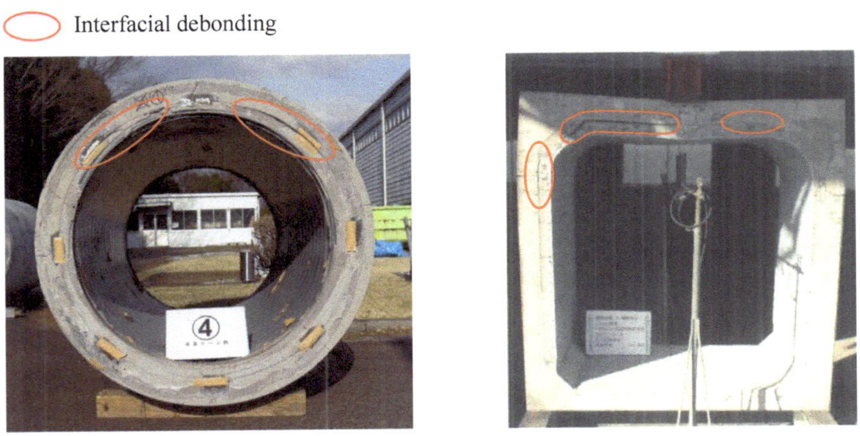

(a) (b)

Photo 5.1: Interfacial debonding of renovated pipe specimens at failure loads: (a) a circular pipe specimen; (b) a box-culvert pipe specimen

The following discussion is based on numerical studies of the fracture tests of two renovated pipe specimens, a box-culvert pipe (1500×1500 mm) and a circular pipe (ϕ1000 mm), as defined in Case 2 of Table 3.2 and Case 4 of Table 3.3. Structural dimensions and renovation details are shown in Fig. 3.2(a) and Fig. 3.3(a), and test conditions are shown in Fig. 3.4. Tables 3.4 and 3.5 summarise the material properties of the original pipe and the renovation layer for each case, respectively.

Figure 5.5 shows schematic illustrations of the structural deformations during the external pressure tests. As seen, under the compressive loads multiple tension zones develop in the renovation layer, which is generally thinner than the plates and the walls of the original pipe and structurally functions like a tension member. In these tension zones tensile stress normal to the axis of the tension member may occur at the interface, and its occurrence can be attributed to the Poisson effect, the interfacial shear transfer, the hardening shrinkage of the fill mortar and its self weight. As mechanical connectors are not used to integrate the original pipe and the renovation layer, the complete integration of structural components required by the Guidelines depends solely on the direct tensile strength of the fill mortar.

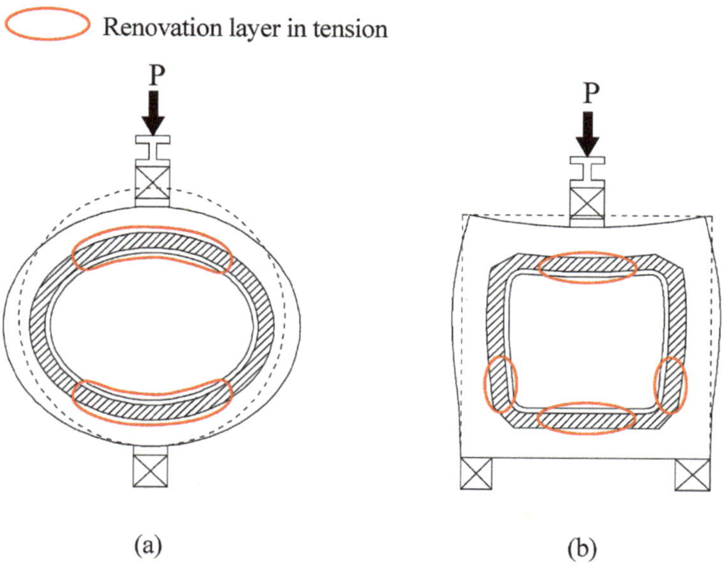

Figure 5.5: Schematic drawings of structural deformation of composite pipe specimens during external pressure test: (a) a circular pipe specimen; (b) a rectangular pipe specimen

Figure 5.6 shows the numerically-obtained maximum tensile stresses that occur in the neighbouring elements of the interfaces in the two renovated pipes at their respective maximum loads, assuming rigid bonds for the interface. As shown, the maximum tensile stress at the interface reaches 3.74 MPa in the box-culvert pipe, and 2.73 MPa in the circular pipe at the peak loads. Compared with the direct tensile strengths of various types of SPR fill mortar (Table 3.18) with the maximum value being obtained at 1.7 MPa, debonding at the interface of a composite pipe seems to be inevitable at high loads.

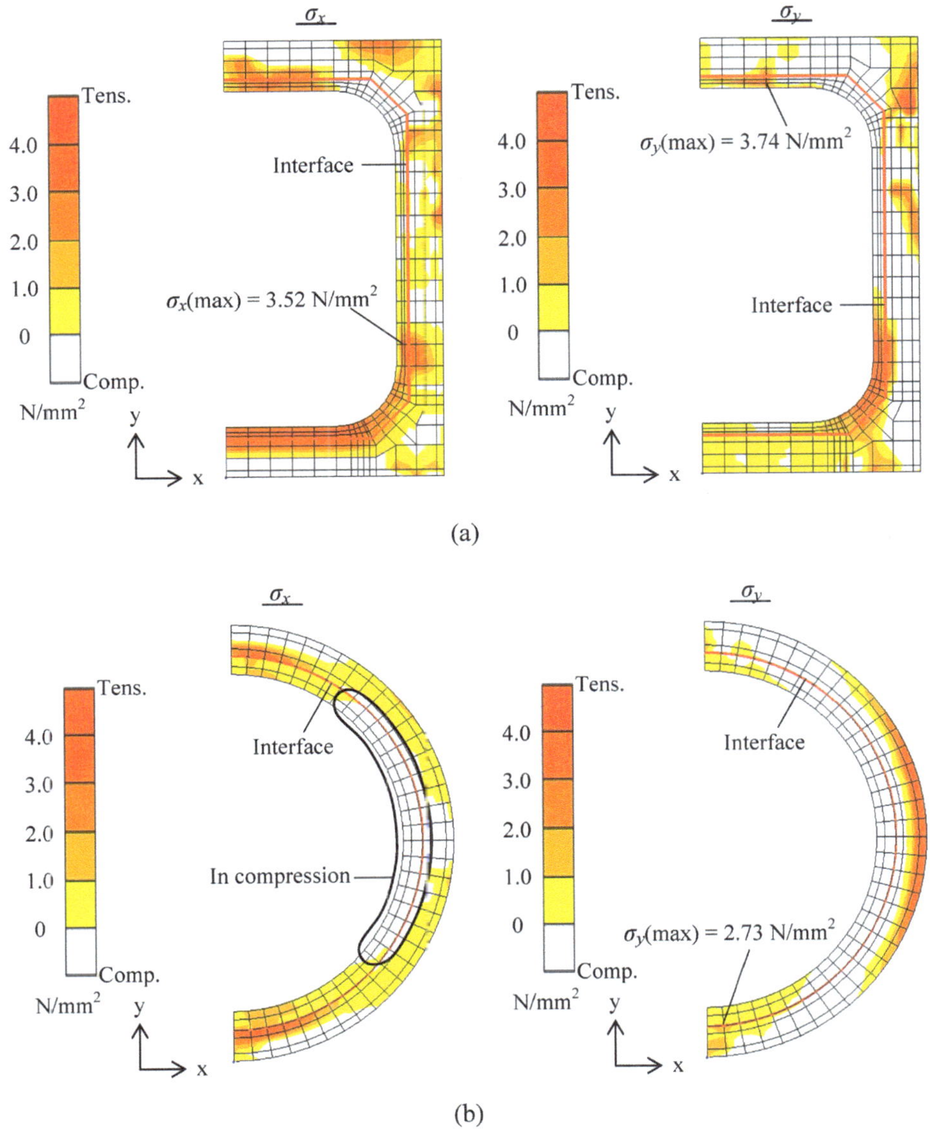

Figure 5.6: Numerical results of maximum tensile stress at the interfaces of renovated pipe specimens with rigid bonds under compressive loading: (a) box-culvert pipe; (b) circular pipe

To fully explore the advantages of the composite pipe method in sewer renovation while keeping safe design in mind, a semi-composite pipe structure is proposed for this renovation method based on the following assumptions:

1. Perfect bond after construction: At the completion of renovation construction, complete integration of the original pipe and the renovation layer is achieved.
2. No-tension interface: When tensile stress occurs at the interface under the external loads, the perfect bond is replaced by free surface modelling, or a no-tension interface model, to terminate stress transfer.
3. Perfect bond under compression: In the compression zones of the renovation layer, a perfect bond for the interface is assumed, which allows a continuous transfer of shear and compressive stresses between the original pipe and the renovation layer.

The resulting structure may be considered as a composite structure with mechanical connectors set in the compression zones that are intercepted by no-tension interface zones along the enclosed renovation layer, and thus is called a semi-composite pipe structure. The no-tension interface zones and the perfect bond interface zones of a semi-composite pipe structure can evolve with the change of load condition. A perfect bond zone can become a no-tension zone when tensile stress arises. However, a no-tension zone is irreversible, though under compression it can still transfer compressive stresses, but without transferring shear.

Figure 5.7 illustrates the corresponding semi-composite pipe structures for the present two cases. The perfect integration of the original pipe and the renovation layer in the compression zones maximises the renovated pipe strength, and the free interface deformation of the structural components in the no-tension zones compromises the integration effect of a composite structure in these regions. It is this potential debonding of the interface in the tension zones of the renovation layer in an otherwise fully-integrated composite pipe that gives rise to the unique semi-composite pipe structure of the composite pipe method. No-tension interface modelling in numerical analysis is illustrated in Fig. 5.8, where the interface is modelled by using dummy elements and dual nodes of the same coordinates. For the rigid bond condition the spring coefficients connecting the dual nodes are assumed to be infinite. When tensile stress arises, the spring connections in the vertical and horizontal directions are removed to allow free deformation of the interfaces.

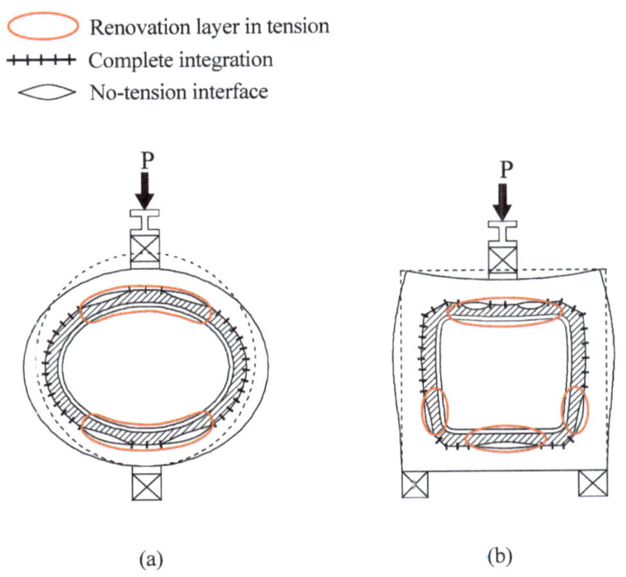

Figure 5.7: Conceptual drawings of the semi-composite pipe structures of composite pipe specimens during external pressure test

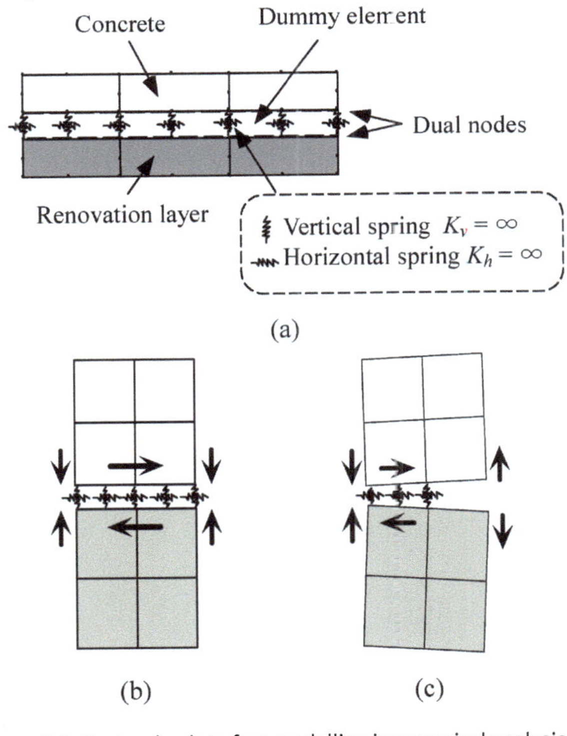

Figure 5.8: No-tension interface modelling in numerical analysis: (a) interface modelling by dummy elements and dual nodes; (b) perfect bond in compression; (c) no-tension interface in tension

5.2.2 Material modelling

(1) Overview

In order to predict the structural behaviour of the semi-composite pipe structure of a renovated sewer pipe up to the failure load by numerical analysis, accurate material modelling of the structural components is essential. The materials involved are the concrete and reinforcing steel bars in the existing pipe, the cementitious grout and the surface material (forming the lining surface) in the renovation layer. In most of the pipe-reforming methods, the renovation layer is usually reinforced by profiled steel sheets or steel frames. In the following, the stress-strain relations of these materials that form the fundamental basis for FE analysis are explained.

(2) Concrete in compression

As the structural members of sewer pipes (i.e., the top and bottom plates, and the sidewalls) are mainly designed to resist bending moments and axial forces under the traffic loads and the surrounding earth pressure, the following code-recommended stress-strain relations are employed (JSCE, 2009):

$$\sigma'_c = k_1 f'_{cd} \frac{\varepsilon'_c}{0.002}\left(2 - \frac{\varepsilon'_c}{0.002}\right) \quad \text{for } \varepsilon'_c \le 0.002 \tag{5.1a}$$

$$\sigma'_c = k_1 f'_{cd} \quad \text{for } \varepsilon'_c > 0.002 \tag{5.1b}$$

with

$$k_1 = 1 - 0.003 f'_{ck} \le 0.85$$

$$\varepsilon'_{cu} = \frac{155 - f'_{ck}}{0.002}\left(2 - \frac{\varepsilon'_c}{0.002}\right); \quad 0.0025 \le \varepsilon'_{cu} \le 0.0035$$

where σ'_c = compressive stress of concrete; ε'_c = compressive strain of concrete; f'_{ck} = characteristic value of compressive strength of concrete (N/mm²); f'_{cd} = design compressive strength of concrete (N/mm²). The corresponding stress-strain curve is illustrated in Fig. 5.9.

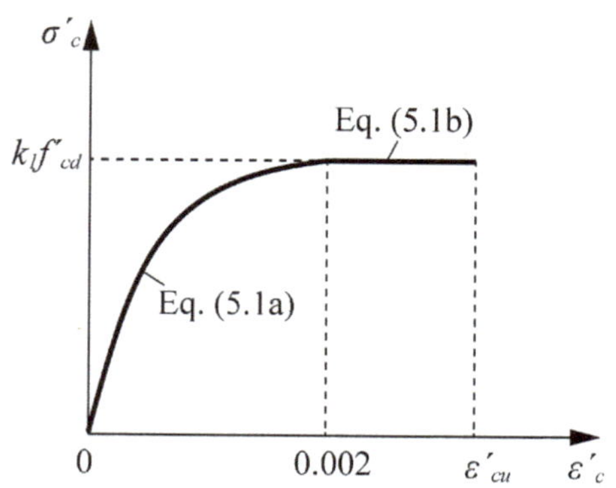

Figure 5.9: Stress-strain relationship of concrete in compression (JSCE, 2009)

(3) Concrete in tension: crack analysis based on fracture mechanics of concrete
When in tension, the stress-strain relation of concrete is assumed to be linear-elastic up to the tensile strength of concrete. After that, crack analysis of concrete is carried out based on the tension-softening law of concrete, which governs the dissipation of fracture energy in cracked concrete. Refer to Chapter 4, Section 4.4 for details.

For reinforced concrete cracks are usually distributed along the tension bars in the tension zones of the structure, and this type of cracking behaviour is best modelled by employing the smeared crack modelling approach discussed in Section 4.6. The stress-strain relation of the cracked concrete can be defined by using the crack band model, the non-orthogonal crack model, or the localised smeared crack model. Respectively, the stress-strain relation is defined by Eq. (4.62), Eq. (4.69), and Eq. (4.81). In this book the localised smeared crack model, which uses the secant modulus of elasticity for strain softening, is employed for crack analysis.

In plain concrete structures such as some types of manhole and some ageing sewer pipes with little steel reinforcement left due to corrosion, cracks tend to localise in the tension zones. This type of cracking behaviour should be modelled by the discrete crack modelling approach discussed in Section 4.5. Discrete crack modelling usually requires the crack location to be identified before carrying out crack analysis (although this can be avoided by presetting multiple crack paths in potential cracking areas).

Unlike the smeared crack approach in which the stress-strain relation of a cracked element is defined based on the continuum assumption, in the discrete crack approach this relation is obtained only after the crack equations for localised cracking behaviour are solved to obtain exact information on the crack or cracks, i.e., the length and orientation of the crack, the cohesive forces acting on the crack surfaces, and the crack-opening width. By solving the boundary value problem including the new crack surfaces as a part of the boundary, the stress-strain relation of a cracked element is obtained. If the crack problem involves only a single crack, the crack equations are given by Eqs. (4.46)-(4.48). If multiple cracks are involved, the crack equations can be obtained by expanding Eqs. (4.49)-(4.59), which define the crack equations for two cracks.

(4) Reinforcing steel bars
It is assumed that the bond between concrete and rebar is rigid. The stress-strain curve of reinforcing steel bars in tension and in compression is assumed to be elastic-perfectly plastic, as shown in Fig. 5.10 with f_{yd} = design yield strength of steel and E_s = elasticity coefficient of steel.

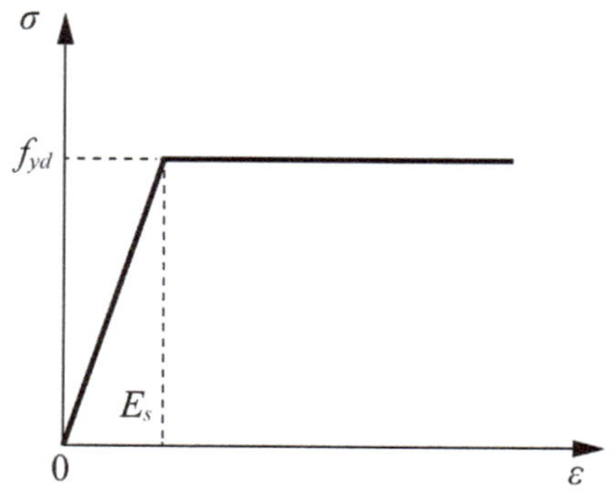

Figure 5.10: Stress-strain relationship of reinforcing steel bars

(5) Renovation layer

a) Profiled steel sheets or steel frames

The stress-strain curve of profiled steel sheets or steel frames used for reinforcing the renovation layer is also assumed to be elastic-perfectly plastic, as shown in Fig. 5.10.

b) Cementitious grout

- When in compression, the stress-strain relation of the cementitious grout is approximated by that of the concrete in compression, as shown in Fig. 5.9.
- When in tension, cracks develop in the cementitious grout as in concrete, and the previous discussion on concrete cracking is also valid for the cementitious grout.

c) PVC profile

The strength of the PVC profile used as the surface material of the renovation layer is not taken into account in evaluating the load-carrying capacity of a renovated sewer pipe in numerical analysis. In buckling analysis of the invert lining under groundwater pressure, however, its material properties are considered in evaluating the equivalent elasticity coefficient of the composite lining. This topic is treated in Section 5.4.

5.3 Numerical Studies of Fracture Behaviours in Renovated Sewer Pipes and Manholes

5.3.1 Numerical analyses of load-carrying capacity tests on real-size pipe specimens using the smeared crack modelling approach

(1) Cases studied

The cases to be studied are chosen from the load-carrying capacity tests discussed in Chapter 3, which are listed in Tables 3.2 and 3.3. Figure 5.11 shows four 1500×1500 mm rectangular pipes selected from the seven cases of Table 3.2. Case 1 is an original pipe with a standard doubly-reinforced cross section, and Case 2 is a standard renovation of Case 1. Case 3 is an original pipe without the inner concrete cover over the tension rebars, and Case 4 is a standard renovation of Case 3. Cases 1 to 4 correspond to Cases 1, 2, 4 and 5 of Table 3.2, respectively. The geometric dimensions and reinforcement details are shown in Fig. 3.2, and the material properties are given in Table 3.4.

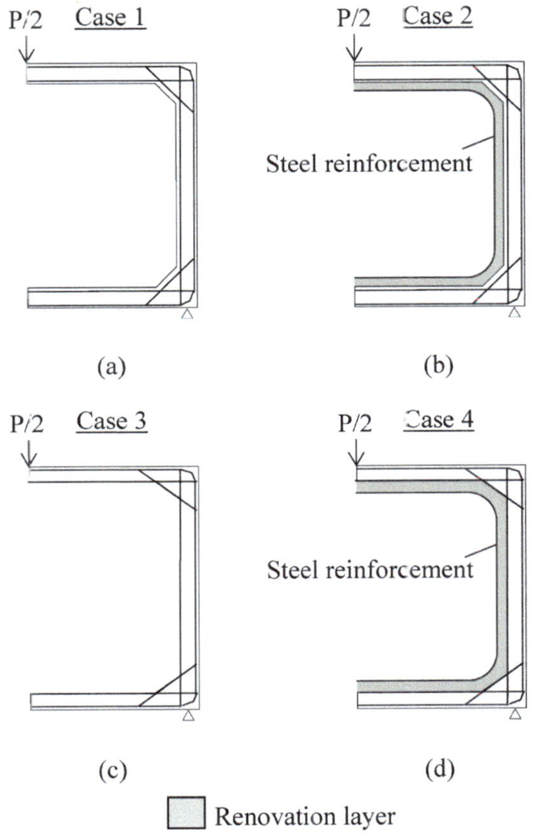

Figure 5.11: Numerical studies on fracture tests of 1500 × 1500 mm rectangular pipes: (a) Case 1: original pipe with standard doubly-reinforced cross section; (b) Case 2: standard renovation of Case 1; (c) Case 3: original pipe without inner concrete cover; (d) Case 4: standard renovation of Case 3

Figure 5.12 shows four ϕ1500 mm circular pipes selected from the five cases of Table 3.3. Case 5 is an original pipe with a standard doubly-reinforced cross section, and Case 6 is a standard renovation of Case 5. Cases 7 and 8 are also standard renovations of Case 5, using non-reinforced PVC profiles. In Case 7 the lining thickness is uniform as in Case 6, and in Case 8 it thins down from top to bottom: at the bottom of the pipe the renovation layer is reduced to the thickness of the PVC profile. Cases 5 to 8 correspond to Cases 1, 2, 4 and 5 of Table 3.3, respectively. The geometric dimensions and reinforcement details are shown in Fig. 3.3, and the material properties are given in Table 3.5. The loading conditions of the tests are shown in Fig. 3.4. Note that these cases are chosen because they represent the common types of existing sewer pipe and renovation condition.

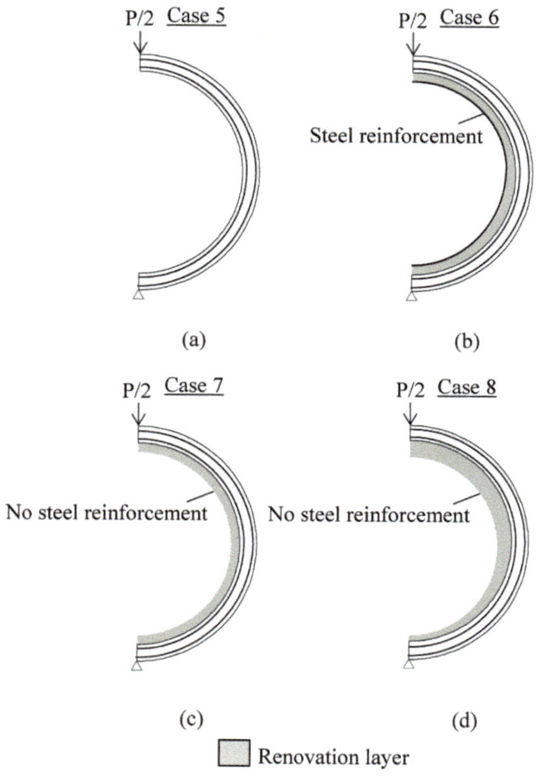

Figure 5.12: Numerical studies on fracture tests of ϕ1000 mm circular pipes; (a) Case 5: original pipe with standard doubly-reinforced cross section; (b) Case 6: standard renovation of Case 5 with steel reinforcement in the PVC profiles; (c) Case 7: standard renovation of Case 5 without steel reinforcement in the PVC profiles; (d) Case 8: standard renovation of Case 5 without steel reinforcement in the PVC profiles which are in contact with the bottom slab

Figures 5.13 and 5.14 show FE models of Cases 1 to 4, and Cases 5 to 8, respectively. As shown, the interface of a renovated pipe is modelled by using dummy elements. Note that these FE models are generated automatically by the SPRana design software, which will be introduced in Chapter 7.

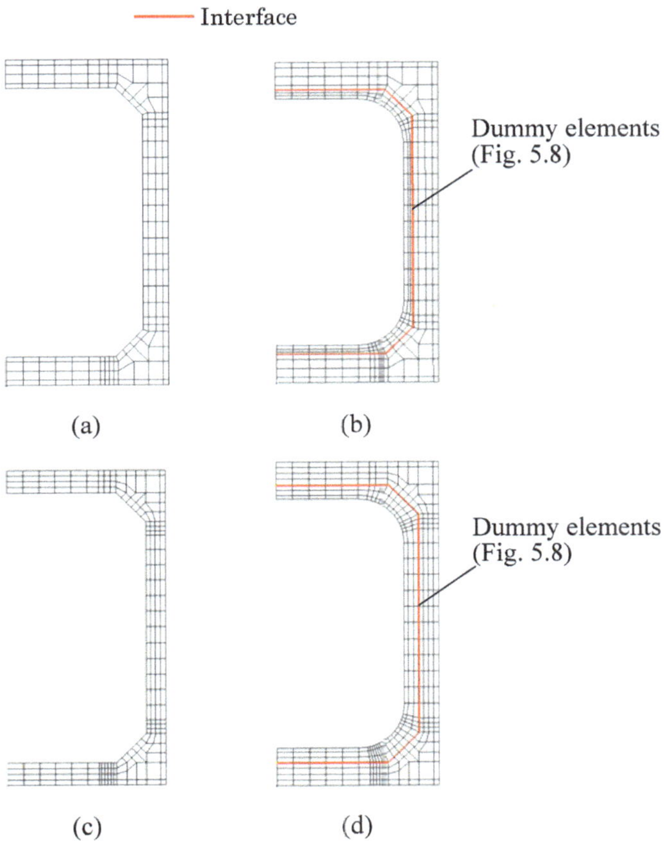

Figure 5.13: FE models for numerical analysis of fracture tests on rectangular pipes: (a) Case 1; (b) Case 2; (c) Case 3; (d) Case 4

(2) Numerical results of rectangular pipes: Cases 1 to 4
Figures 5.15 to 5.18 summarise the numerical results of Cases 1 to 4, including the load-displacement relations, cracking behaviour, stress distribution and debonding areas at two stages of loading. For comparison, the test results of load-displacement relation are also shown.

In Case 1, cracking initiated in the top plate at the load level of 87.0 kN/m, and the maximum load was obtained at 327.4 kN/m, which agreed reasonably well with the average test result of 364.3 kN/m with a difference ratio of 10%. At the maximum load, cracks occurred mainly in the top plate and the wall near the haunch area from the outside. There were also some cracks in the wall near the lower haunch from the inside. In Case 2, after renovation the crack initiation load increased to 205.0 kN/m, and the maximum load reached 552.0 kN/m. Compared with the average test result of 643.4 kN/m, the difference ratio was 14%. At the maximum load, cracks concentrated mainly in the top plate and the upper wall from the outside, with some cracks appearing in the lower wall, near the interface.

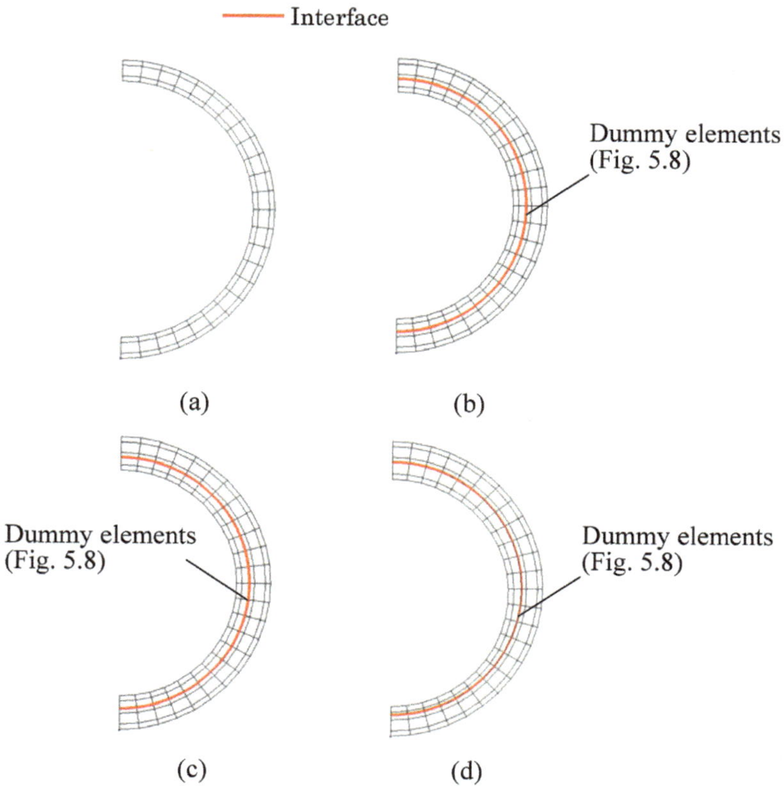

Figure 5.14: FE models for numerical analyses of fracture tests on circular pipes: (a) Case 5; (b) Case 6; (c) Case 7; (d) Case 8

Debonding of the interface is illustrated in the stress contours in Fig. 5.16. As seen, at the crack initiation, debonding occurred around the lower haunch area, with a debonding ratio of 26% over the total length of the interface. At the maximum load, the debonding areas expanded greatly, with a debonding ratio of 73%. Note that these figures do not necessarily reflect the actual debonding ratio occurred in the renovated test specimens. In the no-tension interface model the bonding strength of the cementitious grout is completely ignored when tension occurs in the interface, while in the test specimens debonding occurs only when the interfacial tensile stress reaches the bonding strength. Therefore, the actual debonding ratio in the test specimens should be lower.

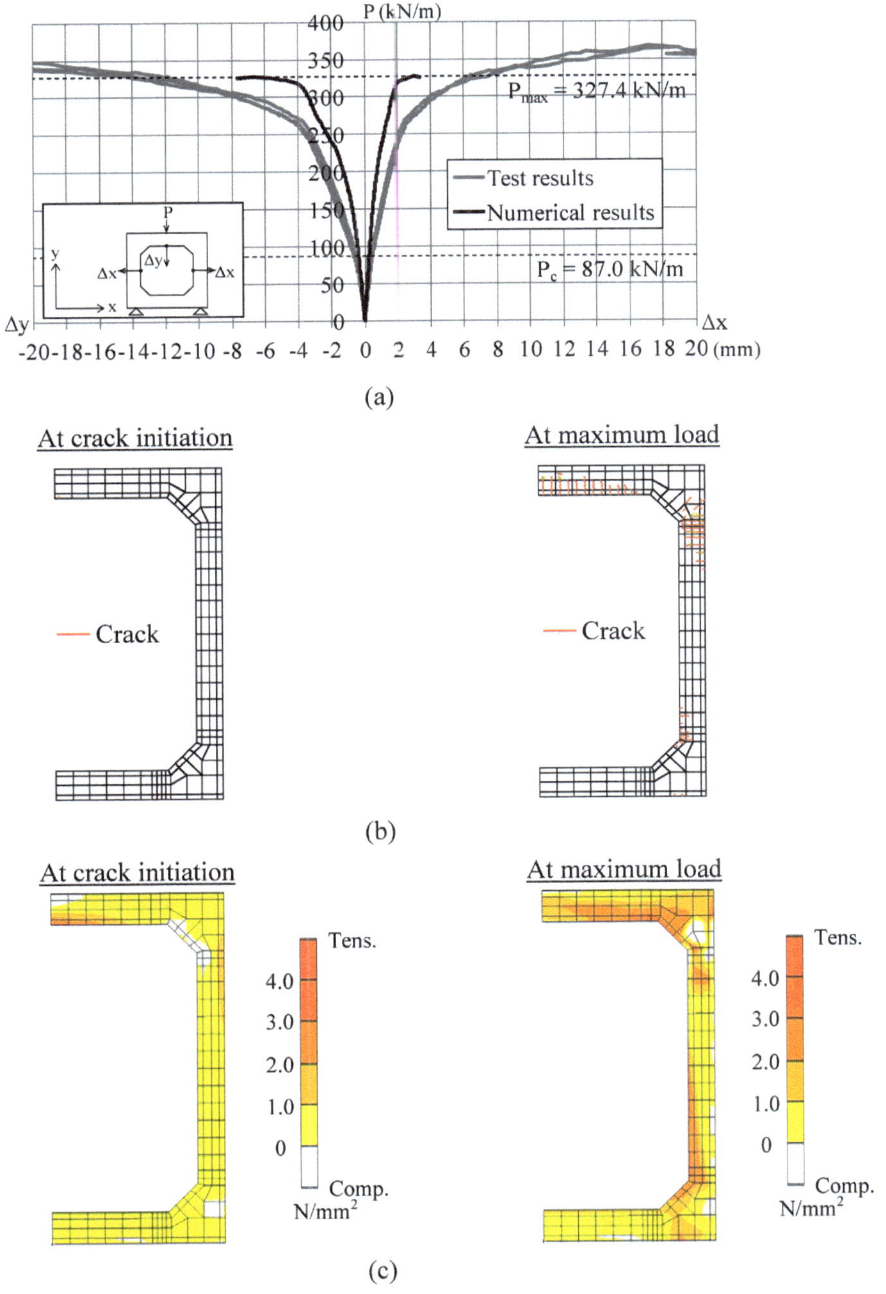

Figure 5.15: Numerical results of Case 1: (a) load-displacement relations; (b) cracking behaviour; (c) maximum principal stress

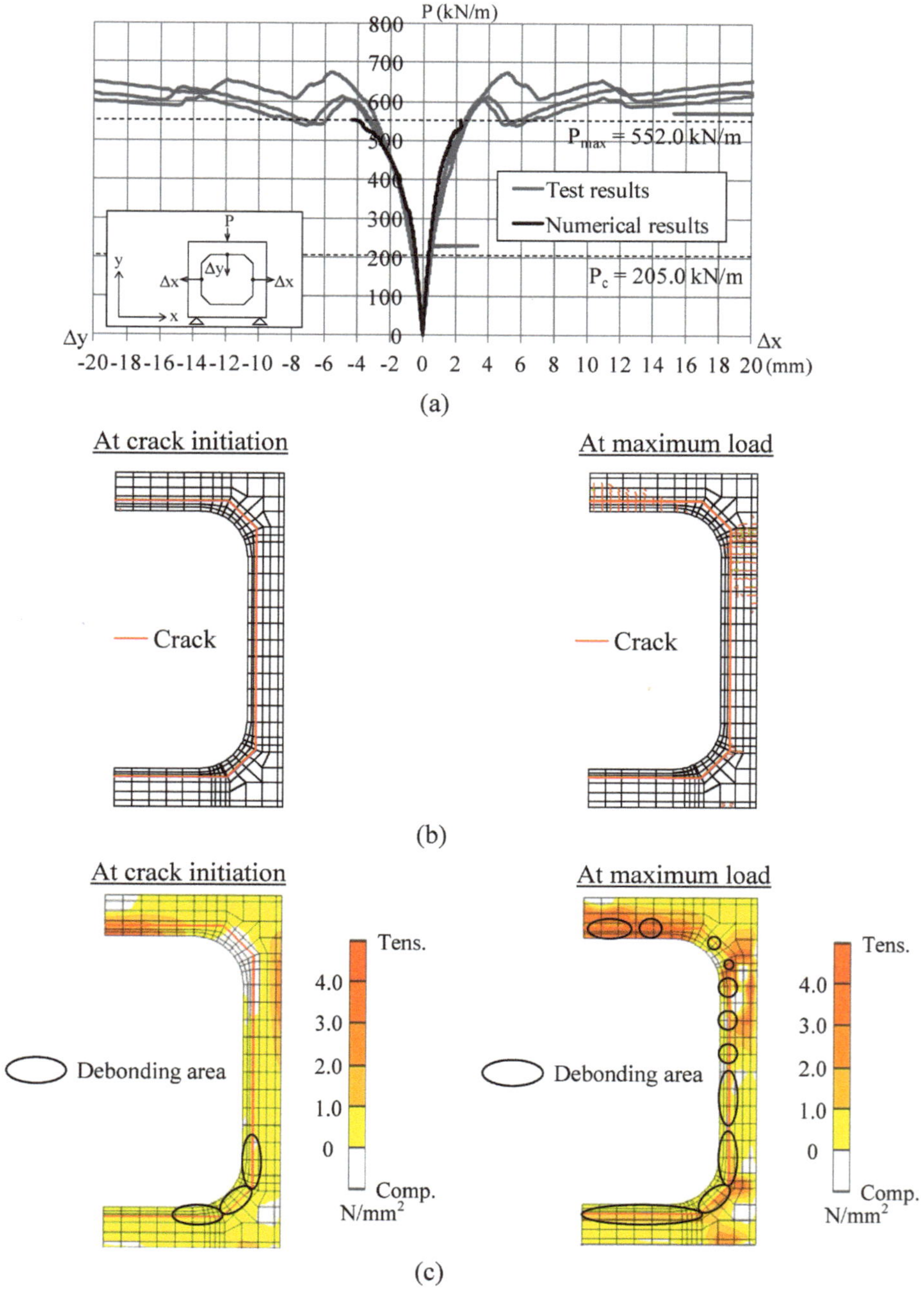

Figure 5.16: Numerical results of Case 2: (a) load-displacement relations; (b) cracking behaviour; (c) maximum principal stress and debonding areas

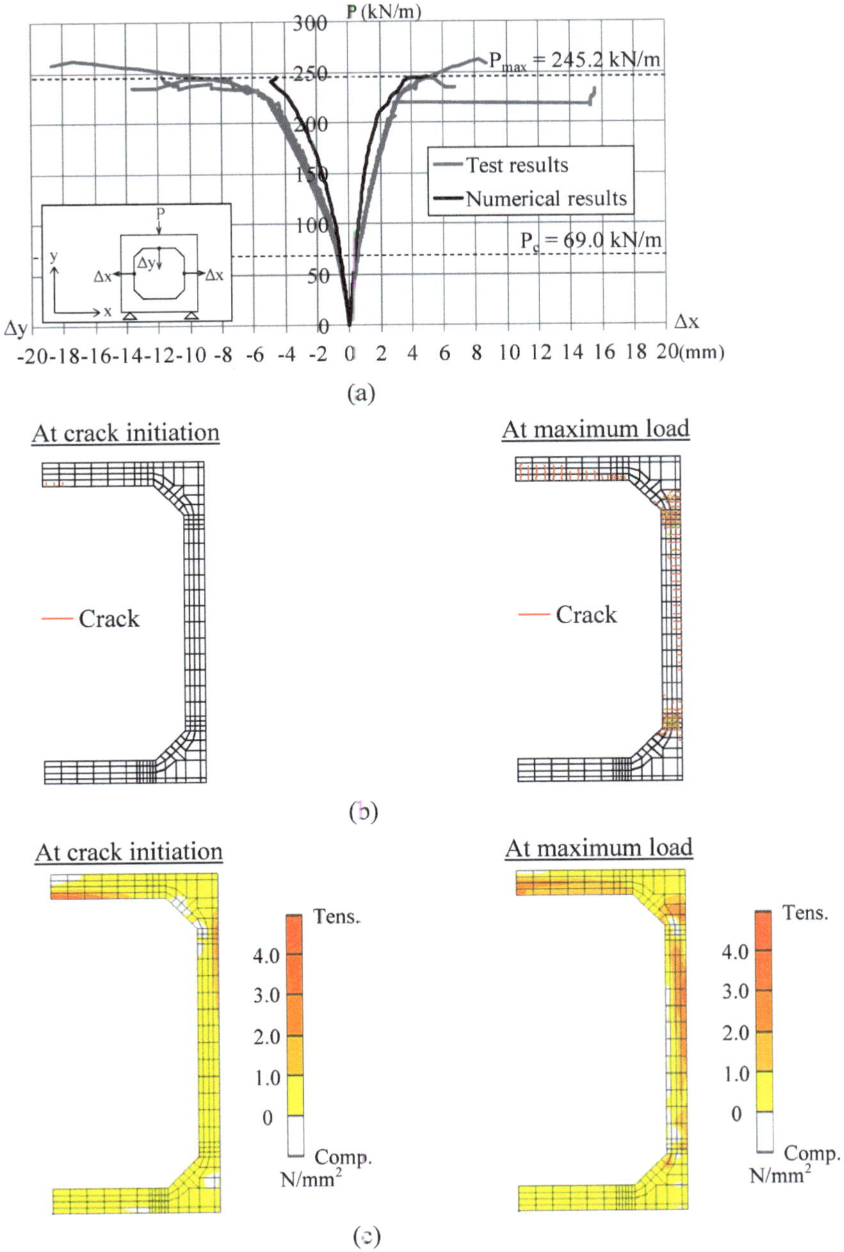

Figure 5.17: Numerical results of Case 3: (a) load-displacement relations; (b) cracking behaviour; (c) maximum principal stress

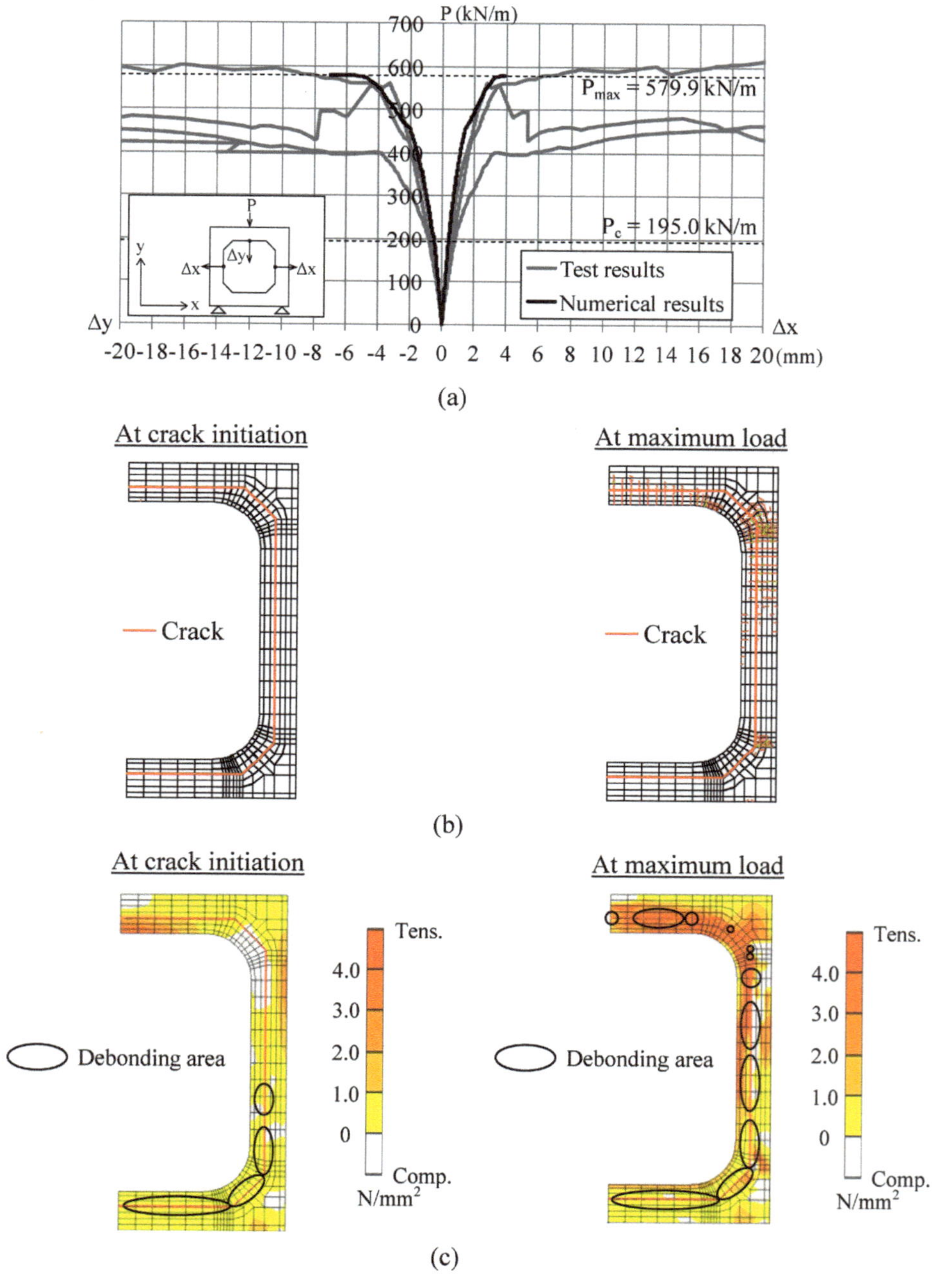

Figure 5.18: Numerical results of Case 4: (a) load-displacement relations; (b) cracking behaviour; (c) maximum principal stress and debonding areas

In Case 3, cracking initiated at 69.0 kN/m, and the maximum load was 245.2 kN/m, which compared well with the test average of 251.2 kN/m. By removing the inner concrete cover over the tension rebars, the maximum load decreased 25% from that of Case 1. At the maximum load, cracks spread to the entire top plate and the wall from the outside, and the areas near the upper and lower haunches were heavily cracked. In Case 4, after renovation the crack initiation load increased to 195.0 kN/m, and the maximum load reached 579.9 kN/m, also comparing relatively well with the test average of 545.1 kN/m. At the maximum load cracks occurred not only in the top plate and the upper wall from the outside, but also in the middle wall from the inside. As shown in Fig. 5.18, debonding at the crack initiation occurred in the bottom plate and the lower wall, and the debonding ratio was 44%. At the maximum load debonding areas spread to the top plate and the upper wall with a debonding ratio of 77%. The interface in the upper haunch area was mainly under compression in the direction normal to the interface, and therefore the bonding condition there remained basically rigid.

(3) Numerical results of circular pipes: Cases 5 to 8

Figures 5.19 to 5.22 show the numerical results of Cases 5 to 8. In Case 5, the crack initiation load was 33.0 kN/m, and the maximum load was 92.4 kN/m, agreeing perfectly with the test average of 92.6 kN/m. In Case 6, after renovation the crack initiation load became 55.5 kN/m, and the maximum load increased to 143.4 kN/m, also comparing well with the test average of 145.5 kN/m. At the crack initiation, debonding of the interface occurred in the top and bottom areas, and also in some areas near the 45-degree lines measured from the horizontal centreline of the pipe. The debonding zones remained the same up to the maximum load, with a debonding ratio of 50%.

In Case 7, cracks initiated at the load level of 36.0 kN/m, and the maximum load reached 132.4 kN/m, slightly greater than the test average of 118.1 kN/m with a difference ratio of 12%. At the crack initiation, debonding occurred in the top and bottom of the pipe and two other locations, with a debonding ratio of 43%. At the maximum load the ratio of debonding increased to 61%. In Case 8 the crack initiation load was 39.0 kN/m, and the maximum load was 125.0 kN/m, comparing relatively well with the test average of 118.6 kN/m. The debonding behaviour was similar to Case 7, and the debonding ratio was 47% at the crack initiation and was 54% at the maximum load. Cracking behaviours in the four circular pipes were more or less similar: cracks started from the top and the bottom of the pipe, and at the maximum load the tension sides of the pipe from the outside were heavily cracked.

(4) General comments

Based on the results of numerical study, it can be concluded that the no-tension interface model and the semi-composite pipe structure combined with the fracture-mechanics-based modelling theories work well for predicting the maximum loads of composite sewer pipes. As discussed above, the difference ratios between the numerical results and the test results are less than 15%, which is considered quite accurate in predicting the load-carrying capacity of reinforced concrete structures.

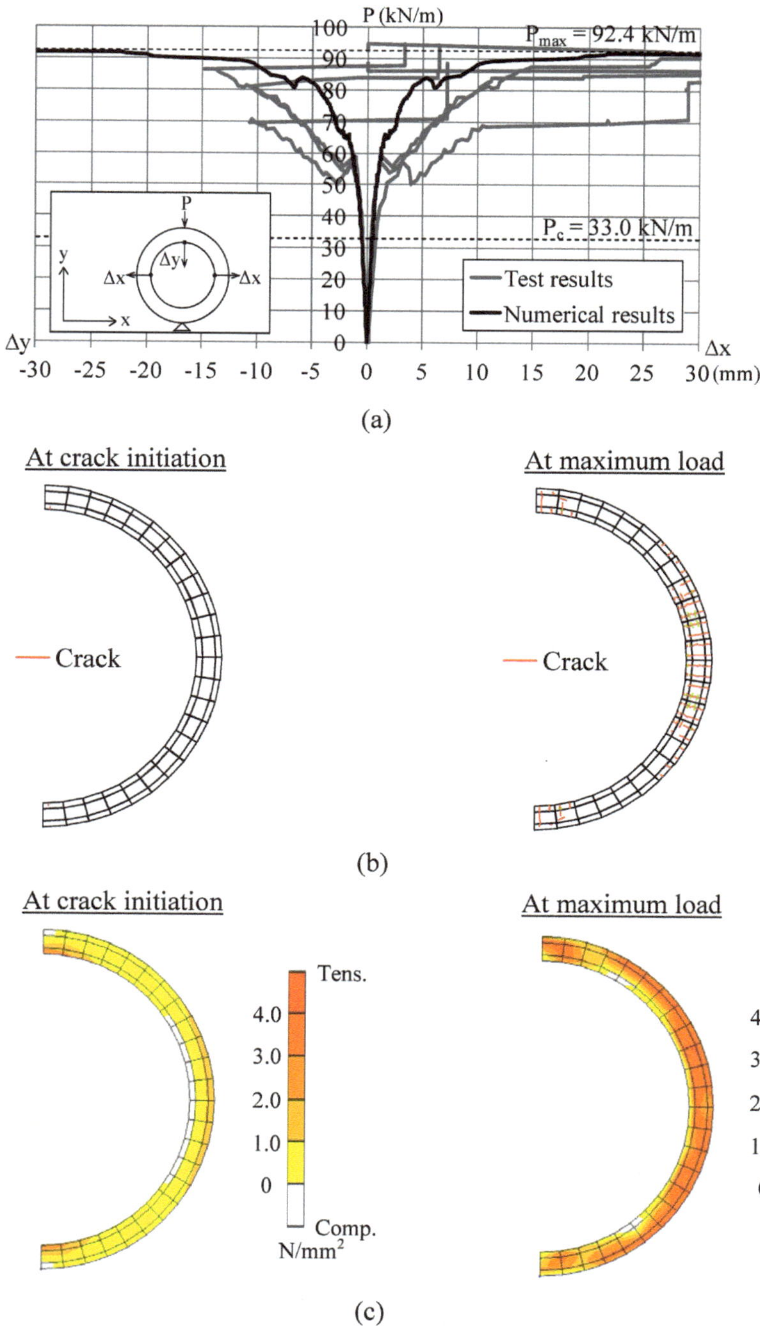

Figure 5.19: Numerical results of Case 5: (a) load-displacement relations; (b) cracking behaviour; (c) maximum principal stress

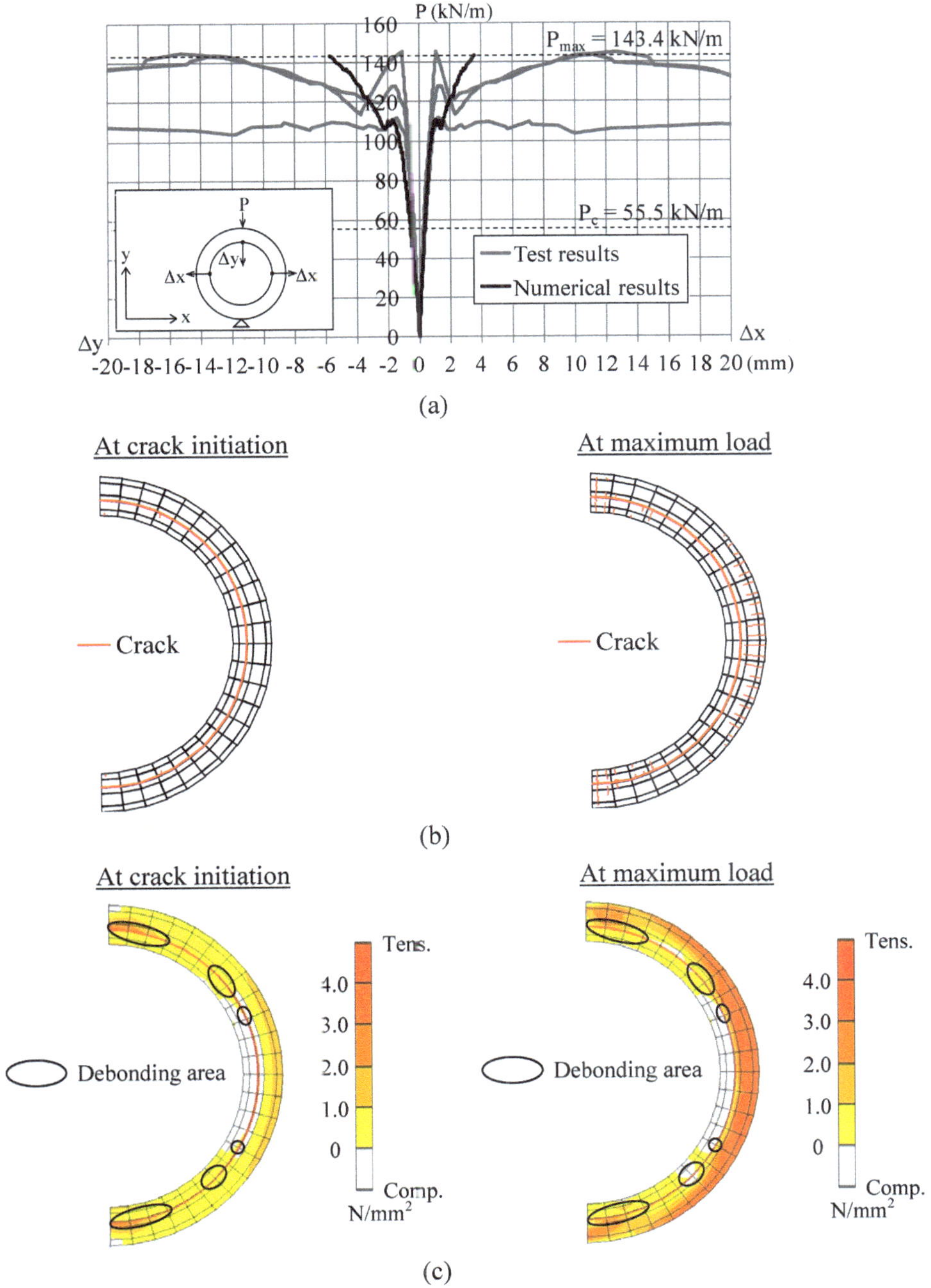

Figure 5.20: Numerical results of Case 6: (a) load-displacement relations; (b) cracking behaviour; (c) maximum principal stress and debonding areas

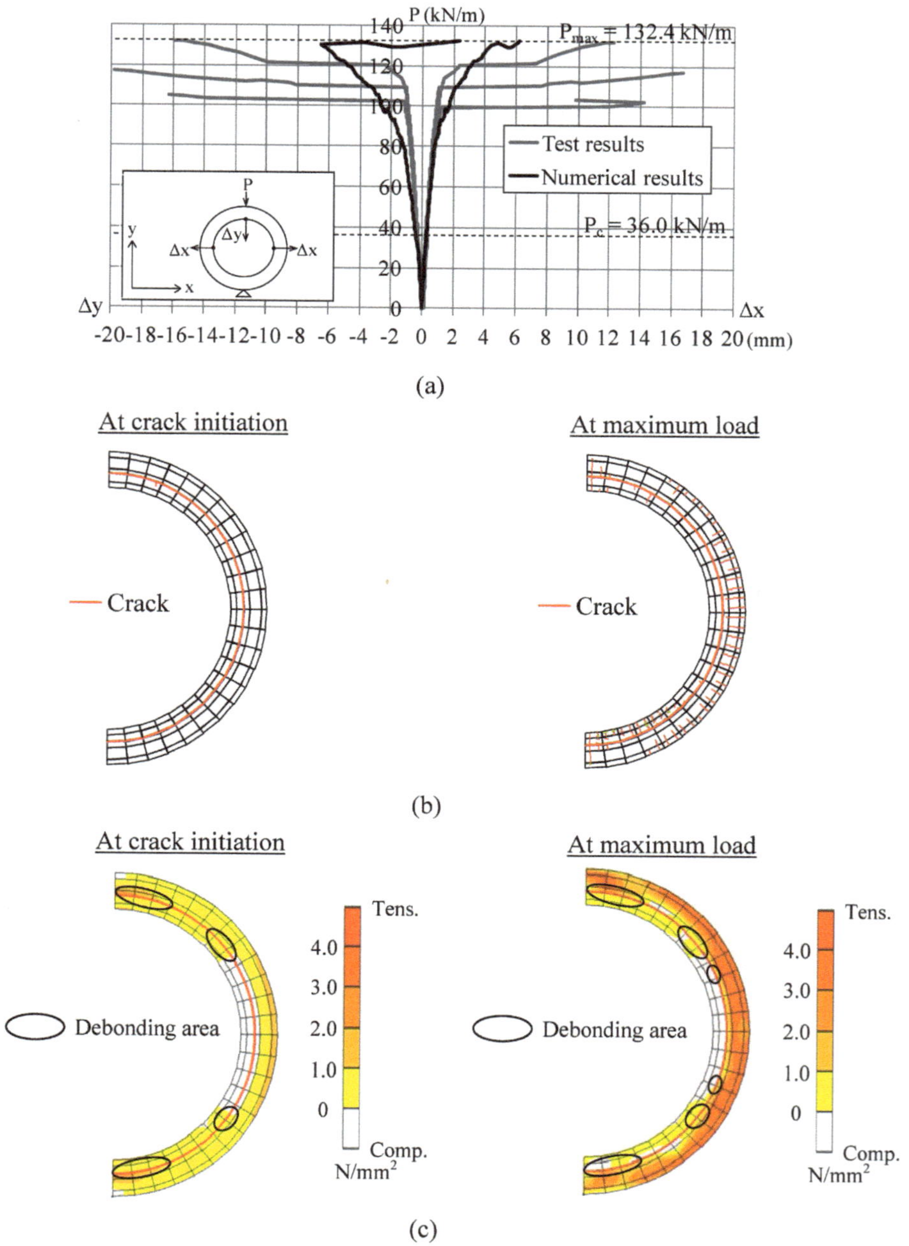

Figure 5.21: Numerical results of Case 7: (a) load-displacement relations; (b) cracking behaviour; (c) maximum principal stress and debonding areas

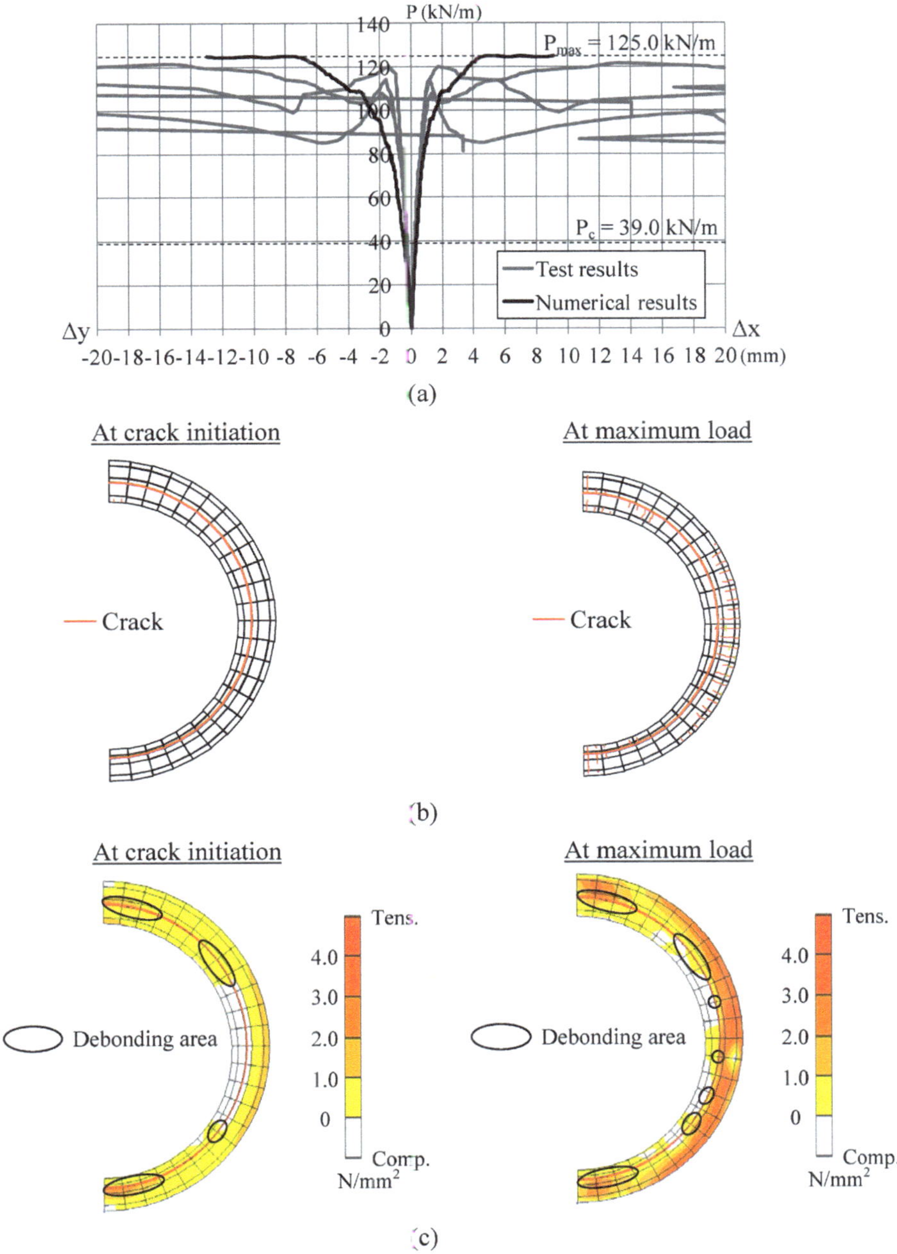

Figure 5.22: Numerical results of Case 8: (a) load-displacement relations; (b) cracking behaviour; (c) maximum principal stress and debonding areas

The no-tension interface model is a conservative modelling approach for evaluating the bonding strength of a composite pipe based on a safe-design consideration. Although it may not reflect the actual debonding process in a renovated pipe under loading, based on the above accuracy assessment the final semi-composite pipe structure with the debonding and various cracking conditions derived from the numerical analyses should compare reasonably well with the actual structural state of the test specimen at its maximum load.

5.3.2 Numerical analyses of load-carrying capacity tests on real-size manhole specimens using the discrete crack modelling approach

(1) Cases studied

A modified epoxy resin as a new corrosion-preventive coating material, which utilises treated incineration ash of sewage sludge as its base compound, has been developed recently in Japan and has found its first application in repairing corroded manholes. A fracture test on plain concrete manhole specimens with their inner surfaces scraped and coated with this new material revealed that the resin greatly strengthens the repaired test specimens, showing its potential usage as a new renovation material in the ongoing reconstruction of ageing sewage facilities in Japan (Kurozumi, 2014).

Table 5.1 shows the material properties of the concrete and the surface coating material (SCM) used in the manhole fracture test. Compared with normal concrete, the elasticity modulus of the SCM is much lower, but its compressive strength and tensile strength are much higher, and its fracture energy is estimated to be ten times that of normal concrete. Also, it possesses a high bonding strength with concrete.

Table 5.1: Material properties of manhole specimens

	Compressive strength	**Tensile strength**	**Elasticity**	**Poisson's ratio**	**Fracture energy**	**Bond strength**
			E	v	G_F	
	(N/mm²)	**(N/mm²)**	**(kN/mm²)**		**(N/m)**	
Concrete	24.0	2.40	2.57	0.2	85.5	
Surface coating material (SCM)	50.0	9.5	6.5	0.3	951.87	2.70

Figure 5.23 shows the fracture test on a plain concrete manhole specimen, which is laid on its side and the load is applied from above. Figure 5.24 shows three types of full-size manhole specimen with the same outer diameter of 1300 mm. Case 1 is

an original manhole pipe with a wall thickness of 200 mm. Case 2 was prepared by first scraping the inner surface of an original pipe by 10 mm to represent the surface corrosion in ageing manholes (mainly caused by the sulphide in sewage) and then repairing it with the SCM by 10 mm, so the wall thickness of Case 2 remained the same as that of Case 1. Case 3 was prepared by scraping the inner surface of an original pipe by 20 mm and repairing it with the SCM by 10 mm, so the wall thickness of Case 3 was reduced by 10 mm to investigate the cost-effective coating thickness. Photo 5.2 shows the preparation procedures of the test specimens, and Photo 5.3 shows a fracture test in progress. The four cracks that occurred at the crown and the bottom from the inside, and at the two end points at the horizontal centre line from the outside are shown in Photo 5.4 for three loading stages: at crack initiation, at maximum load and at failure. Photo 5.5 shows the four broken pieces of the specimen after the test.

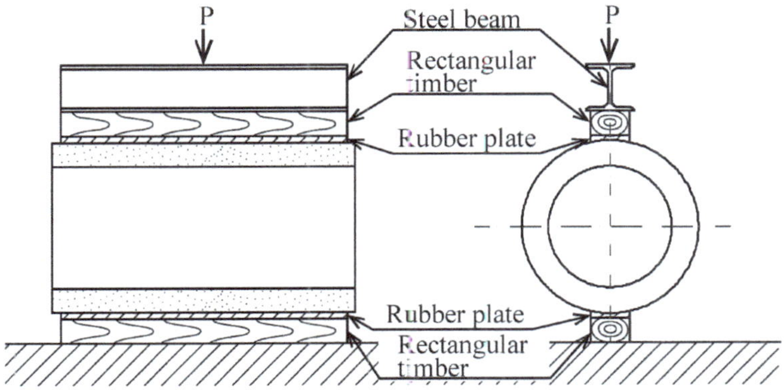

Figure 5.23: Fracture test of manhole specimens

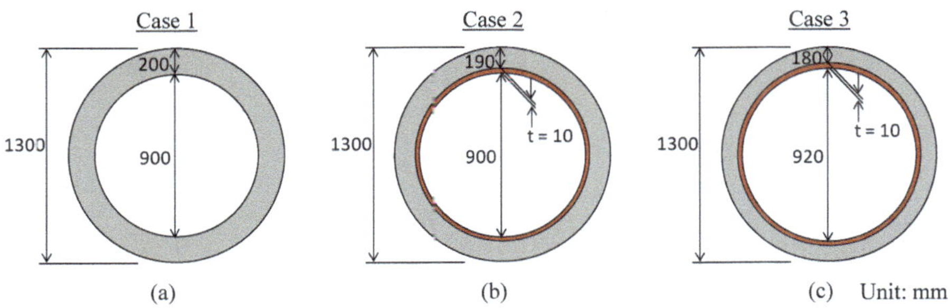

Figure 5.24: Test cases in fracture test on manhole specimens: (a) Case 1: original pipe; (b) Case 2: repaired pipe of unchanged thickness with 10 mm scraping and 10 mm coating; (c) Case 3: repaired pipe of reduced thickness with 20 mm scraping and 10 mm coating

Photo 5.2: Preparation of manhole specimens

Photo 5.3: Fracture test on manhole specimens

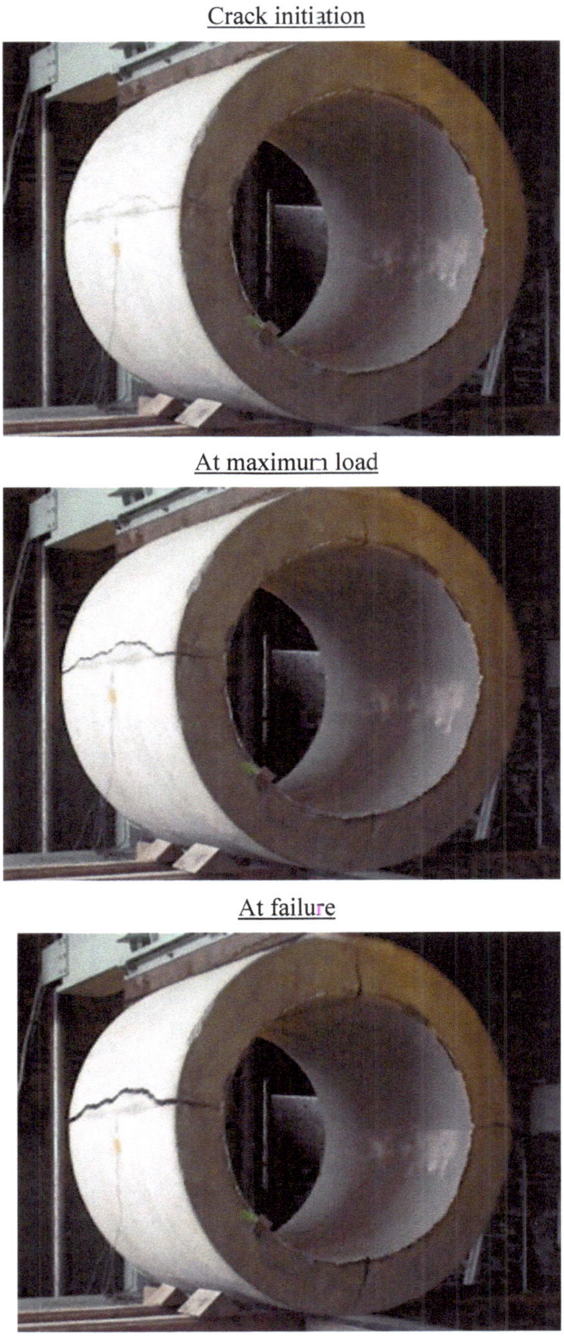

Photo 5.4: Crack propagation in manhole specimens during fracture test

Photo 5.5: Broken manhole specimen in four pieces

Figure 5.25 shows FE models of Cases 1 to 3 for crack analysis using the discrete crack modelling approach. Due to symmetry, only half models of manhole specimens are employed with three discrete cracks being modelled for each case, i.e., two vertical cracks, Crack A and Crack C, and one horizontal crack, Crack B. As shown, the crack path is modelled by using dummy elements. In numerical analysis the bilinear tension softening relation of concrete shown in Fig. 4.15(b) is employed, and the material properties of Table 5.1 are assumed.

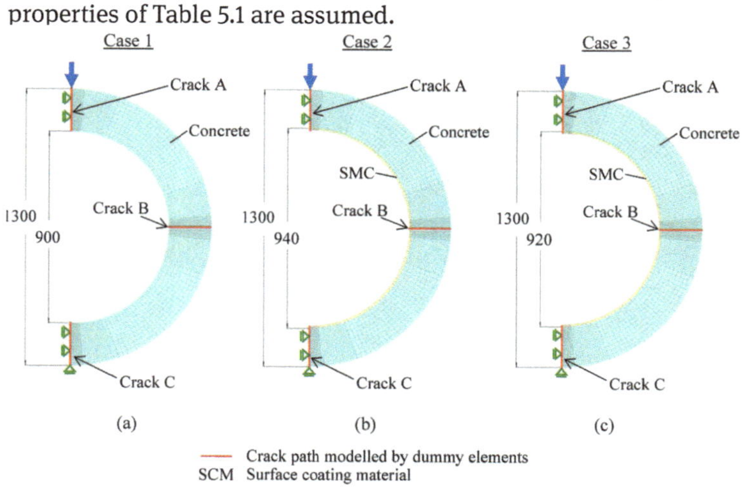

Figure 5.25: FE models for numerical analysis of fracture tests on manhole specimens: (a) Case 1: original pipe; (b) Case 2: repaired pipe of unchanged thickness with 10 mm scraping and 10 mm coating; (c) Case 3: repaired pipe of reduced thickness with 20 mm scraping and 10 mm coating

(2) Numerical results of Cases 1 to 3

Table 5.2 shows the maximum loads obtained from the fracture tests and numerical analyses, and the ratios of load comparison as well. For each case three tests were carried out, and the average value of the three obtained maximum loads was used for comparison with the numerical result. Based on the test results, the maximum loads of Cases 2 and 3 increased by 74% and 54% from Case 1, respectively. Figures 5.26 to 5.28 summarise the numerical results of crack analysis, including the load-displacement relation at the loading point, the stress distribution at the maximum load, and the crack propagation with the crack opening width (COW) along the wall thickness at the designated points of pipe deformation.

Table 5.2: Maximum loads of fracture test and numerical analysis

Case	Test	Average	/Case 1	Numerical analysis		
	P_{max} (N/mm)			P_{max} (N/mm)	Numer. /Test	/Case 1
1	110	108.7		137.9	1.27	
	100					
	116					
2	192	189.3	1.74	196.5	1.04	1.42
	193					
	183					
3	174	167.0	1.54	177.0	1.06	1.28
	171					
	156					

In Case 1, crack initiation began with the vertical cracks from the inside in the pre-peak region at the load level of 93.2 N/mm. The maximum load was obtained at the load level of 137.9 N/mm, which is greater than the test result by 27%. The horizontal crack initiated in the post-peak region at 104.4 N/mm, from the outside. Two target points in pipe deformation are selected for presenting the results of stress distribution and crack propagation: Δ = 0.29 mm at the maximum load and Δ = 0.70 mm before the final failure. The minimum and maximum principal stresses at Δ = 0.29 mm are shown. For crack propagation, Crack A and Crack C grew to 70 mm at the maximum load, reaching approximately one third of the wall thickness with a COW of about 0.03 mm at the wall surface. At Δ = 0.70 mm, the vertical cracks almost penetrated through the wall thickness, and the horizontal crack extended to one third of the wall thickness from the outside. The COWs of the vertical and horizontal cracks at the wall surface were 0.27 mm and 0.02 mm, respectively.

In Case 2, the vertical cracks initiated in concrete in the pre-peak region at the load level of 89.1 N/mm, and the crack initiation in the SCM next to the cracked concrete did not occur until the load reached a high level of 163.0 N/mm, which was clearly the result of the combined effects of the SCM's large plasticity, high tensile strength and high bonding strength. Unlike Case 1, the horizontal crack initiated in the pre-peak region at 180.7 N/mm, which was soon followed by the crack penetration in the SCM at 185.5 N/mm. The maximum load reached 196.5 N/mm, which compares well with the test result of 189.3 N/mm. Note that before reaching the maximum load, the stiffness of the manhole specimen was much reduced and the pipe deformation greatly increased. Compared with the numerical result of Case 1, the maximum load increased by 42%. The three selected target points for illustration are $\Delta = 0.34$ mm at the crack initiation in the SCM, $\Delta = 0.54$ mm at the maximum load, and $\Delta = 0.70$ mm before the final failure. The minimum and maximum principal stresses at $\Delta = 0.54$ mm are shown.

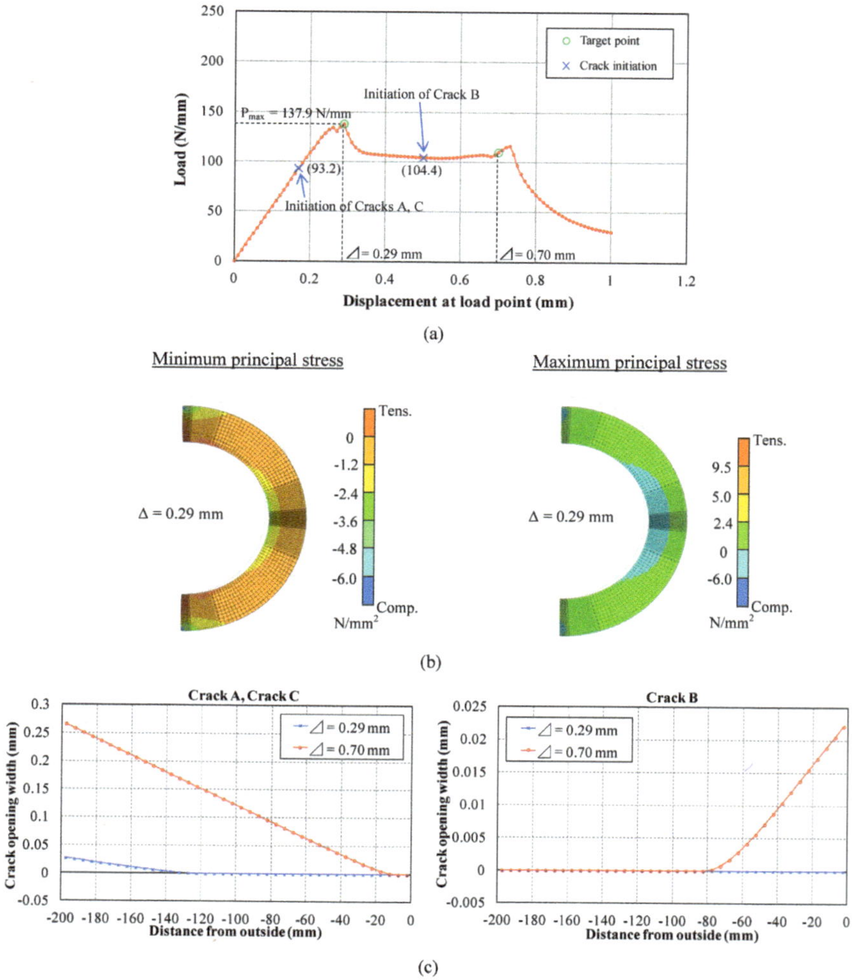

Figure 5.26: Numerical results of Case 1: (a) load-displacement relation; (b) principal stresses at Δ = 0.29 mm; (c) crack opening widths along the wall thickness

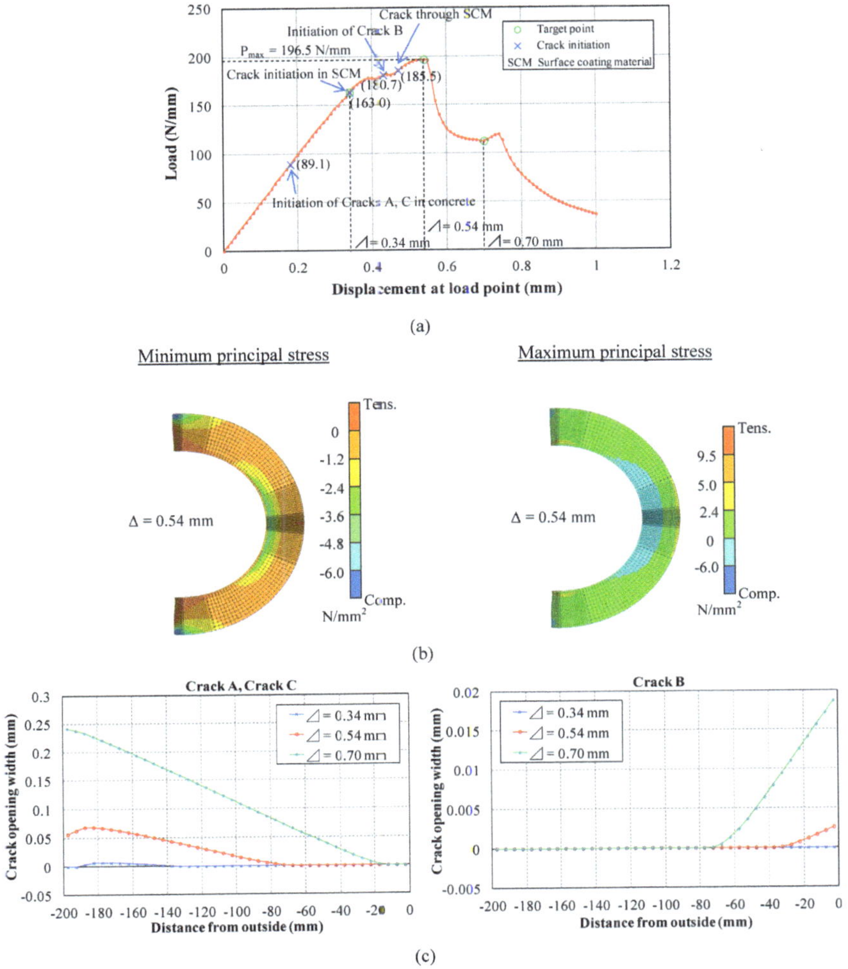

Figure 5.27: Numerical results of Case 2: (a) load-displacement relation; (b) principal stresses at Δ = 0.54 mm; (c) crack opening widths along the wall thickness

For crack propagation, at Δ = 0.34 mm Crack A and Crack C were 50 mm in length, and the COW at the wall surface was merely 0.01 mm. Compared with the cracking behaviour in Case 1 at the maximum load of 137.9 N/mm, it is clear that the crack propagation at Δ = 0.34 mm in Case 2 was greatly restrained to a shorter propagation length and a much smaller COW, even though under a much greater load of 163.0 N/mm. This demonstrates the crack restraining effect of the SCM on its host material, and consequently, its strength increasing effect as well. At the maximum load Crack A and Crack C extended to 130 mm, while Crack B was about 30 mm. At Δ = 0.70 mm, the vertical cracks almost penetrated through the wall thickness, and the horizontal crack reached one third of the wall thickness. The COWs of the vertical and horizontal cracks at the wall surface were 0.25 mm and 0.02 mm, respectively.

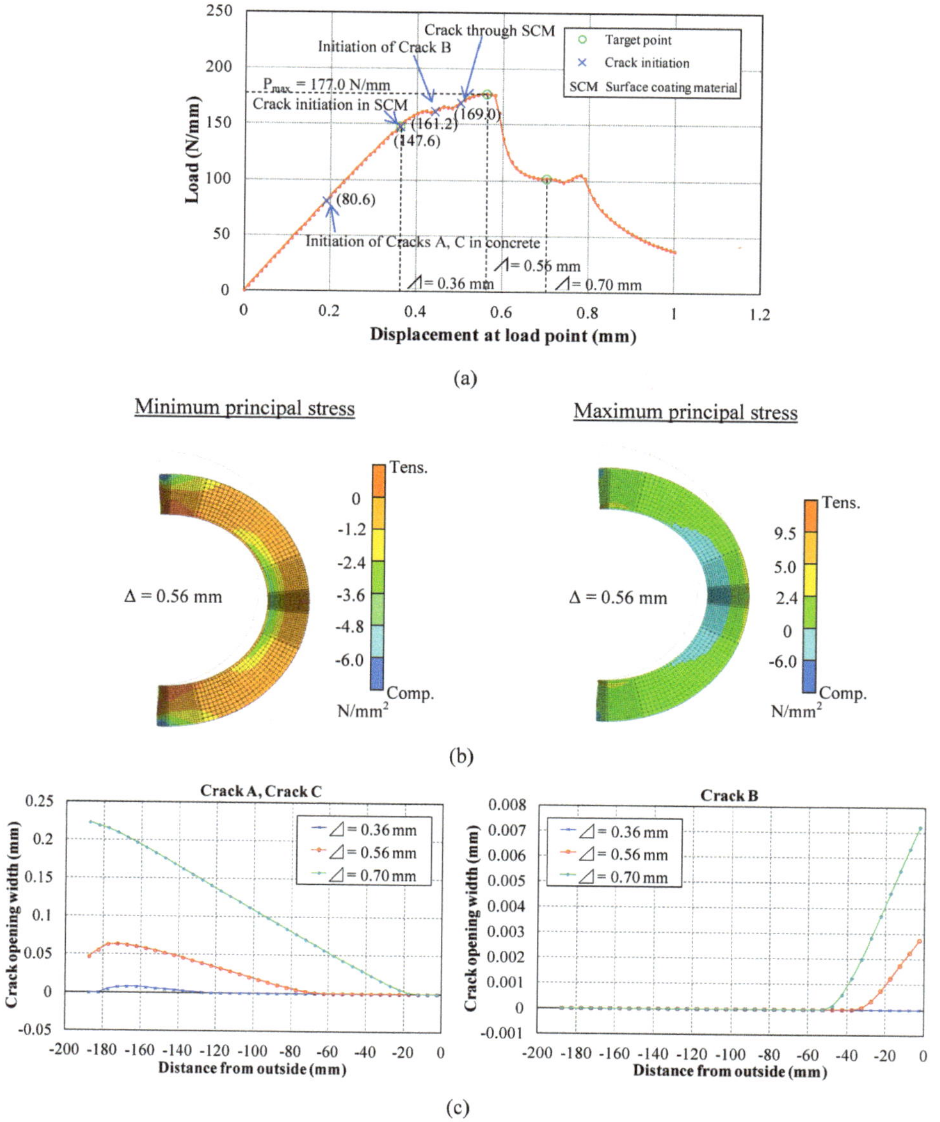

Figure 5.28: Numerical results of Case 3: (a) load-displacement relation; (b) principal stresses at Δ = 0.56 mm; (c) crack opening widths along the wall thickness

In Case 3, the vertical cracks initiated in the concrete in the pre-peak region at the load level of 80.6 N/mm, and the cracking of the SCM occurred when the load reached 147.6 N/mm. Like Case 2, the horizontal crack initiated in the pre-peak region at 161.2 N/mm, and the crack penetration through the SCM took place at 169.0 N/mm. The calculated maximum load was 177.0 N/mm, which also compares well with the test result of 167.0 N/mm. Compared with the numerical result of Case 1, the maximum load increased by 28%, even though the wall thickness in Case 3 was 10 mm thinner than the original pipe. This demonstrates the possibility of rehabilitating severely corroded manholes with a thin layer of coating by the SCM to restore the manhole strength to its original level. The three selected target points for illustration are $\Delta = 0.36$ mm at the crack initiation in the SCM, $\Delta = 0.56$ mm at the maximum load, and $\Delta = 0.70$ mm before the final failure. The minimum and maximum principal stresses at $\Delta = 0.56$ mm are shown.

For crack propagation, at $\Delta = 0.36$ mm the vertical cracks were about 50 mm in length. At the maximum load Crack A and Crack C extended to 120 mm, while Crack B was about 30 mm from the outside. At $\Delta = 0.70$ mm, Crack A and Crack C almost penetrated through the wall thickness, and Crack B reached one fourth of the wall thickness. The COWs of the vertical and horizontal cracks at the wall surface were 0.23 mm and less than 0.01 mm, respectively. Under the given load condition, the two vertical cracks in the manhole specimens are expected to be much more active than the two horizontal cracks because the bending moment at the load point and the reaction point is much larger than the bending moment at the two horizontal points, with a ratio of approximately 1.75.

(3) General comments

The results of the numerical study show that the discrete crack modelling approach is an effective tool for studying the cracking behaviour and the load-carrying capacity of plain concrete manholes repaired by the SCM, where the fracture behaviour of the structure is dominated by a few localised cracks. As discussed above, the accuracy of numerical predictions on the maximum loads shows some variations, from fair in Case 1 to very good in Cases 2 and 3. In general, these results are sufficiently accurate for predicting the load-carrying capacity of plain concrete structures. This is because compared with reinforced concrete structures, fracture behaviour in a plain concrete structure tends to be more brittle and more influential on the load-carrying capacity of the structure, thus requiring sound discrete crack models and accurate material data because both can greatly influence the results of crack analysis.

Based on the verification study, the discrete crack modelling approach can be used for evaluating the structural strength of ageing manholes of different sizes and shapes that suffer surface corrosions and need to be repaired by the SCM. To determine the cost-effective coating thickness, the strengths of an original manhole structure, the structure after removing the corroded surface and the repaired structure with different coating thicknesses can all be studied numerically. With appropriate safety factors included, a safe and economic renovation approach for ageing manholes is possible.

In general, when cracking behaviour in a sewage structure is dominated by a few localised cracks, the problem should be studied by discrete crack models. Examples include sewers with fractured bottom plates or sidewalls, shear fracture between the top plate and the sidewall in some sewers, and some localised large cracks in lightly reinforced sewers due to severe corrosion of rebars, etc. These problems should be treated individually, using adequate numerical models and the results of crack analysis should be carefully studied. If necessary, fracture tests should be performed to verify the accuracy of numerical modelling.

5.4 Buckling Theory of Invert Lining under Groundwater Pressure

5.4.1 Buckling of invert lining under groundwater pressure

In sewer renovation, the thickness of the renovation layer varies in the circumferential direction of the existing pipe, and at the bottom plate it is frequently reduced to its smallest possible value, i.e., the thickness of the PVC profile as in the SPR method (Fig. 5.29). Furthermore, the invert lining is usually built with a flat arc shape to meet the hydraulic requirement, inadvertently lowering its buckling strength further. Therefore,

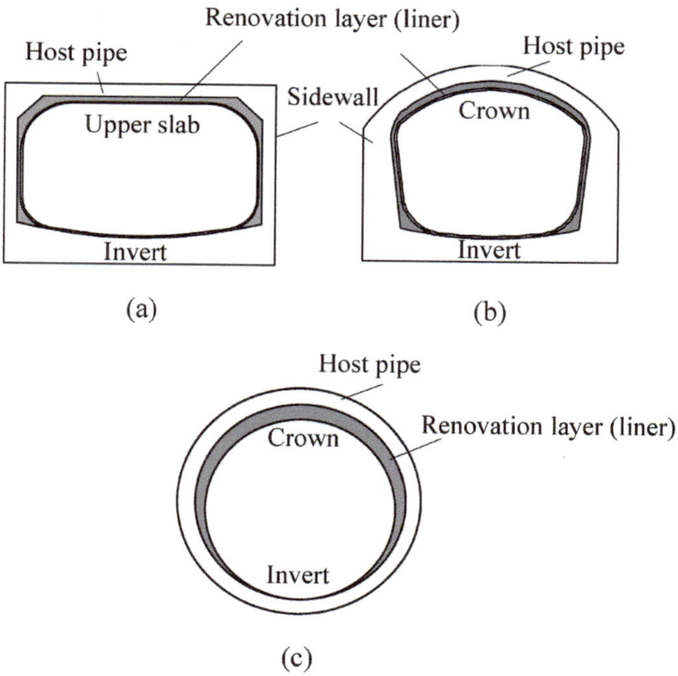

Figure 5.29: Thickness variations in renovated sewer pipes: (a) rectangular pipe; (b) horseshoe pipe; (c) circular pipe

when subjected to external groundwater pressure as in the case of a bottom plate with through-thickness cracks, the possibility of local buckling of the liner should be taken into consideration in renovation design. In practice, there have been reports of local buckling of the invert lining in sewer pipes renovated by the composite pipe method, where the invert lining buckled inward with a single-lobe shape.

Buckling phenomena of encased linings with various cross-sectional shapes have been well studied (Amstutz, 1970; Jacobsen, 1974; Glock, 1977; Thepot, 2001), and theoretical equations have been derived based on various ring models of the entire cross section of the lining. As these equations cannot be directly applied to solving the local buckling of invert lining, relevant buckling equations are derived below, focusing on the thin renovation layer at the bottom plate.

5.4.2 Derivation of buckling equation for invert lining

(1) Basic assumptions

Figure 5.30 shows the geometric shapes and load conditions of non-circular sewer pipes, and the modelling concept for the local buckling of invert lining. As shown, the invert lining is clamped by the thick grout at the bottom corners, which are simplified as fixed ends. Thus, the local buckling problem can be treated as a built-in arc under pressure. No tension and free sliding between the host pipe and the liner are assumed. The buckling wave is simulated using a quadratic cosine equation, which satisfies the boundary condition at the two ends. Variables of radial displacement (w) and one-half of the centre angle (ϕ_0) are used to define the buckling wave, and its trial expression is given by

$$w = w_0 \cos^2 \frac{\pi\phi}{2\phi_0} \quad \text{with} \quad (-\phi_0 \le \phi \le \phi_0) \tag{5.2}$$

Note that the annular gap is disregarded as it is filled with cementitious grout as in the composite pipe method.

According to Flügger's cylindrical shell theory (Flügger, 1999), the general strain-displacement relation is expressed as

$$\varepsilon = \frac{w}{R} - \frac{\mathrm{d}u}{\mathrm{d}s} - \frac{1}{2}\left(\frac{\partial w}{\partial s}\right)^2 \tag{5.3}$$

$$\chi = \frac{w}{R^2} + \frac{\mathrm{d}^2 w}{\mathrm{d}s^2} \tag{5.4}$$

where $s = R\Theta$. Refer to Fig. 5.31 for the local axes at a point in the middle surface of a shell, denoting the components of the displacement at that point by u, v, and w. Note that compression is defined as positive. The hoop force and bending moment are calculated by

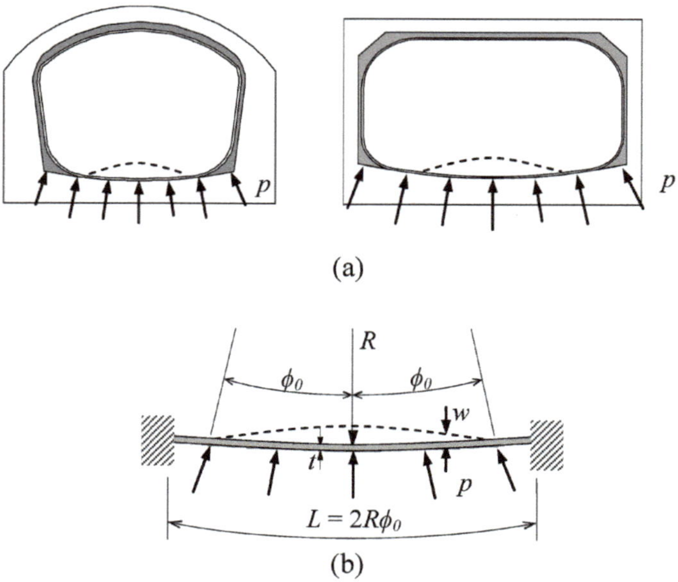

Figure 5.30: Schematic drawings of local buckling of liner in non-circular renovated sewers: (a) geometric shape and load condition; (b) theoretical model of local buckling

$$N = EA\varepsilon \tag{5.5}$$

$$M = EI\chi \tag{5.6}$$

where, E = elastic modulus; A = cross-sectional area; I = moment of inertia of area; ε = hoop strain; χ = change of curvature.

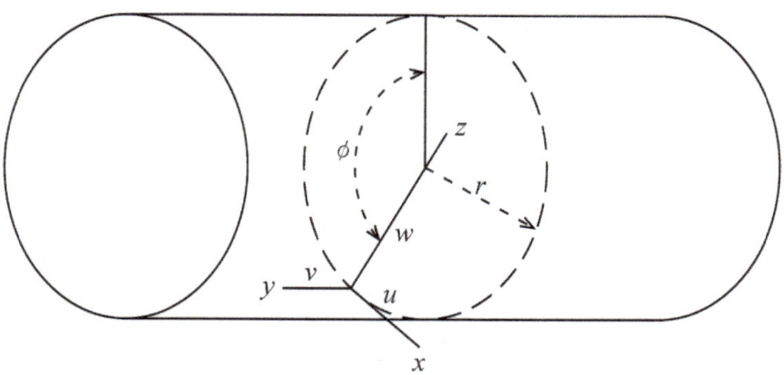

Figure 5.31: Local axes of a point in the middle surface of a shell

By substituting $\beta = \frac{\pi}{2\phi_0}$ and Eq. (5.2) into Eq. (5.3) and Eq. (5.4), and integrating the hoop stress for half the length of the buckled lining, the average hoop thrust force of the buckled lining can be evaluated from:

$$\frac{L}{2} \cdot \frac{\overline{N}}{EA} = \int_0^{L/2} \varepsilon ds = R\int_0^{\phi_0} \left(\frac{w_0 \cos^2 \beta\phi}{R} - \frac{{w_0}^2 \beta^2}{2R^2} \sin^2 2\beta\phi \right) d\phi$$

which leads to

$$\overline{N} = \frac{EA}{L}\left(w_0\phi_0 - \frac{{w_0}^2\beta^2}{2R}\phi_0 \right) \tag{5.7}$$

Similarly, the bending moment is obtained as

$$M = EI\chi = \frac{EI}{2R^2\phi_0^2}\left[w_0\phi_0^2 + w_0\left(\phi_0^2 - \pi^2\right)\cos 2\beta\phi \right] \tag{5.8}$$

(2) Buckling equation

For an elastic buckling problem the buckling equation can be derived by the energy method. The total potential energy (Π) is expressed by the strain energy of hoop force (U_ε), the strain energy of bending (U_χ) and the external work (W), as

$$\Pi = U_\varepsilon + U_\chi - W \tag{5.9}$$

where

$$U_\varepsilon = \frac{1}{2}\int_0^L \frac{N^2}{EA} \mathrm{d}s = \frac{\overline{N}^2 L}{2EA} \tag{5.10}$$

$$U_\chi = \frac{1}{2}\int_0^L \frac{M^2}{EI} \mathrm{d}s = \frac{EI}{4R^3\phi_0^3} \cdot \pi^4 \left[{w_0}^2 \left(\frac{\phi_0}{\pi} \right)^4 + {w_0}^2 \left(\frac{\phi_0^2}{\pi^2} - 1 \right)^2 / 2 \right] = \frac{EI\pi^4}{8R^3\phi_0^3} {w_0}^2 \tag{5.11}$$

$$W = \int_0^L wp\mathrm{d}s = pRw_0\phi_0 \tag{5.12}$$

In deriving Eq. (5.11), the higher order term $(\phi_0/\pi)^4$ is neglected. Finally, the total potential energy of the deformed lining is obtained as

$$\Pi = \frac{1}{2}\frac{\overline{N}^2}{EA}L + \frac{EI\pi^4}{8R^3\phi_0^3}{w_0}^2 - pRw_0\phi_0 \tag{5.13}$$

By applying the principle of minimum total potential energy ∂Π=0, the partial differential equations with respect to ϕ_0 and w_0 are obtained as

$$w_0\left(\overline{N}\frac{\pi^2}{8R\phi_0^2} - \frac{3EI\pi^4}{8R^3\phi_0^4} \right) = pR - \overline{N}$$

$$w_0\left(-\overline{N}\frac{\pi^2}{4R\phi_0^2}+\frac{EI\pi^4}{4R^3\phi_0^4}\right)=pR-\overline{N} \tag{5.14}$$

By solving the above simultaneous equations, the average hoop force is obtained as

$$\overline{N}=\frac{5EI\pi^2}{3R^2\phi_0^2} \tag{5.15}$$

Comparing Eq. (5.15) with Eq. (5.7) leads to

$$\frac{EA}{L}\left(w_0\phi_0-\frac{w_0^{\,2}\pi^2}{8R\phi_0}\right)=\frac{5EI\pi^2}{3R^2\phi_0^2} \tag{5.16}$$

By solving the above equation the maximum displacement (w_0) is obtained as

$$w_0=\frac{R}{\beta^2}\left[1-\sqrt{1-\frac{80IL}{3R^3\pi A}\beta^5}\right] \tag{5.17}$$

The maximum displacement (w_0) can also be obtained by substituting Eq. (5.15) into Eq. (5.14):

$$w_0=\frac{5R}{2\beta^2}-\frac{3R^4}{8EI\beta^4}p \tag{5.18}$$

By comparing Eqs. (5.17) and (5.18), a relation between the load p and the angle variable β is obtained as

$$\frac{3R^3p}{8EI}=\beta^2\left(\frac{3}{2}+\sqrt{1-\frac{80IL}{3R^3\pi A}\beta^5}\right) \tag{5.19}$$

Differentiating p with respect to β and substituting $\beta=\dfrac{\pi}{2\phi_0}$, the critical subtended angle is obtained as

$$\phi_{cr}=\pi\left(\frac{15}{4(1+\sqrt{6})}\frac{IL}{R^3A\pi}\right)^{1/5} \tag{5.20}$$

Finally, by substituting Eq. (5.20) into Eq. (5.19) the critical pressure load for buckling is derived as

$$\frac{p_{cr}}{E}=\frac{8}{3}\left(\frac{1+\sqrt{6}}{120}\right)^{2/5}\left(\frac{9+2\sqrt{7-2\sqrt{6}}}{6}\right)\left(\frac{I}{R^3}\right)^{3/5}\left(\frac{\pi A}{L}\right)^{2/5} \tag{5.21}$$

or

$$p_{cr}=1.28E\left(\frac{I}{R^3}\right)^{3/5}\left(\frac{\pi A}{L}\right)^{2/5} \tag{5.22}$$

Taking the longitudinal length of a sewer pipe into account, Eq. (5.22) is corrected for the effect of Poisson's ratio (ν) in a plane strain situation and is rewritten as

$$p_{cr} = \frac{1.28E}{1-\nu^2}\left(\frac{I}{R^3}\right)^{3/5}\left(\frac{\pi A}{L}\right)^{2/5} \tag{5.23}$$

By substituting the parameters $L = 2\pi R$, $A = t$ and $I = t^3/12$ into Eq. (5.23), the buckling equation of an encased circular pipe is obtained as

$$p_{cr} = \frac{E}{1-\nu^2}\left(\frac{t}{2R}\right)^{2.2} \tag{5.24}$$

This is exactly Glock's buckling equation for a close-fit circular pipe (Glock, 1977).

5.4.3 Verification study

The local buckling equation derived was examined by a numerical study of the invert buckling in two renovated sewer pipes. using FE software package MSC Marc (2009). Figure 5.32 shows two non-circular sewers, a rectangular type and a horseshoe type, and their respective geometric dimensions. At the bottom plate the renovation layer is reduced to the composite lining, as shown in Fig. 5.32(c). Nonlinear analysis was required to consider the effects of geometric nonlinearity, contact behaviour and non-conservative pressure. For the contact analysis, free sliding and no-tension conditions were assumed in the contact surface between the lining and the host pipe.

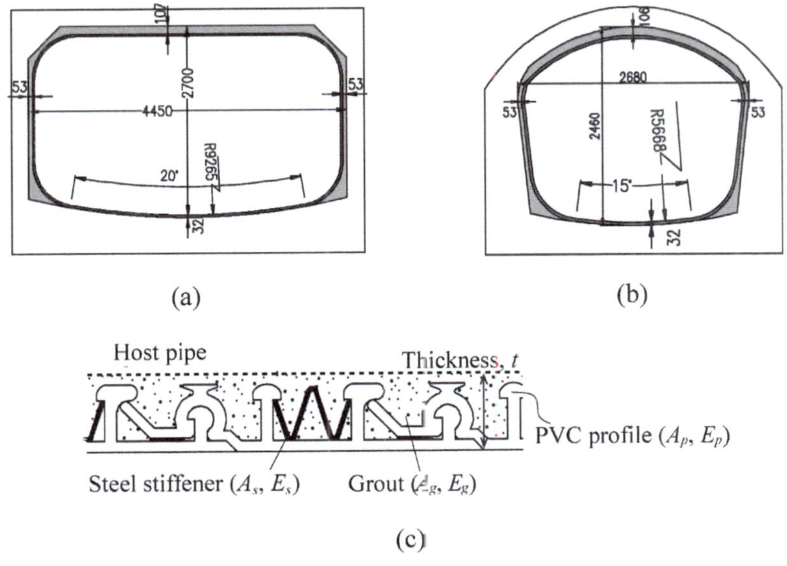

Figure 5.32: Dimensions of two non-circular sewers: (a) rectangular type; (b) horseshoe type; (c) composite lining

Buckling of the composite lining under groundwater pressure was studied as a two-dimensional problem in the plane strain condition. Figure 5.33 shows the FE models and boundary conditions of the rectangular pipe and the horseshoe pipe under two types of load condition: full loading along the cross section of the pipe and partial loading at the bottom plate. Due to symmetry, only one half of the cross section was modelled. In numerical analysis the host pipe was assumed to be rigid, and the renovation layer was modelled using linear elastic elements. To trace the buckling behaviour and obtain the buckling load, the full Newton–Raphson numerical procedure was employed in conjunction with an arc-length-based automatic increment method. Table 5.3 shows the dimensions and material properties used in the study.

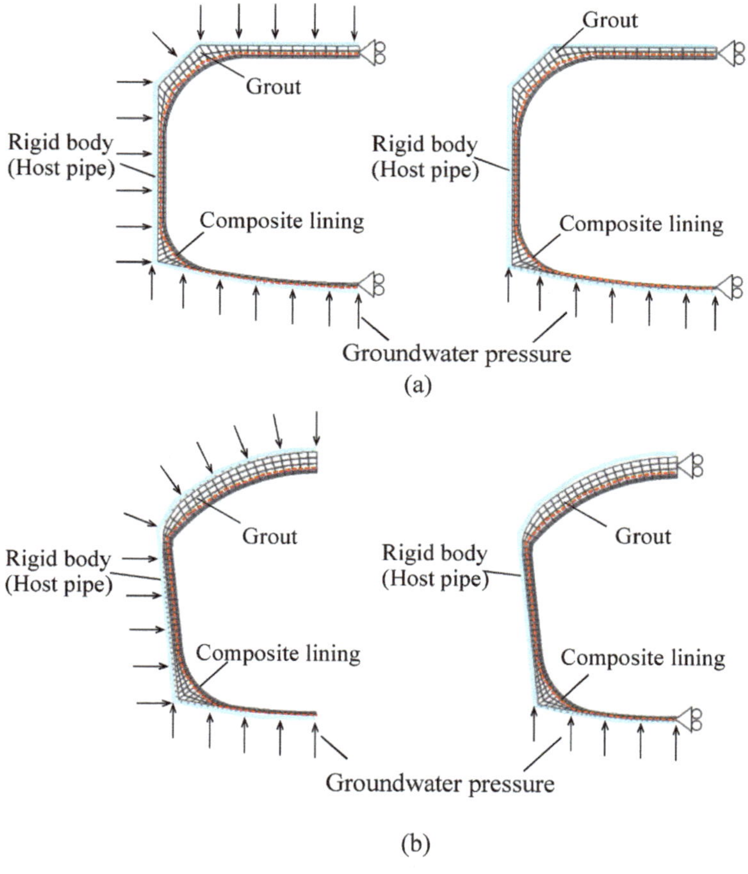

Figure 5.33: Finite element models and boundary conditions: (a) rectangular pipe; (b) horseshoe pipe

Table 5.3: Dimensions and material properties of invert lining

Item	Rectangular pipe	Horseshoe pipe	Remarks
R (mm)	5668	9265	
t (mm)	31.7	31.7	
R/t	179	292	
ϕ_0 (°)	7.5	10.0	
L (mm)	1483.9	3234.1	
I (mm^4)	2654.6	2654.6	
A (mm^2)	31.7	31.7	
E (MPa)	6300	6300	Equivalent elasticity coefficient
E_g (MPa)	22,000	22,000	Elasticity coefficient of grout
v	0.35	0.35	

Table 5.4 shows the buckling loads obtained by Eq. (5.23) and by numerical computation. While the theoretical buckling load for the rectangular pipe is 0.0187 MPa, just 5% smaller than the numerical result of 0.0196 MPa, the theoretical value of 0.0620 MPa for the horseshoe pipe is slightly larger than the numerical result of 0.0561 MPa by 10%. In either case the difference is considered reasonably small to justify Eq. (5.23). Note that in the case of the horseshoe pipe, the critical subtended angle ϕ_{cr} obtained by Eq. (5.20) is slightly larger than the centre angle ϕ_0 of the invert perimeter. To account for the uncertainty of various assumptions in deriving the buckling equation, the actual buckling length may be considered a little longer than that of the invert. Herein, one and half perimeters of the invert is assumed to be the limit of the potential buckling length: i.e., buckling may occur when the critical angle $\phi_{cr} < 1.5\phi_0$, otherwise invert buckling is disregarded.

Numerical results on sewer deformation at buckling loads in the two pipes are shown in Fig. 5.34. As seen, the invert lining buckled into a single-lobe shape with the maximum deformation at the centre of the bottom plate, which justifies the assumed buckling wave used in deriving Eq. (5.23). As the numerical results did not show any differences between the buckling loads obtained under the full and partial loading conditions, it should be concluded that the local buckling of invert lining in a renovated sewer pipe is not affected by the sidewalls, again justifying Eq. (5.23) for the study of invert buckling.

Table 5.4: Comparison of buckling loads obtained by Eq. (5.23) and numerical computation

	Item	Rectangular pipe	Horseshoe pipe	Remarks
Theoretical results				
	ϕ_{cr} (°)	5.57	8.37	Eq. (5.20)
	ϕ_{cr}/ϕ_o	0.6	1.1	
	p_{cr} (MPa)	0.0187	0.0620	Eq. (5.23)
Numerical results				
	p_{cr} (MPa)	0.0196	0.0561	Full loading
	p_{cr} (MPa)	0.0196	0.0561	Partial loading

Note: Disregard buckling if $\phi_{cr} > 1.5\phi_o$; otherwise calculate the buckling length $L = R\phi_o$ for $\phi_{cr} > \phi_o$ and $L = R\phi_{cr}$ for $\phi_{cr} < \phi_o$.

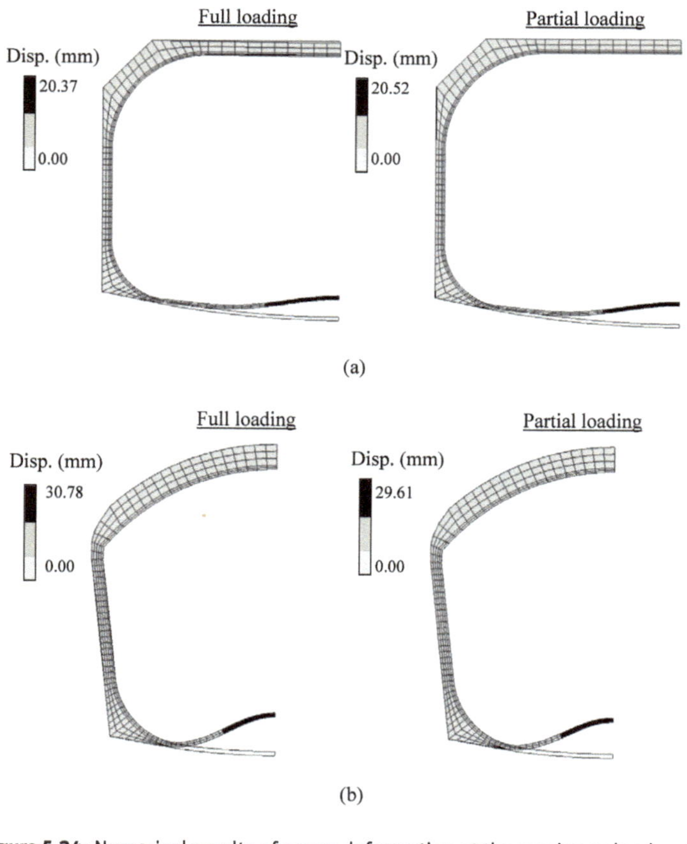

Figure 5.34: Numerical results of sewer deformation at the maximum load under full loading and partial loading: (a) rectangular pipe; (b) horseshoe pipe

5.4.4 Buckling design

The equivalent elasticity coefficient E for the composite lining in Table 5.3 is evaluated based on the equivalence principle of axial stiffness, as

$$E = \frac{E_P A_P + E_G A_G + E_s A_s}{t} \tag{5.25}$$

where E_p, A_p, E_g, A_g and E_s, A_s are the elasticity coefficient and the cross-sectional area per unit length of pipe for the PVC profile, grout and profiled steel sheets, respectively.

Figure 5.35 shows a flowchart for verifying the buckling strength of invert lining in renovation design by the composite pipe method. The verification procedure includes the following steps:

1. Determine the sewer conditions including the thickness t of the composite lining, radius R, invert length L, centre angle ϕ_o of invert lining, and the cross-sectional areas (A_p, A_g, A_s) and elasticity coefficients (E_p, E_g, E_s) of the PVC profile, grout and steel stiffener;
2. Calculate the equivalent elasticity coefficient of composite lining by Eq. (5.25);
3. Calculate the critical centre angle of buckling lobe ϕ_{cr} by Eq. (5.20) and compare it with ϕ_o: if $\phi_{cr} > 1.5\phi_0$ buckling is disregarded, otherwise calculate the buckling length $L = R\phi_o$ for $\phi_{cr} \geq \phi_o$ and $L = R\phi_{cr}$ for $\phi_{cr} < \phi_o$;
4. Calculate the buckling load p_{cr} by Eq. (5.23);
5. Compare the buckling strength p_{cr} with the design groundwater pressure p_{wd}: if $p_{cr} > f_s \cdot p_{wd}$, where f_s is a safety factor, the invert lining is judged safe against buckling; otherwise, local buckling may occur and countermeasures should be taken by reexamining the renovation design.

Verification studies of invert buckling in renovated sewer pipes are mainly concerned with non-circular pipes, usually disregarding circular pipes. In general, circular pipes possess much stronger buckling strengths with a minimum value at 0.5 MPa for the SPR method. In other words, buckling will not occur in a circular pipe unless it is buried more than 50 m below the groundwater level.

Note that when the bottom plate is flat such as in a rectangular pipe with lower haunches, the problem should not be treated as a buckling problem. Instead, it is the bending of a thin bottom lining under groundwater pressure. In situations like this, efforts should be made to seal off the existing cracks in the bottom plate to prevent the groundwater from entering the sewer, or mechanical connectors should be used to rigidly connect the lining with the bottom plate. These measures help prevent local buckling or large deformation of the bottom lining under groundwater pressure. Therefore, if one of these measures is taken, the verification study on invert buckling can be omitted.

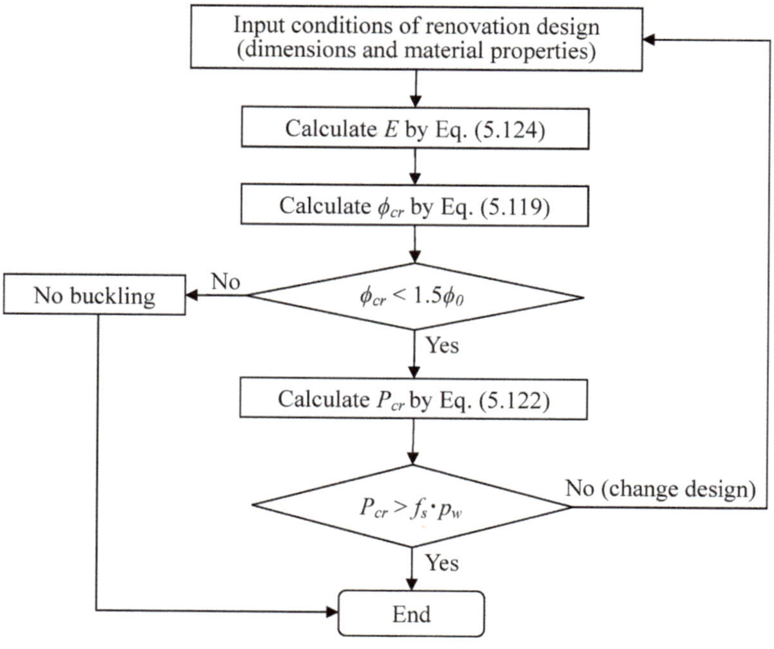

Figure 5.35: Flowchart for verifying the buckling strength against local buckling of invert lining in non-circular sewer pipes

References

Amstutz, E. (1970). Buckling of pressure-shaft and tunnel linings. Water Power, 12, 391-400.

Flugge, W. (1932) Die Stabilit/it der Kreiszylinderschalen. Ingenieur-Archiv 3, 463-506.

Glock, D. (1977). Uberkritisches verhalten eines starr ummantelten kreisrohres bei wasserdruck von aussen und temperaturerhohung (English translation: Post-critical behaviour of a rigidly encased circular pipe subject to external water pressure and temperature rise). Der Stahlbau, 46(7), 212-217 (in German).

Jacobsen, S. (1974). Buckling of circular rings and cylindrical tubes under external pressure. Water Power, 26, 400-407.

Johnson, R. P. (2004). Methods of shear connection. In: Composite Structures of Steel and Concrete. 3rd ed., Blackwell Publishing: 26-29.

JSCE (2009). Standard Specifications for Hybrid Structures – 2009. Tokyo, Japan Society of Civil Engineers.

JSWA (2003). JSWAS A-1, Reinforced Concrete Sewerage Pipes. Tokyo, Japan Sewage Works Association.

JSWA (2011). Design and Construction Guidelines for Sewer Pipe Rehabilitation. Tokyo, Japan Sewage Works Association.

Kurozumi, M. (2014). Private communication.

MSC Marc (2009). User's Manual Release, 09.17, MSC. Software Co., U.S.A.

Thépot, O. (2001). Structural design of oval-shaped sewer linings. Thin-walled structures, 39(6), 499-518.

Masaaki Nakano

6 Renovation Design of Ageing Sewers as Composite Pipes by the Limit State Design Method

6.1 Application of the Limit State Design Method

Ageing in-service sewers requiring renewal tend to be in a structurally unsound condition because of concrete cracking, rebar corrosion and other anomalies arising from their long service history. Though these reinforced concrete structures were designed by the allowable stress method, which was the prevailing design theory at the time of construction, the critical stresses caused by cracks and other damages in existing sewers are likely to have already exceeded the allowable stress. For this reason, when renovating ageing sewers using the composite pipe method in which the existing sewer also serves as an important structural component, the allowable stress method may not be suitable for structural design because it focuses on the critical stress.

In contrast, the limit state design method focuses on the limit states of damage or failure that a structure may experience during its design life, and the criterion for a safe design is to ensure that such a limit state is not reached. For reinforced concrete structures, the limit states can be classified into

1. The ultimate limit state:
   This involves the collapse of the whole structure or its elements under the design loads.
2. Serviceability limit states:
   The structure is unfit to serve its normal function due to excessive deformation, cracking or vibration.
3. Special limit states:
   This type of limit state mainly concerns damage or failure due to abnormal conditions such as strong earthquake, structural effects of corrosion or deterioration, and structural effects of other extreme conditions. Obviously, the relevance of a special limit state to structural design must be justifiable.

As such, limit state design is a performance-verification design method. In the design process, partial safety factors are employed with respect to load, material, structural analysis, structural member, and structural importance, and the influence of these factors on the required performance concerning a limit state can be evaluated separately. Therefore, important design considerations can be clearly identified so that a comprehensive evaluation can be carried out with respect to multiple performance requirements. Table 6.1 shows typical performance requirements, limit states and check items. In the case of sewer renovation, structural safety can be verified by evaluating the possibility of cross-sectional failure of the renovated structure under design loads, and serviceability can be checked by evaluating its functional sustainability and trafficability. Seismic safety can be verified by performing dynamic structural analysis under Level 1 and Level 2 earthquake loading.

Table 6.1: Examples of performance requirements, limit states and check items

Performance requirements	Limit states	Check item
Safety	Cross-sectional failure	Cross-sectional force
	Fatigue failure	Stress, cross-sectional force
	Structural instability	Deformation, deformation due to foundation structure
Serviceability	Appearance	Crack width, stress
	Noise, vibration	Noise/vibration level
	Trafficability, etc.	Displacement, deformation
	Watertightness	Permeability of structure, crack width
	Functional damage	Force, deformation, etc.
Restorability	Repairability	Force, deformation, etc.

6.2 Basic Concept of Performance Verification

Performance verification of renovated sewer pipes by the limit state design method involves two types of loading condition, normal loading and earthquake loading. This section outlines the basic concept of limit state design, using the sectional forces of a renovated sewer (moment, shear, axial force) as design indices.

The performance of the structure of interest in the ultimate limit state is checked by comparing the response value under the design load with the capacity value under the failure load. Specifically, the basic criterion is that the sectional force under the design load multiplied by a structure factor must not exceed the sectional capacity determined from the ultimate failure analysis of the structure, which is expressed as

$$\gamma_i \frac{S_d}{R_d} \leq 1.0 \tag{6.1}$$

where y_i: structure factor which measures the degree of importance of the structure; S_d: design sectional force; and R_d: design sectional capacity.

The design sectional force S_d is expressed as

$$S_d = \gamma_{ap} S(\gamma_{fp} F_p) + \gamma_a S(\gamma_f F_{kt}) = S(F_p) + \gamma_a \gamma_f S(F_{kt}) \tag{6.2}$$

where $S(\ )$: sectional force obtained by structural analysis under the load shown in parentheses; F_p: characteristic value of sustained load; F_{kt}: characteristic value of primary variable load; y_a: structural analysis factor for primary variable load; y_f: load factor for primary variable load; y_{ap}: structural analysis factor for sustained load (= 1.0); and y_{fp}: load factor for sustained load (= 1.0). In general, the design sectional force is obtained by linear structural analysis, and therefore, the load factor y_f can be applied after structural analysis, as shown in Eq. (6.2).

The design sectional capacity R_d can be expressed as

$$R_d = R(f_k / \gamma_m) / \gamma_b \cong R(f_k) / \gamma_m / \gamma_b \tag{6.3}$$

where $R()$: sectional capacity obtained by structural analysis at failure; f_k: characteristic value of material strength; y_m: material factor; and y_b: member factor. As the design sectional capacity or ultimate strength is usually obtained by nonlinear structural analysis, when the nonlinearity is weak the practice of applying the material factor y_m after structural analysis requires an approximately-equal sign, as shown in Eq. (6.3).

The flowchart of performance verification for renovation design of ageing sewers is shown in Fig. 6.1. As seen, the verification process is composed of two parts, normal loading analysis and earthquake loading analysis. In each analysis, a serviceability limit state and an ultimate limit state are specified, and the renovation cross section has to pass the verification check based on a specific design criterion in each limit state.

6.3 Performance Requirements for Renovated Sewer

6.3.1 Under normal loading

(1) Serviceability limit state

Sewer pipes are required to maintain watertightness under normal loading, and sewer renovation must not accelerate the deterioration of the existing pipe. New cracking in the renovated sewer must therefore be prevented.

(2) Ultimate limit state

Sewer pipes are required to maintain their discharge function even under rare loading conditions encountered during their service life. Therefore, the renovated sewer must not fail even under a large load exceeding the design load and must maintain sufficient load-carrying capacity as a structure.

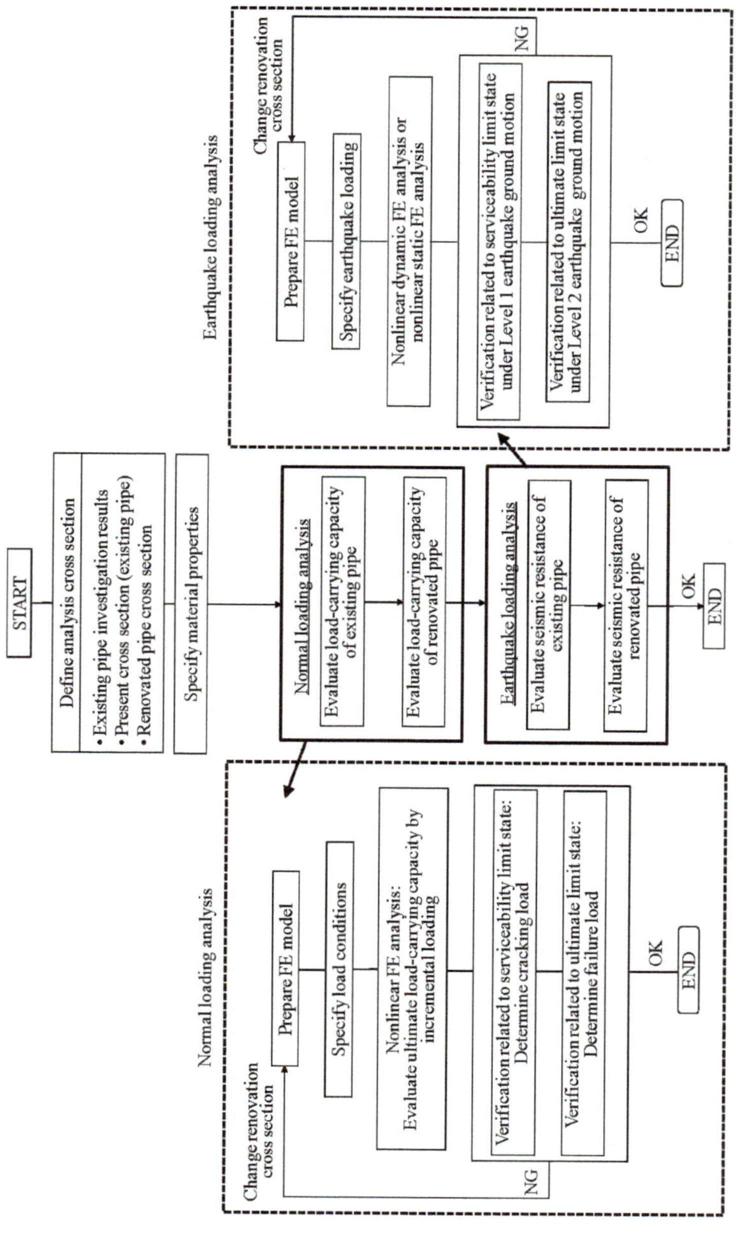

Figure 6.1: Flow of performance verification of composite pipe based on limit state design method

(3) Fatigue limit state
Because sewer pipes are underground structures, live loads do not act directly on them, so it is highly unlikely that sewer pipes will be endangered by fatigue failure caused by cyclic loading. Accordingly, fatigue strength is not taken into consideration as a performance requirement for renovation design.

6.3.2 Under earthquake loading

Earthquake resistance requirements for existing and renovated sewers are as specified in the Japan Sewage Works Association's *Guidelines for Seismic Design and Retrofit of Sewerage Facilities* (hereafter referred to as the JSWA *Seismic Guidelines*) (JSWA 2014). As shown in Table 6.2, performance requirements for sewerage pipelines under earthquake loading are broadly classified according to earthquake ground motion levels.

Table 6.2: Concept of seismic design of pipelines (JSWA, 2014)

Pipeline		Design ground motion		Seismic performance required	
		Level 1	Level 2	Level 1	Level 2
Existing	Important trunk line, etc.	R	R	Design discharge capacity	Discharge function
	Other pipeline	R	NR	Design discharge capacity	
New	Important trunk line, etc.	R	R	Design discharge capacity	Discharge function
	Other pipeline	R	NR	Design discharge capacity	

Note: R = required; NR = not required

Level 1 earthquake ground motion refers to earthquake ground motion of a magnitude that is likely to occur several times during the service life of a structure. Level 2 earthquake ground motion refers to very strong earthquake ground motion, and its probability of occurrence is very small during the service life of a structure.

(1) Under Level 1 earthquake ground motion: serviceability limit state
Sewer pipes are required to maintain their design discharge capacity even after earthquake ground motion (Level 1 earthquake ground motion) that is likely to be encountered several times during the service life of the pipelines. Therefore, performance verification should ensure that no large structural deformation of renovated sewers occurs under such earthquake conditions.

(2) Under Level 2 earthquake ground motion: ultimate limit state
Sewer pipes are required to maintain their discharge function even after very strong earthquake ground motion (Level 2 earthquake ground motion) that can occur during their service life, although its probability of occurrence is low. Therefore, performance verification should ensure that no structural failure of renovated sewers occurs under such earthquake conditions.

6.4 Performance Verification under Normal Loading

This section describes in detail the methods for verifying the safety of renovated sewers under normal loading. Nonlinear structural analysis of composite pipe is performed to verify structural performances at the serviceability limit state and the ultimate limit state, using the no-tension interface modelling approach introduced in Chapter 5.

The limit states defined under normal loading and their verification criteria are as follows (JSWA, 2011):

For serviceability limit state: New cracking must not occur anywhere in the renovated sewerage structure under the design load.

For ultimate limit state: The critical sectional force of the renovated sewerage structure under the design load must not exceed the sectional capacity or ultimate strength of the structure.

6.4.1 Verification for serviceability limit state

With respect to the serviceability limit state of a renovated sewer, the design sectional moment under the sustained load and traffic load must not exceed the cracking moment of the cross section. Specifically, the maximum tensile stress occurring in the renovated structure must not exceed the tensile strength of the material.

6.4.2 Verification for ultimate limit state

Sectional forces which are taken into consideration regarding the ultimate limit state of a renovated sewer are bending moment and shear force. This means that the design sectional force S_d and the design sectional capacity R_d expressed by Eqs. (6.2) and (6.3), respectively, can be rewritten as:

$$S_d : \begin{cases} M_d = \gamma_a \gamma_f M \\ V_d = \gamma_a \gamma_f V \end{cases} ; R_d : \begin{cases} M_{rd} = M_u / \gamma_m / \gamma_b \\ V_{rd} = V_u / \gamma_m / \gamma_b \end{cases} \tag{6.4}$$

where M, M_d: bending moment under the design load, design bending moment; V, V_d: shear force under the design load, design shear force; M_u, M_{rd}: maximum bending moment at failure, design moment capacity; and V_u, V_{rd}: maximum shear at failure, design shear capacity.

6.4.3 Safety factors

Table 6.3 shows typical safety factor values indicated in the JSCE *Standard Specifications for Concrete Structures: Design* (hereafter referred to as the JSCE *Standard Specifications*) (JSCE, 2012):

Table 6.3: Standard safety factors (JSCE, 2012)

Safety factors	Material factor γ_m		Member factor	Structural analysis factor	Load factor	Structure factor
	Concrete γ_c	Steel γ_s	γ_b	γ_a	γ_f	γ_i
Serviceability limit state (serviceability)	1.0	1 0	1.0	1.0	1.0	1.0
Ultimate limit state (cross-sectional failure)	1.3	1 0 or 1.05	1.1–1.3	1.0	1.0–1.2	1.0–1.2

(1) Material factor γ_m:
The material factor is to be determined by taking into consideration factors such as a change in an undesirable direction from the characteristic value of material strength, differences in material properties between test specimens and structure materials, the influence of material properties on the limit state and changes over time in material properties. Such an evaluation needs to be made for the materials used in each renovation method.

(2) Member factor γ_b:
The member factor is to be determined by taking into consideration factors such as the uncertainty of the member strength calculation, the effect of variability of member dimensions, and the degree of importance of members, that is, the influence that the member of interest has when it reaches a limit state on the entire structure. The evaluation needs to take into consideration the fact that a composite pipe is a renovated structure consisting partially of existing sewer members that are not in a sound condition.

(3) Structural analysis factor γ_a:
The structural analysis factor is to be determined by taking into consideration the uncertainty of the calculation of the design sectional force and the design sectional

capacity through nonlinear analysis. The evaluation needs to take into consideration the accuracy of analytical results relative to past test results.

(4) Load/action factor γ_f:
The load factor is to be determined by taking into consideration factors such as a change in an undesirable direction from the characteristic value of load (action), the uncertainty of the load calculation, changes in load during the design service life, and the influence of load characteristics on the limit state. Unlike in the case of new construction, the primary load acting on the renovated sewer is a load acting through soil compacted to a certain degree. Accordingly, the evaluation needs to take into consideration the uncertainty of fact finding and the variability of obtained data.

(5) Structure factor γ_i:
The structure factor is to be determined by taking into consideration factors such as the degree of importance of the structure, the socio-economic impact anticipated when the limit state is reached, and economic factors including reconstruction or repair cost. Since there is no established method for renovating these composite pipes in future, the evaluation needs to take into consideration the necessity of reliably rehabilitating sewer pipes.

6.4.4 Loads to be considered

Table 6.4 lists the loads to be taken into consideration at the design stage.

Table 6.4: Loads to take into consideration when designing composite pipes

Dead load	Weight of members	R
	Weight of water in pipe	CR
Live load	Live load from above	R
	Impact	R
Earth pressure	Vertical earth pressure	R
	Horizontal earth pressure	R
	Earth pressure due to live load	R
Groundwater pressure		CR
Buoyancy		CR

Note: R (required): load that must be taken into consideration; CR (conditionally required): load that does not need to be taken into consideration except in cases where its influence is expected to be significant

6.4.5 Structural analysis model

As discussed in Chapter 5, the cross section of a renovated sewer pipe is modelled by using two-dimensional plane elements. As for materials, the nonlinearity of concrete, steel and lining materials is taken into account. By using the nonlinear finite element analysis based on a fracture mechanics model, the behaviour of the pipe from the occurrence of cracking to the ultimate failure is analysed.

6.4.6 Nonlinear structural analysis

The design of a composite pipe is characterised by calculation of the ultimate limit state using nonlinear structural analysis for accurately evaluating the load-carrying capacity of the pipe. Figure 6.2 shows the differences in the safety verification methods with respect to sectional or structural failure applied to a typical reinforced concrete structure and a composite pipe. As seen, the fundamental difference between the two is the method of capacity calculation. While for the design of reinforced concrete structures, member capacity can be determined from theoretical equations based on idealised structural mechanics models, in the renovation design of ageing sewers it is obtained through numerical analysis.

The principal loads which act on the structure are classified as described below according to overburden thickness:

(a) For overburden thickness of less than 4 m
Assume that primary loads are live loads (traffic loads) and increase the primary loads by increasing the load weighting factor of the live loads till structural failure. In the following, the load weighting factor is referred to as the load coefficient. The load coefficient obtained at failure shows the ratio of the failure load to the design load.

- Load step 1: Apply dead loads, which include earth and groundwater pressures and self-weight.
- Load step 2: Apply live loads incrementally until the design load is reached.
- Load step 3: While keeping dead loads constant, apply live loads by gradually increasing the load coefficient until structural failure is reached.

(b) For overburden thickness of 4 m or greater
Assume that primary loads are the earth and groundwater pressures and the live loads, and increase the primary loads by increasing the load coefficient till structural failure. The load coefficient obtained at failure shows the ratio of the failure load to the design load.

- Load step 1: Apply self-weight.
- Load step 2: Apply primary loads incrementally until the design load is reached.

- Load step 3: While keeping the self-weight of the pipe constant, apply the primary loads by gradually increasing the load coefficient until structural failure is reached.

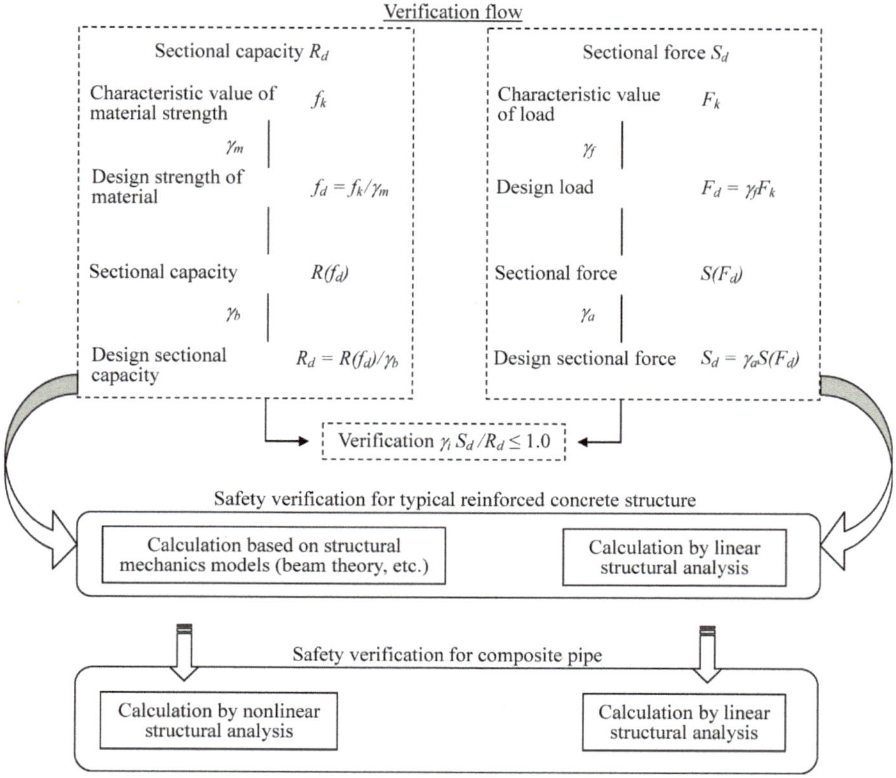

Figure 6.2: Comparison of computational methods for ultimate limit state between reinforced concrete structures and composite pipes

In the verification with respect to the serviceability limit state, checks are made, by following the steps described above, to determine whether cracking occurs under the design load. In the verification with respect to the ultimate limit state, the design sectional capacity and the design sectional force are calculated as follows:

1. The maximum sectional force (= design sectional capacity) of each member (such as the top plate, the bottom plate, or the sidewalls) at structural failure is calculated, and its location in that member is recorded.
2. The sectional force (= design sectional force) at the abovementioned location under the design load is calculated.

Figure 6.3 shows the analysis procedure for evaluating the ultimate load-carrying capacity by incremental loading. It should be pointed out that the maximum sectional force obtained as described above may not necessarily be the actual capacity of that member, because under the designated primary loads the critical failure mode of that member may not occur.

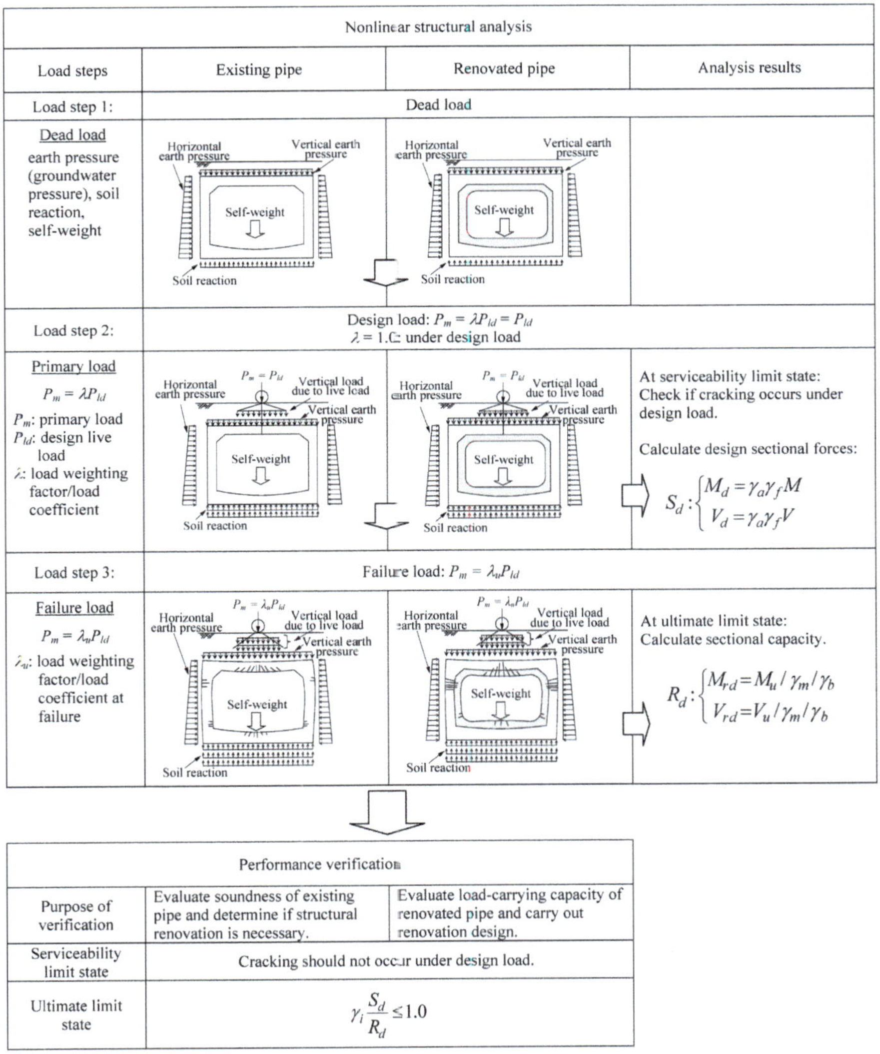

Figure 6.3: Flow of performance verification for renovation design of ageing sewers (example: overburden thickness less than 4 m)

6.4.7 Performance evaluation in terms of load coefficients

(1) Basic concept

Conventional practice when designing a reinforced concrete pipe for sewers is to determine the maximum bending moment occurring in the pipe under the design load based on the type of foundation and to select a pipe type so that the moment remains on the safe side relative to the allowable bending moment for the pipe (JSWA, 2003). The moment capacity for a pipe can be calculated by adding the bending moment due to the self-weight of the pipe to the maximum bending moment occurring under the cracking load determined according to the code-specified values. Thus, the design approach uses the margin of safety of the cracking load relative to the design load as a safety factor, which is comparable to the load coefficient introduced above. Therefore, performance evaluation based on the margin of safety relative to the design load is a suitable method for evaluating the structural safety of sewer pipes.

In the design of composite pipes, the design load is increased incrementally through the load coefficient, and structural strength is evaluated by nonlinear structural analysis. Consequently, the margin of safety relative to the design load with respect to the cracking load and the failure load can be evaluated directly. Therefore, besides the verification method based on sectional forces, the verification method based on the load coefficient has also been in use.

The JSWA *Design and Construction Guidelines for Sewer Pipe Rehabilitation* (hereafter referred to as the JSWA *Guidelines*) shows examples of load-carrying capacity evaluation using the load coefficient as a safety factor when designing composite pipes by the limit state design method.

To illustrate this load coefficient design concept, consider the design of a member subjected to axial force by the limit state design method. The ultimate limit state is verified by

$$\gamma_i \frac{\gamma_a(\gamma_f F_k)}{(Af_k/\gamma_m)/\gamma_b} \leq 1.0 \tag{6.5}$$

which leads to

$$\frac{Af_k}{F_k} \geq \gamma_i\gamma_a\gamma_b\gamma_m\gamma_f \quad \text{or} \quad \frac{F_u}{F_k} \geq \gamma_i\gamma_a\gamma_b\gamma_m\gamma_f \tag{6.6}$$

where F_k: the characteristic value of load; f_k: the characteristic value of material strength; A: the effective cross-sectional area of the member; and F_u: the ultimate sectional capacity or failure load.

Clearly, Eqs. (6.5) and (6.6) are equivalent so the verification of the ultimate limit state can be performed by checking the ratio of the failure load to the characteristic value of the load. It can be shown that for linear elastic materials, this statement is also true for the design of members under bending and shear loads. If nonlinearity in the stress-strain relation is weak, it then becomes approximately true.

Figure 6.4 shows the simple relationship between the sectional force-based verification method and the load coefficient-based verification method. Using the load coefficient, the performance requirements for renovation design of ageing sewers can be expressed as

For serviceability limit state: $\lambda_c > 1.0$

For ultimate limit state: $\lambda_u \geq 2.5$

Note that λ_c is the load coefficient for cracking, and λ_u is the ultimate load coefficient at structural failure.

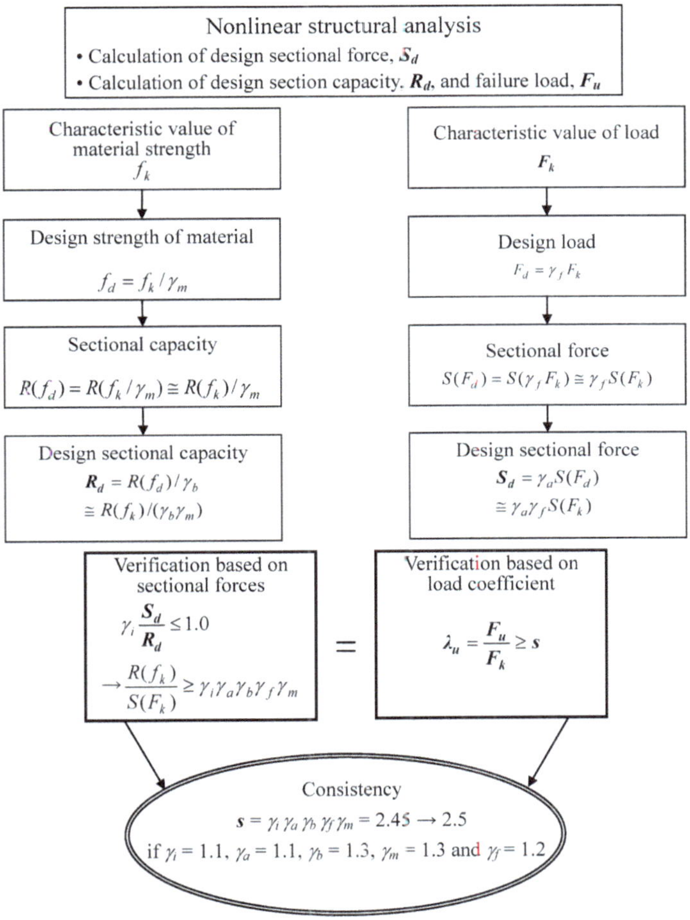

Figure 6.4: Relationship between the load coefficient-based verification method and the sectional force-based verification method for renovation design of ageing sewers

(2) Safety factors used in performance evaluation by load coefficients

Table 6.5 shows the recommended values of safety factors that have been used for performance evaluation based on load coefficients for renovation design of ageing sewers. In determining these safety factors for the ultimate limit state, the emphasis is placed on safe design. Figure 6.5 shows the analysis procedure for evaluating the ultimate load-carrying capacity using the load coefficient.

Table 6.5: Values of standard safety factors and safety factors used for renovation design of ageing sewers

	Material factor γ_m		Member factor	Structural analysis factor	Load factor	Structure factor
Safety factors	**Concrete γ_c**	**Steel γ_s**	γ_b	γ_a	γ_f	γ_i
Serviceability limit state (serviceability)	1.0	1.0	1.0	1.0	1.0	1.0
For composite pipe	**1.0**	**1.0**	**1.0**	**1.0**	**1.0**	**1.0**
Ultimate limit state (sectional failure)	1.3	1.0 or 1.05	1.1–1.3	1.0	1.0–1.2	1.0–1.2
For composite pipe	**1.3 (for both γ_c and γ_s)**		**1.3**	**1.1**	**1.2**	**1.1**

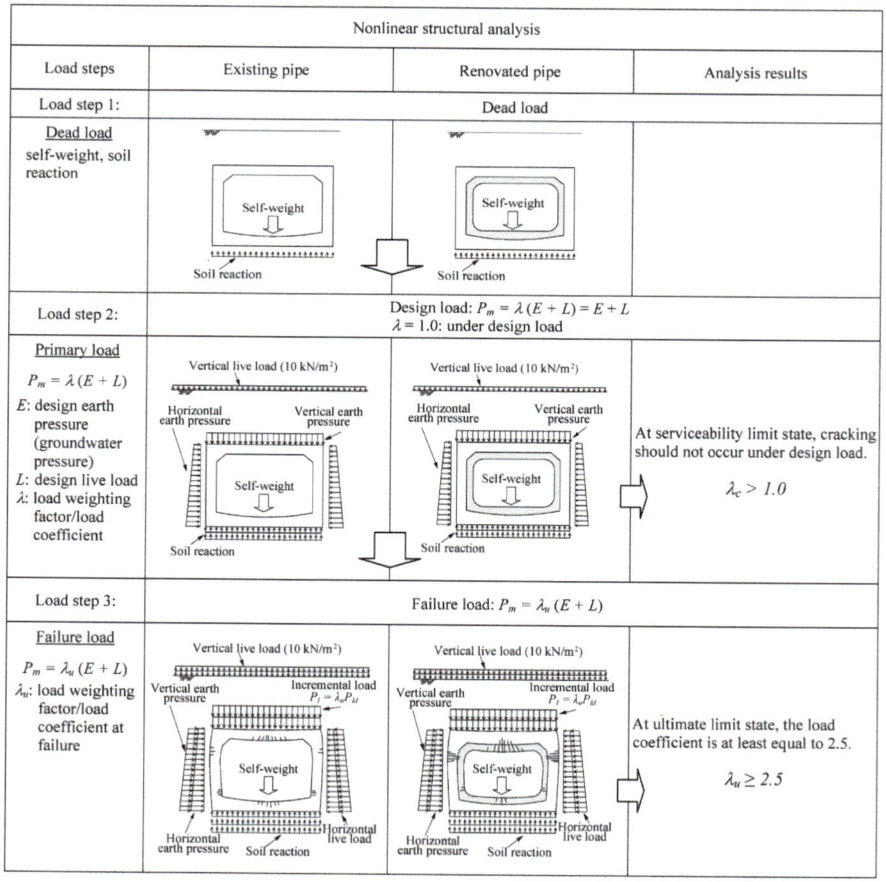

Figure 6.5: Flow of performance verification for renovation design of ageing sewers based on load coefficients (example: overburden thickness of 4 m or greater)

6.5 Performance Verification under Earthquake Loading

6.5.1 Seismic performance requirements

As set forth in the JSWA *Seismic Guidelines*, seismic performance requirements for renovated sewer pipes are classified according to the earthquake ground motions concerned. The earthquake resistance is evaluated according to the following verification criteria:

- Verification criterion for Level 1 earthquake ground motion

The verification criterion corresponds to the serviceability limit state.

- Verification criterion for Level 2 earthquake ground motion

The verification criterion corresponds to the ultimate limit state.

The seismic performance requirements specified in the JSCE *Standard Specifications* meet the requirements specified in the JSWA *Seismic Guidelines*. Current practice, therefore, is to use the verification indicators specified in the JSCE code. As shown in Table 6.6, in the seismic performance verification of a composite pipe in the cross-sectional direction, the occurrence of rebar yielding is used as the indicator for the serviceability limit state, and the ultimate displacement and the shear capacity are used as the indicators for the ultimate limit state. The flow of seismic performance verification for renovation design of ageing sewers is shown in Fig. 6.6.

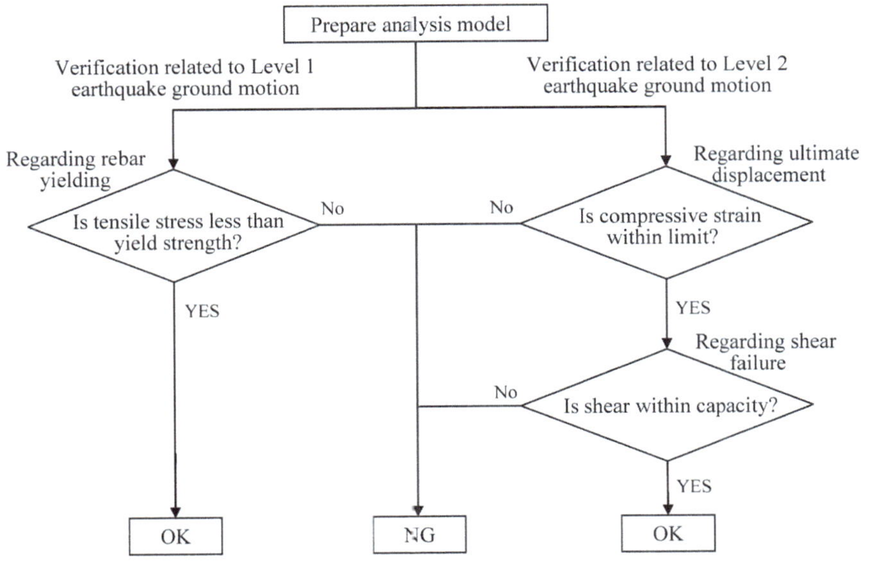

Figure 6.6: Flow of verification on seismic performance for renovation design of ageing sewers

Table 6.6: Comparison of guidelines for performance requirements under earthquake loading

Design earthquake ground motion	Seismic performance required	
	JSWA Guidelines for Seismic Design and Retrofit of Sewerage Facilities (2014)	**JSCE Standard Specifications for Concrete Structures (2012)**
Level 1	Retaining design discharge capacity	Level 1 seismic performance
	Ability to maintain the discharge capacity shown on the flow calculation sheet	Able to remain in a functionally sound and usable condition even in the event of an earthquake Post-earthquake residual displacement is sufficiently small. **Steel reinforcement does not yield.**
Level 2	Retaining discharge function	Level 2 seismic performance
	Design discharge capacity is difficult to maintain, but the pipeline is still able to pass wastewater downstream until corrective measures such as repair and reconstruction are taken.	Functions can be restored shortly after an earthquake, and there is no need for structural reinforcement. The load-carrying capacity of the structure does not decrease because of an earthquake. **Response displacement is within ultimate displacement.** **Shear failure does not occur.**

Note: As a Level 3 seismic performance requirement under Level 2 earthquake ground motion, the JSCE code specifies the ability of a structure to withstand an earthquake without collapsing. From the viewpoint of the ability to maintain the discharge function, that can be regarded as a synonymous requirement. It was considered appropriate, however, to check on Level 2 seismic performance as shown above in order to be on the safe side.

In the longitudinal direction of the pipe, the basic rule is to make sure the lining members (push-fit connectors) in the pipe are joined together to be watertight. Common practice is to conduct experiments under the conditions described below to check the seismic performance.

1. In the verification of push-fit connector performance, checks are made to make sure push-fit joints remain connected and watertight in the event of pullout due to the permanent strain (1.5%) assumed in view of the data obtained from the Hyogoken Nanbu Earthquake.
2. In the verification of push-fit connector performance in the event of land subsidence due to liquefaction, checks are made to make sure push-fit joints remain connected and watertight in the event of deformation corresponding to an average span of 30 m and a subsidence of 30 cm.

6.5.2 Verification for serviceability limit state

The performance requirement in the serviceability limit state, which is the verification criterion for Level 1 earthquake ground motion, is the ability of a renovated sewer to remain fully functional without needing repair in the event of an earthquake. Therefore, checks are made to make sure the residual displacement of the structure after an earthquake is sufficiently small, and the criterion is the non-occurrence of rebar yielding.

6.5.3 Verification for ultimate limit state

The performance requirement in the ultimate limit state, which is the verification criterion for Level 2 earthquake ground motion, is the ability of a renovated sewer to restore functionality within a short period of time following an earthquake without needing structural reinforcement. Therefore, checks are made to make sure the load-carrying capacity of the structure does not decrease, that response displacement does not exceed the ultimate displacement and that shear failure does not occur.

Specifically, the criterion that response displacement does not exceed ultimate displacement can be checked by making sure that in-plane compressive strain does not reach or exceed two times the strain corresponding to the compressive strength.

$$\gamma_i \gamma_a \gamma_{m(response)} \varepsilon_c < 2\varepsilon_{peak} / \gamma_b / \gamma_{m(\lim it)} \tag{6.7}$$

where ε_c: in-plane compressive strain; ε_{peak}: strain corresponding to compressive strength (usually 0.002); y_b: member factor; y_i: structure factor; y_a: structural analysis factor; and y_m: material factor (= response value: 1.0, limit value: 1.3). The non-occurrence of shear failure can be verified by verifying that in-plane shear force does not exceed the design shear capacity of the member.

$$\gamma_i \gamma_a V < V_d \tag{6.8}$$

where V: in-plane shear force and V_d: design shear capacity.

6.5.4 Safety factors

The safety factors used in this study are shown in Table 6.7. Basically, they are in accordance with the JSCE *Standard Specifications* and have been set on the assumption that a nonlinear analysis method with fully proven accuracy is used. In the verification, it is desirable that different material factors and member factors be used for response analysis in the case where response displacement is calculated and in the case where the non-occurrence of shear failure is verified. In practice, however, these two tasks are often performed in one response analysis with all safety factors being set at 1.0.

Safety factors (material factor y_m, member factor y_b, structural analysis factor y_a, load factor y_f and structure factor y_i) for Grade 2 seismic performance required against Level 2 earthquake ground motion have been determined according to the concepts described below.

Table 6.7: Safety factors considered for seismic performance verification (JSCE, 2012)

		Safety factor					
		Material factor γ_m Concrete	**Steel**	**Member factor**	**Structural analysis factor**	**Load factor**	**Structure factor**
Seismic performance		γ_c	γ_s	γ_b	γ_a	γ_f	γ_i
Grade 1 seismic performance	Response value and limit value	1.0	1.0	1.0	1.0	1.0	1.0
Grade 2 seismic performance	Response value	1.0	1.0	1.0	1.0 (1.0–1.2)	1.0 (1.0–1.2)	1.0 (1.0–1.2)
	Limit value	1.3	1.0	1.0* 1.3 or 1.56** (1.1–1.3)			1.0

* For displacement limit value

** For shear force calculation

1. Material factor y_m: The material factor is determined by taking into consideration factors such as a change in an undesirable direction from the characteristic value of material strength, differences in material properties between test specimens and structure materials, the influence of material properties on the limit state and changes over time in material properties. The response value and limit value used were 1.0 and 1.3, respectively.
2. Member factor y_b: It is generally known that the shear capacity of a structural member subjected to reversed cyclic loading decreases in the large deformation range, but the rate of such decrease has not yet been evaluated quantitatively. The JSCE *Standard Specifications* require that the shear force to be resisted by the concrete be increased by a factor of 1.2 or so in the shear capacity calculation formula (i.e. $1.3 \times 1.2 = 1.56$). The limit value, therefore, is 1.56 in the case where flexural yielding occurs and 1.3 in the case where flexural yielding does not occur.
3. Structural analysis factor y_a: The structural analysis factor is 1.0.
4. Load factor y_f: The load factor is 1.0.
5. Structure factor y_i: The structure factor is 1.0.

6.5.5 Analysis method used for verification

Records of observation of underground pipelines, immersed tube tunnels and other underground structures during earthquakes and the results of model vibration experiments have shown that earthquake-induced vibration properties of underground structures have the following characteristics:

1. The apparent unit weight of underground structures is smaller than or similar to that of the surrounding ground. Therefore, during an earthquake, underground structures do not vibrate differently, but vibrate with the surrounding ground.
2. Seismically induced structural behaviour is governed not by seismic inertia force but by the relative displacement of the surrounding ground (ground strain).

Hence, methods for calculating the earthquake resistance of underground structures have been developed on the basis of the abovementioned vibration characteristics. Those methods can be broadly classified into static analysis methods and dynamic analysis methods.

Static analysis methods include the response displacement method and the response seismic coefficient method. In these methods, seismically induced dynamic external forces are replaced with static forces, and structural behaviour under earthquake loading (e.g. response displacement, seismically induced stress) is calculated by applying those forces on the structure of interest.

In dynamic analysis methods, the object to be analysed is modelled as a dynamic mechanics model with dynamic response characteristics, and the time history response of the structure of interest is calculated by directly inputting the time history waveform of ground motion. The response behaviour of an underground structure can be determined by solving an equation of motion formed by taking into account its mass, stiffness and damping characteristics.

The JSWA *Seismic Guidelines* describe a number of methodologies for evaluating the earthquake resistance of existing pipes and renovated pipes. Among those methods, the Tokyo Metropolitan Government has adopted a verification method based on nonlinear dynamic response analysis, aiming at accurately reflecting the condition of buried pipes. This verification method is discussed in the following section.

6.5.6 Earthquake resistance verification based on nonlinear dynamic analysis

(1) Overview of the nonlinear dynamic analysis method

The dynamic response of a structure subjected to earthquake ground motion varies depending on mass, stiffness and damping. This section briefly describes the time history response analysis method as a technique for analysing the response of a vibrating system to dynamic disturbances such as earthquake ground motion.

The equation of motion of a single-degree-of-freedom elastoplastic system can be expressed as shown below. If given the input $-m\ddot{z}$ at a given time, this equation of motion can be solved by using direct integration.

$$m\ddot{x} + c\dot{x} + Q(x) = -m\ddot{z} \tag{6.9}$$

where m: mass; c: damping factor; x: relative displacement; and $\ddot{z}$: the acceleration of ground motion. $Q(x)$, called the restoring force characteristic, is a function of the deformation history. Within the small structural deformation range, structural behaviour can be deemed more or less elastic. As deformation increases, however, damage resulting from cracking, plasticisation, etc. accumulates locally so that the restoring force–deformation curves form hysteresis loops. When trying to determine the response properties of a structure through nonlinear dynamic analysis, it is important to set such elastoplastic restoring force characteristics or hysteresis characteristics appropriately.

Various nonlinear hysteresis models for nonlinear dynamic analysis have been proposed for use under different material and structural conditions. Figure 6.7 shows three types of basic restoring force model. Type (a), the bilinear model, is widely used as a model of the characteristics of a highly deformable (i.e. tough) structure. A model of this type is often used to model a steel structure, and is characterised by relatively large energy consumption due to the hysteresis effect. Type (b), the stiffness reduction type model, does not follow the same loops and shows gradual stiffness reduction after the occurrence of plastic deformation. A model of this type is often used to model a reinforced concrete structure. Type (c), the slip type model, shows low unloading stiffness in the small restoring force range and is characterised by relatively small energy consumption due to the hysteresis effect. Thus, if a nonlinear hysteresis model to be used is to be determined, its skeleton curve and hysteresis rule are necessary. Since an ageing sewer to be renovated is a reinforced concrete structure, a Type (b) model is used for nonlinear hysteresis modelling of a composite pipe.

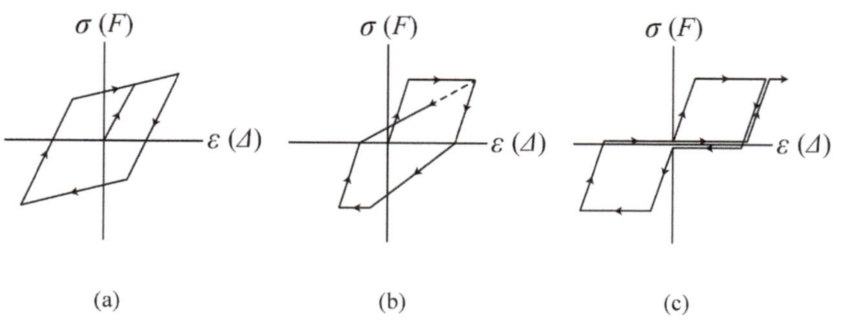

Figure 6.7: Examples of restoring force models: (a) bilinear type; (b) reduced stiffness type; (c) slip type.

(2) Structural analysis model

In view of the nonlinearity of materials, the basic rule is to perform verification on the basis of response values obtained through time history response analysis. Another rule is to conduct structure–soil coupling analysis by using reliable nonlinear hysteresis models of materials and members. Cracking of concrete is modelled by using a smeared crack model. For dynamic modelling, path-dependent hysteresis models are used for both concrete and reinforcing steel. For ground modelling, the nonlinearity of ground during a large-scale earthquake is considered. As an example, UC-win/WCOMD, a commercial programme that meets the requirements mentioned above, is selected for dynamic response analysis (Okamura and Maekawa, 1991; FORUM 8, 2013).

In cases where a coupling analysis of a soil–structure system is conducted, the vibration of the structure and the energy of scattered waves generated by irregularities of the ground are confined in the system. It is therefore necessary to use a finite element model covering a sufficiently large area of ground. It is also necessary to use artificially introduced imaginary boundaries to absorb wave energy. Such boundaries that are widely used include viscous boundaries, energy-transmitting boundaries and superposition boundaries. Superposition boundaries, which are used in UC-win/WCOMD, are briefly explained below.

When a scattering wave reaches an artificial boundary, the wave is reflected in the same phase if the boundary is free. If the boundary is fixed, the wave is reflected in the opposite phase. The reflected wave can then be eliminated by adding the two waves together. If, however, the reflection is repeated at two or more boundaries, the reflected waves cannot be eliminated. Another drawback of the superposition boundary approach is that the entire region needs to be analysed. In view of these problems, Cundall *et al.* proposed a method of superimposing the solutions obtained from constant-velocity and constant-strain boundary conditions, in place of fixed and free boundaries (Cundall, Kunar, Carpenter, et al., 1978). In the proposed method, waves can be cancelled in near-boundary zones. Since two-layer finite elements are defined only in the near-boundary zones as shown in Fig. 6.8, reflected waves can be cancelled with good accuracy with only a small number of degrees of freedom.

(3) Limit values in verification

Limit values used in the verification for Level 2 earthquake ground motion are as follows. In the verification for ultimate displacement, based on Eq. (6.7) a limit value of 0.004, which is two times the strain corresponding to typical concrete compressive strength (0.002), is used. In shear-related verification, design shear capacity is calculated according to the design shear capacity formula for linear members (JSCE, 2012):

$$V_{yd} = V_{cd} + V_{sd} + V_{ped} \tag{6.10}$$

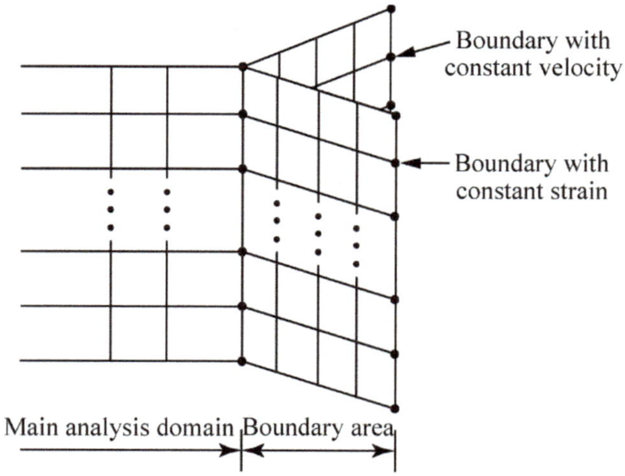

Figure 6.8: Concept of superposition boundary

where V_{cd} is the design shear capacity of a linear member without shear reinforcement, calculated as

$$V_{cd} = \beta_d \cdot \beta_p \cdot \beta_n \cdot f_{vcd} \cdot b_w \cdot d / \gamma_b \tag{6.11}$$

and

$$f_{vcd} = 0.20\sqrt[3]{f'_{cd}}\,(N/mm^2) \text{ where } f_{vcd} \le 0.72(N/mm^2)$$

$$\beta_d = \sqrt[4]{1/d}\,(d:m) \text{ where 1.5 is used if } \beta_d > 1.5$$

$$\beta_p = \sqrt[3]{100P_v} \text{ where 1.5 is used if } \beta_p > 1.5$$

$$\beta_n = 1 + 2M_0/M_{ud} \text{ (if } N'_d \ge 0\text{) where 2 is used if } \beta_n > 2$$

$$= 1 + 4M_0/M_{ud} \text{ (if } N'_d < 0\text{) and } \beta_n = 0 \text{ if } \beta_n > 0 \tag{6.12}$$

where N'_d: design compressive force; M_{ud}: pure flexural capacity; M_0: bending moment necessary to cancel stress caused by axial force at extreme tension fibre corresponding to design bending moment M_d; b_w: midsection width; d: effective height; $P_v = A_s/(b_w \cdot d)$; A_s: cross-sectional area of steel in tension zone; f'_{cd}: design compressive strength of concrete, in N/mm^2; γ_b: usually, 1.3 may be used (1.56 if flexural yielding occurs); V_{sd}: design shear capacity provided by shear reinforcement; and V_{ped}: shear component of effective tension in axial tendon.

The formula shown above cannot be applied directly to lining members because they are composite members. It is therefore necessary to make effective use of the advantages of each renovation construction method and to evaluate shear capacity on the basis of relevant test results. For example, for the SPR method, the shear strength of lining members shown in Table 3.35 is calculated from the results of single shear tests conducted on composite members, and shear capacity is calculated on the basis of Eq. (6.13) and Eq. (6.14):

$$V_{spr} = \sigma_{spr} H \tag{6.13}$$

where V_{spr}: shear capacity of SPR lining member (contribution by SPR lining member); σ_{spr}: shear strength of SPR lining member (from Table 3.35); and H: thickness of SPR lining member. From the above, the design shear capacity of an SPR composite pipe can be calculated from:

$$V_d = V_{cd} + V_{spr} / \gamma_m / \gamma_b \tag{6.14}$$

where V_d: design shear capacity of SPR composite cross section; V_{cd}: design shear capacity of linear member without shear reinforcement; and V_{spr}: shear capacity of a SPR lining member.

(4) Ground motions considered

The ground motions used to verify the earthquake resistance of renovated sewers are Level 1 and Level 2 earthquake ground motions. The correct way in this case would be to decide on ground motions to be considered in view of factors such as the seismic activity level at and near the construction site, earthquake source characteristics and characteristics of the propagation of ground motion from the source to the construction site. Since, however, deriving time history acceleration waveforms by this approach would be laborious, guidelines such as the JSCE *Standard Specifications* and the *Specifications for Highway Bridges* (JRA, 2012) allow the use of simulated ground motions containing vibration components that increase the influence on the structure. Those guidelines state that for Level 2 earthquake ground motion, it is necessary to take into consideration two types of ground motion: inland near-field and oceanic/interplate.

When conducting seismic verification, the Tokyo Metropolitan Government uses the simulated time history waveforms of ground motion shown in the JSCE *Standard Specifications*. The JSCE code shows acceleration response spectra to specify earthquake loads and gives four sample waveforms of Level 2 earthquake ground motion as time history waveforms, as shown in Fig. 6.9. Figure 6.10 shows the acceleration response spectra corresponding to those seismic waveforms. The following four sample seismic waveforms were derived:

1. The "inland type 1" earthquake ground motion waveform was defined on the basis of a past inland strong motion record.
2. The "inland type 2" earthquake ground motion waveform was taken from the Kobe Port Island strong-motion seismograph record of the Hyogoken Nanbu Earthquake of 1995.

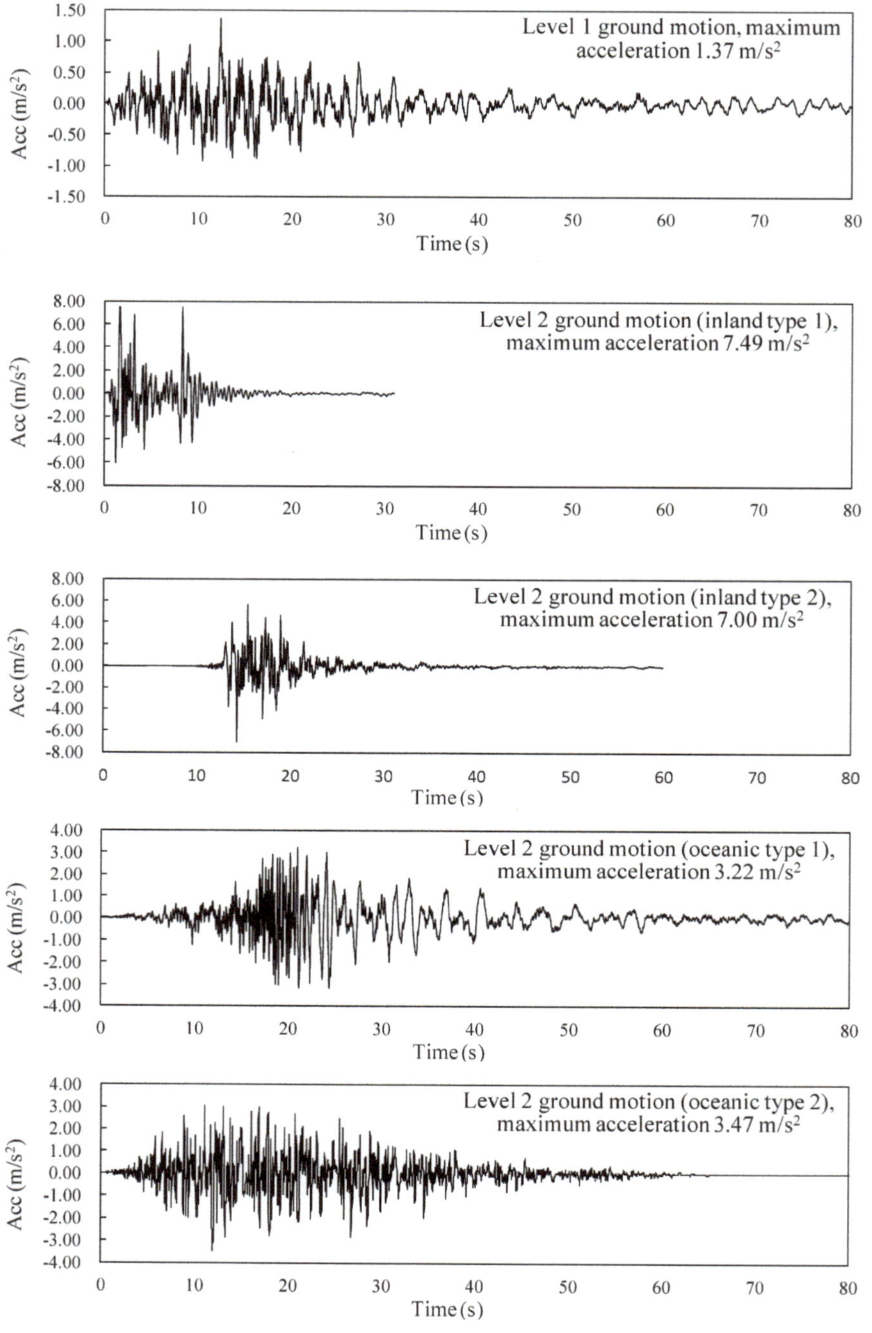

Figure 6.9: Simulated ground motion waveforms (JSCE, 2012)

3. The "interplate (oceanic) type 1" earthquake ground motion waveform was obtained by making corrections to an observation record.
4. The "interplate (oceanic) type 2" earthquake ground motion waveform was derived by calculation from the fault model of the scenario Tokai Earthquake proposed by the Central Disaster Management Council.

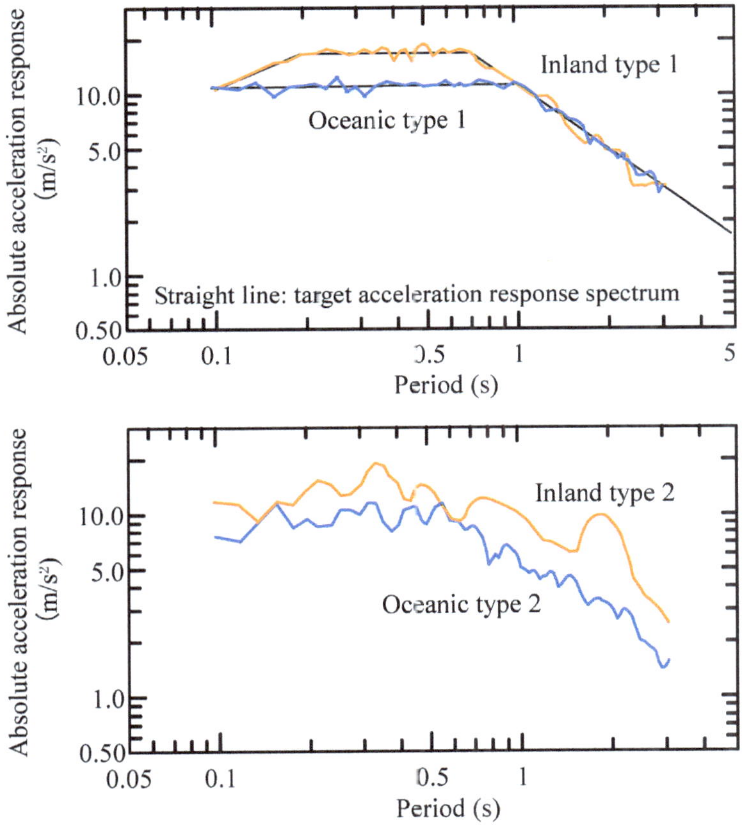

Figure 6.10: Acceleration response spectra of inland and oceanic earthquakes (damping factor = 5%) (JSCE, 2012)

(5) Ground conditions

Ground conditions are extracted from available records of boring surveys conducted near the representative cross section of the sewer line to be renovated. The bedrock surface assumed, at which ground motion is input, must be the upper surface of a ground layer that is as extensive as the project site, is firm and has sufficiently higher shear wave velocity compared with the ground surface. The bedrock surface must be selected by collecting not only geological survey data obtained at the project site but also information on the surrounding areas and comprehensively evaluating the information thus obtained.

The bedrock surface may be chosen as the upper surface of a continuous layer having an *N*-value (SPT blow count) of 50 or more and a shear wave velocity V_s of not lower than about 300 m/s. If detailed geotechnical survey results are available, a bedrock surface can be determined appropriately. It is therefore necessary to take site-specific ground conditions into consideration.

6.5.7 Earthquake resistance verification based on response displacement method

(1) Position of response displacement method
The JSWA *Seismic Guidelines* state that it is good practice to consider the use of a more accurate method such as nonlinear dynamic analysis that takes into consideration the nonlinearity of ground and structural members, but it also permits the use of the response displacement method in view of its simplicity in engineering practice. Although verification by the response displacement method has not been recognised as a sufficiently rational verification approach applicable to renovation design of ageing sewers, it has been widely used in the seismic design of ordinary underground structures and sewer systems. Therefore, the response displacement method will also be used for calculating seismically induced loads. Concepts concerning the use of the response displacement method for earthquake resistance verification are described in the subsequent sections.

(2) Earthquake-induced ground displacement
1) Calculation of ground displacement based on design response velocity spectrum
The JSWA *Seismic Guidelines* and the JSWA *Guidelines* show response velocity spectra for Level 1 and Level 2 earthquake ground motions for use at the design stage. The response displacement of the surface layer of ground is calculated by the so-called response spectrum method on the basis of the natural period of ground. Types of ground motion, however, are not distinguished.

In the response displacement method, the influence of ground motion is applied as an external force calculated from the response displacement of ground, so the relations concerning design ground motion and the response of ground are important. Basically, the methods shown in the JSWA *Seismic Guidelines* and the JSWA *Guidelines* assume that stiffness is uniform in the depth direction and that the first mode of shear vibration is the only response vibration mode. Therefore, the response displacement of the surface layer of ground can be calculated by using a simple formula:

$$U_h(z) = \frac{2}{\pi^2} \cdot S_v \cdot T_s \cdot \cos\left(\frac{\pi z}{2H}\right) \tag{6.15}$$

where $U_h(z)$: displacement amplitude in horizontal direction at depth z from ground surface; S_v: design response velocity; T_s: natural period of surface layer of ground; and H: thickness of surface layer of ground. Figures 6.11 and 6.12 show response velocity spectra for design use, and Fig. 6.13 shows the seismic areal-division map of Japan for classification of seismic areas.

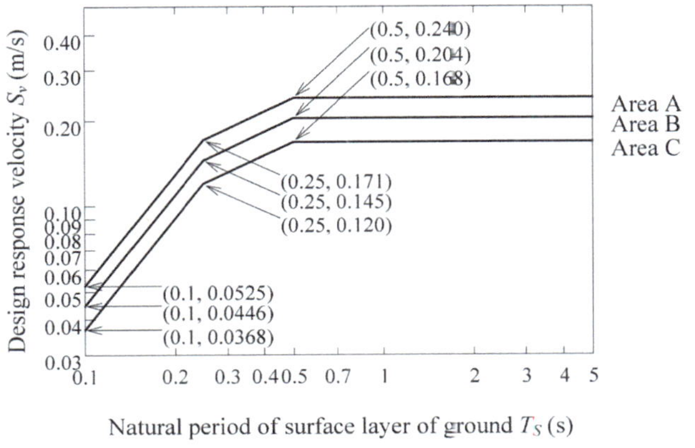

Figure 6.11: Design response velocity (Level 1 earthquake ground motion) (JRA, 2012)

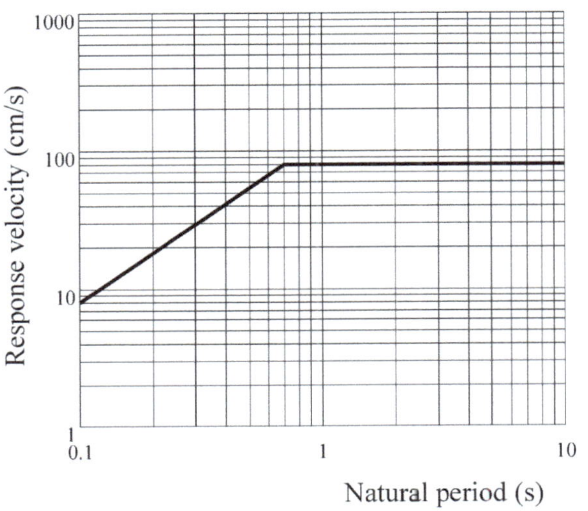

Figure 6.12: Design response velocity (Level 2 earthquake ground motion) (JRA, 2012)

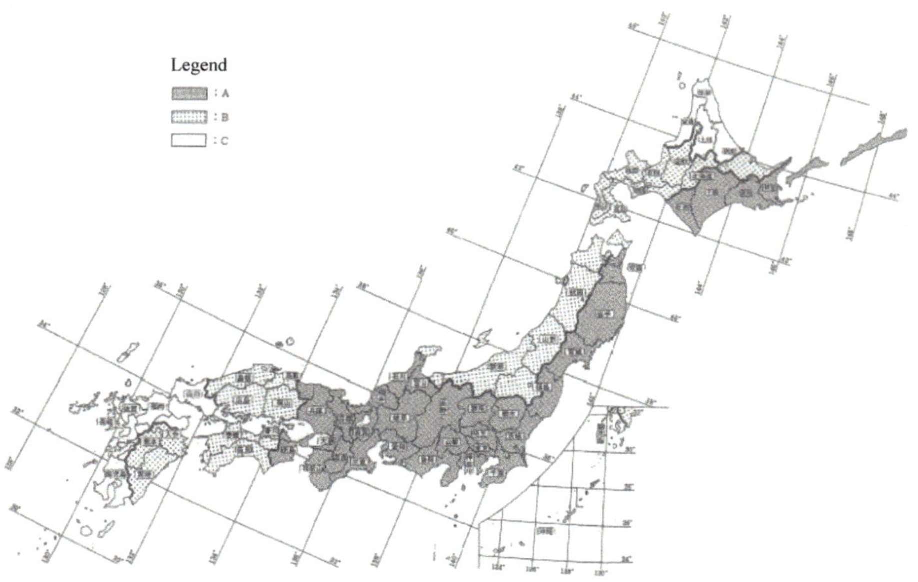

Figure 6.13: Seismic areal-division in Japan (JRA, 2012)

2) Calculation of ground displacement based on ground response analysis
Sewer pipes are often buried in multi-layer ground, and the influence of similar design response spectra on a structure varies depending on the type of ground motion. If, therefore, seismic performance verification is to be made with higher accuracy, it is desirable that the structural response be calculated by inputting the ground response calculated through a time history response analysis of ground alone. The seismic behaviour of an underground structure is greatly affected by the rate of change in horizontal displacement in the vertical direction, instead of the magnitude of absolute displacement of ground at the location of the structure. Thus, the displacement used in the response displacement method is the displacement of ground at the time when the relative displacement of ground between the upper and lower ends of the structure is maximised.

The equivalent linearisation method, a method of response analysis in the frequency domain, is widely used for analysing the time history response of ground. The equivalent linearisation method is used in cases where the shear strain amplitude of ground during an earthquake is relatively small (about 10^{-3} or less). An equivalent linear model expresses soil as a visco-elastic body with stiffness and viscosity and sets equivalent shear moduli and viscous damping factors during the duration of ground motion according to the amplitude of earthquake-induced shear strain (effective strain). Linear analyses are repeated by using the equivalent linear model until the shear strain amplitude converges. This method is practical and useful because stable solutions can be obtained and the calculation time is short.

(3) Inertia force
Since pipeline facilities are usually lighter than the soil of the same cross-sectional area, inertia force is ignored in many cases. For renovated pipes, however, it was decided to take inertia force into consideration to make the renovation design of ageing sewers consistent with the design of other civil engineering structures. As a basic rule, inertia force was calculated by multiplying the weight of the structure by the design seismic coefficient. In FEM analysis, inertia force was made to act as an external force by multiplying each nodal force corresponding to self-weight by the design horizontal seismic coefficient. The design seismic coefficient was determined in accordance with the JSWA *Seismic Guidelines*. The influence of the design vertical seismic coefficient is often ignored because it is usually small.

(4) Soil springs
Soil springs defined along the perimeter of the structure of interest in the response displacement method are positioned normal and tangential to the outer circumference of the pipe. Spring values are determined from moduli of soil reaction, taking into account node-dominated areas in the FEM analysis model. There are a number of methods for defining soil springs, and it was decided to use the method employed in the JSWA *Examples of Seismic Design Calculation for Sewer Facilities* (JSWA, 2001). Although there are various approaches to dealing with loading width, it was decided, in principle, to use product length for ready-made products such as circular pipes, precast box culverts and arch culverts. Cast-in-place pipes require separate definitions. For example, a single span length may be used for tunnels.

(5) Earthquake-induced skin shear force
The outer surface of an underground structure in contact with the surrounding soil is acted upon by skin shear force in the event of an earthquake, and so such skin shear force is taken into consideration. Although there are cases in which earthquake-induced skin shear force is ignored for circular pipes, in view of common practice, such skin shear force should be taken into consideration. A commonly used method is to calculate earthquake-induced skin shear force by using the equation shown below on the basis of the design response velocity spectrum. Another common approach is to use the shear force at the time when the maximum ground displacement occurs obtained from one-dimensional ground response analysis.

$$\tau = \frac{G_d}{\pi H} \cdot S_v \cdot T_s \cdot \sin\left(\frac{\pi z}{2H}\right) \tag{6.16}$$

where τ: earthquake-induced skin shear force per unit area at depth z (m) from ground surface; S_v: design response velocity; T_s: natural period of surface layer of ground; H: thickness of surface layer of ground; and z: depth from ground surface.

(6) Verification method based on nonlinear FEM analysis
Figures 6.14 and 6.15 show the flows of analysis and verification. For Level 1 earthquake ground motion, seismically induced external force is increased and made to act in the initial stress state under normal loading (dead load + static earth pressure + overburden load) to confirm that steel reinforcement does not yield or cracking does not occur until the design ground motion is reached.

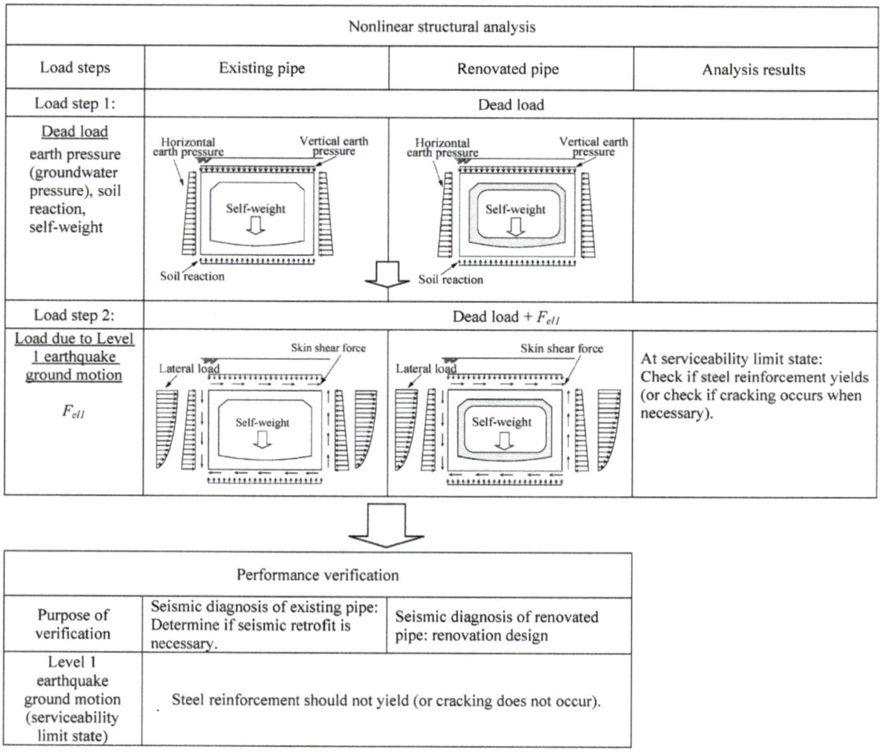

Figure 6.14: Flow of performance verification under Level 1 earthquake ground motion

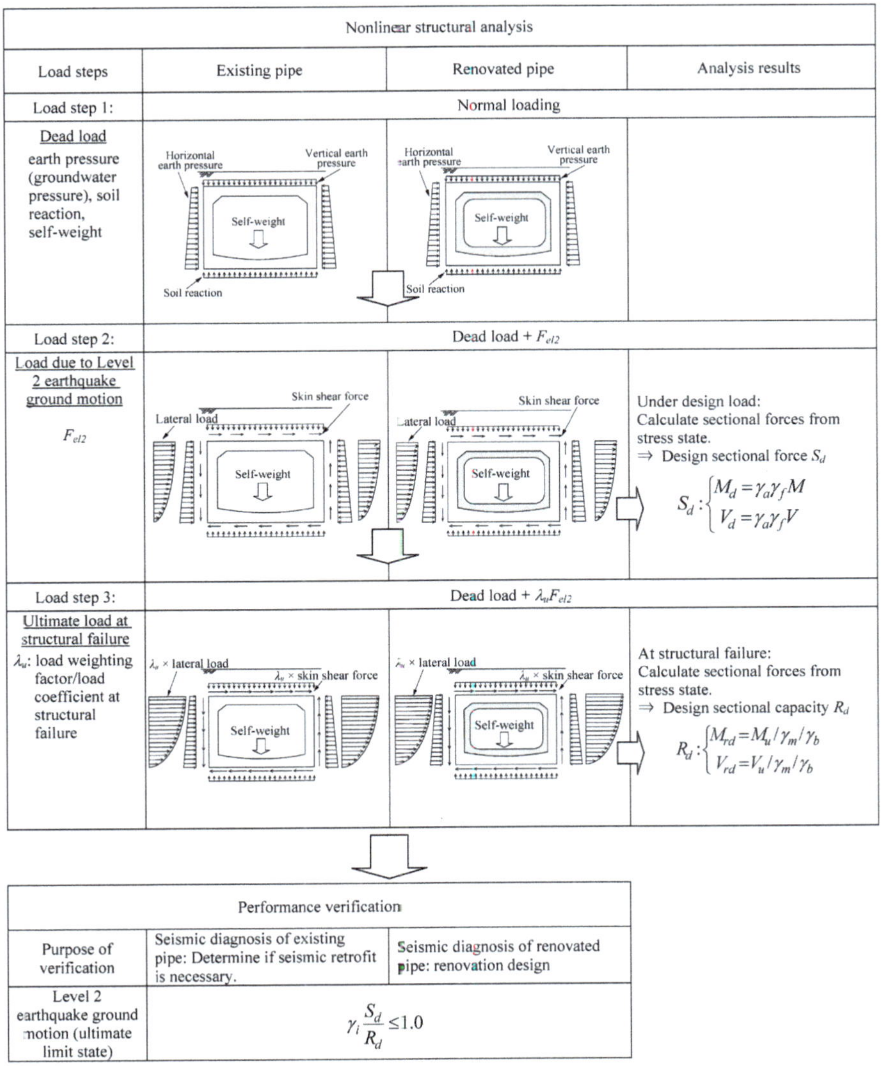

Figure 6.15: Flow of performance verification under Level 2 earthquake ground motion

For Level 2 earthquake ground motion, seismically induced external force is increased and made to act in the initial stress state under normal loading (dead load + static earth pressure + overburden load) and the applied load is increased until the ultimate state is reached. In the verification concerning the ultimate limit state, the design sectional capacity and design sectional force used for verification are calculated as follows:

1. Calculation of the maximum sectional force (= design sectional capacity) at ultimate failure and determination of the location where that occurs
2. Calculation of sectional force (= design sectional force) at the abovementioned location under Level 2 earthquake ground motion

Design sectional force is calculated by applying the specified safety factor to the sectional force calculated from the stress state under design Level 2 earthquake ground motion. Design sectional capacity is calculated by applying the specified safety factor to the maximum sectional force that has occurred at or before failure. In safety verification, checks are made to confirm that design sectional force does not exceed design sectional capacity.

References

Cundall, P. A., Kunar, R. R., Carpenter, P. C. et al. (1978). Solutions of infinite dynamic problems by finite element modeling in the time domain, Proc. 2nd Int. Conf. Appl. Num. Modeling, Madrid, 339-351.

FORUM 8 (2013). UC-win/WCOMD Ver. 2.

JRA (2012). Specifications for Highway Bridges: V. Seismic Design. Tokyo, Japan Road Association.

JSCE (2012). Standard Specifications for Concrete Structures: Design. Tokyo, Japan Society of Civil Engineers.

JSWA (2001). Examples of Seismic Design Calculation for Sewer Facilities. Tokyo, Japan Sewage Works Association.

JSWA (2003). Reinforced Concrete Sewerage Pipes. Tokyo, Japan Sewage Works Association.

JSWA (2011). Design and Construction Guidelines for Sewer Pipe Rehabilitation. Tokyo, Japan Sewage Works Association.

JSWA (2014). Guidelines for Seismic Design and Retrofit of Sewerage Facilities. Tokyo, Japan Sewage Works Association.

Okamura, H. and Maekawa, K. (1991). Reinforced Concrete: Nonlinear Analysis and Constitutive Law. Tokyo, Gihodo.

Yukari Nakamura

7 Development of the Composite Pipe Design Support System

7.1 Design Support Programme

In renovation design of ageing sewers by the SPR method, the limit state design method is used for safety verification. In order to verify safety in the ultimate limit state, it is necessary to determine the ultimate strength and sectional force of a renovated pipe by a nonlinear FE analysis approach and fracture mechanics modelling. The composite pipe design support system, SPRana (CSD, 2015), is a software package that enables a user who may not be familiar with nonlinear structural analysis and fracture mechanics to reach a sound renovation design that satisfies the safety requirements of relevant building codes.

7.2 Overview of the System

SPRana, developed for application to existing reinforced concrete sewer pipes and pipes renovated by the SPR method, makes it possible to verify performance based on nonlinear FEM analysis under normal and earthquake loading. SPRana conforms to the relevant building codes and guidelines (JRA, 1999; JSWA, 2003; JSWA, 2011).

SPRana reduces the time required for data preparation by simplifying structural settings and employing a system for automatically generating FEM analysis models. The system can handle four types of sewer configuration: rectangular, covered, circular and horseshoe-shaped, and so can model typical sewer configurations.

7.3 Programme Structure

Figure 7.1 shows the programme structure of SPRana. The analysis conditions to be input include cross-sectional dimensions, steel reinforcement conditions, material conditions and liner conditions (profile and fill mortar to be used) matching the dimensions of the existing pipe. When the analysis programme is executed, an FE mesh for the reinforced concrete structure to be analysed is automatically created, and input data necessary for the analysis are created at the same time. By using these data, a nonlinear structural analysis is carried out to calculate element stresses and load ratios to be used for verification based on load coefficients. Then, sectional forces are calculated by using the sectional force calculation programme.

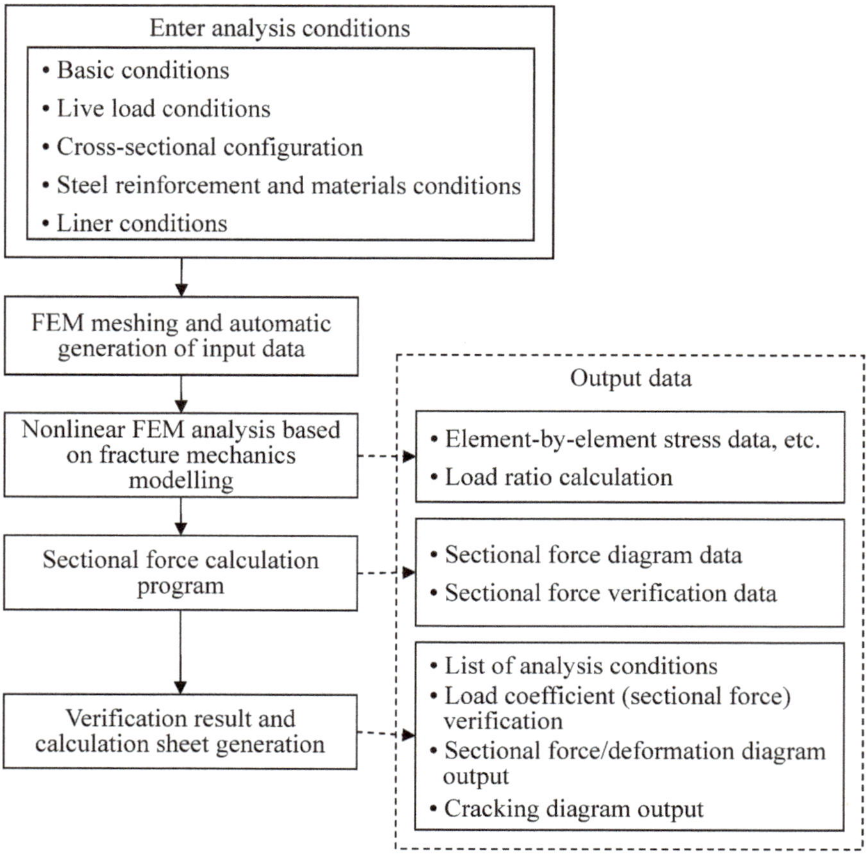

Figure 7.1: Program structure and output data

Upon completion of the analysis, a calculation sheet showing the analysis conditions, results of performance verification and analysis results including sectional forces and crack diagrams is created. The programme is Windows-based, and the computation time depends on the environment.

7.4 Input/Output Functions and Scope of Application

Table 7.1 lists the conditions to be entered into the SPRana system. Table 7.2 shows the items and results output from the system. The output items include not only verification results but also deformation diagrams, sectional force diagrams and cracking diagrams that can be used for validating calculation results. Tables 7.3 and 7.4 show the scope of application of the SPRana system.

Table 7.1: Input items

Conditions			Sub-item	Remarks
Burial conditions		Overburden thickness		
		Unit weight of soil		
		Groundwater level		Only when taken into consideration
		Soil reaction		Only when taken into consideration
		Support angle		Only for circular sewer
Structural conditions	Existing pipe	Shape and dimensions	• Inside width, inside height • Member thickness (loss of concrete cover) • Haunch height • Invert height, radius • Inside radius at crown • Inside radius at side	Sub-items are specific to cross-sectional shapes
		Steel reinforcement conditions	• Concrete cover • Rebar spacing • Rebar diameter (corrosion ratio)	
		Material conditions	• Concrete • Steel reinforcement	
	Renovated pipe	Shape and dimensions	• Inside width and height of renovated pipe • Renovated pipe alignment • Inside radius at invert • Inside radius at crown • Inside radius at side	Sub-items are specific to cross-sectional shapes
		Lining material conditions	• Profile specifications • Fill mortar materials * Profiled steel sheets conditions	
		Slip at interface between existing member and liner		Only when taken into consideration

Table 7.1: (continued)

Conditions			Sub-item	Remarks
Load conditions	Earth pressure	Vertical earth pressure		
		Horizontal earth pressure		Only when taken into consideration
	Live load	Vertical	• Vehicle load • Railway load (EA load) • Overburden load	
		Horizontal		Only when taken into consideration
	Arbitrary load		• Self-weight • Earth pressure • Live load	Only when taken into consideration
Verification		Verification in terms of sectional force		
		Verification in terms of load coefficient		
Safety factors			• Material factor γ_m • Member factor γ_b • Structure factor γ_i • Structural analysis factor γ_a • Load factor γ_f	

Table 7.2: Output items

Main item	Sub-item		Remarks
Design conditions	Cross-sectional view		
	Existing pipe conditions	• Dimensional data (loss of concrete cover) • Rebar arrangement (corrosion ratio) • Concrete material properties • Steel reinforcement material properties	
	Renovation conditions		
	Burial conditions		
	Load conditions		
	Safety factor		
Load calculation and load diagram	Dead load		
	Live load		
	Soil reaction		
Calculation result	Displacement diagram	• Under design load • Under maximum load	
	Sectional force and sectional force diagram	Under design load and under maximum load • Moment diagram • Shear force diagram • Axial force diagram	
	Cracking diagram	• Under design load • At cracking • Under maximum load	
Verification result	Sectional force verification	• Whether cracking occurs under design load • Bending moment • Shear force • Axial force	For verification in terms of sectional force
	Load verification	• Cracking load coefficient • Maximum load coefficient	For verification in terms of load coefficient

Table 7.3: General scope of application

General conditions				**SPRana**	**Remarks**
Structural type				• Ageing sewer • Renovated sewer by SPR method	
Type of sewer	Rectangular			A	*1 Not applicable in case of special shape *2 Not applicable in case of non-uniform sidewall thickness
	Horseshoe-shaped			A*1	
	Covered			A*2	
	Circular			A	
	Covered masonry			NA	
	Oval			NA	
Load conditions	Primary load	Dead load	Self-weight*3	A	*3 Load that must be taken into consideration
			Water in sewer	NA	
		Live load	Overburden load*3	A	
			Live load in sewer	NA	
			Lateral pressure due to live load	A	
			Impact*3	A	
		Earth pressure	Vertical*3	A	
			Horizontal*3	A	
		Groundwater		A	
		Buoyance or uplift pressure		A	
		Influence of drying shrinkage		NA	
		Soil reaction		A	
	Secondary load	Influence of temperature change		NA	
		Influence of earthquake		A	
Ground conditions	Overburden thickness			No restriction	
	Soil layer		Single layer	A	
			Multiple layers	A	
	Coefficient of soil reaction			No restriction	
Material	Concrete			A	
	Rebar			A	
	SPR lining material: mortar			A	
	SPR lining material: profiled steel sheet			A	
	Other (e.g. new material)			NA	
Fracture mode				No restriction	

Note: A: applicable; NA: not applicable

Table 7.4: Scope of application on existing sewers

		SPRana	Remarks
Support conditions		Predetermined (only "circular" is selectable)	
Damage conditions of existing sewer	Partial loss of member	A	
	Partial loss of member not limited to internal concrete cover	NA	
	Rebar exposure/corrosion	A	
	Complete corrosion of rebar	NA	
	Deterioration of material	A	
	Severe local fracture	NA	
Fracture mode		No restriction	

Note: A: applicable; NA: not applicable

7.5 Automatic Meshing Function

FEM analysis models are automatically constructed from the pre-entered information on burial condition, structural dimensions, state of deterioration and material conditions. This automation of FE model building is a major feature of the SPRana system.

The analysis model consists of solid elements representing concrete members and mortar members, and bar elements representing rebars and profiled steel sheets of the renovation layer. The PVC profile in the renovation layer is not modelled. As the structure is symmetrical, a half model is used.

7.6 Basic Rules in Building FE Models

7.6.1 Element types

Table 7.5 shows the elements constituting an analysis model created by the SPRana system and the constitutive rules applied. Joint elements and dummy elements are employed to enable no-tension interface modelling (where the breakage of bonding is enforced to represent the termination of stress transfer in the normal and tangential directions on the interface when tensile stress arises), which is one of the main characteristics of the analysis model used in the system.

Table 7.5: Analysis model elements

Element	Corresponding member	Formation rule
Concrete element	Existing pipe concrete and lining mortar	Rectangular element (8-node configuration) Counterclockwise numbering (Fig. 7.2)
Steel reinforcement element	Rebars in existing pipe and profiled steel sheets in lining	One-dimensional bar element
Joint element (dual nodes)	Bond at interface between existing member and liner	Defined by connecting dummy element nodes (Fig. 7.3)
Dummy element	Interface between existing member and liner	Element used to define joint element Rectangular element (8-node configuration)

7.6.2 Basic rules for meshing

In generating the mesh components shown in Fig. 7.4, the rules of Table 7.6 are followed for the existing pipe and the rules of Table 7.7 are followed for the renovation layer or liner. Figures 7.5 and 7.6 show examples of mesh generation, and Fig. 7.7 illustrates the interface defined by using dummy elements.

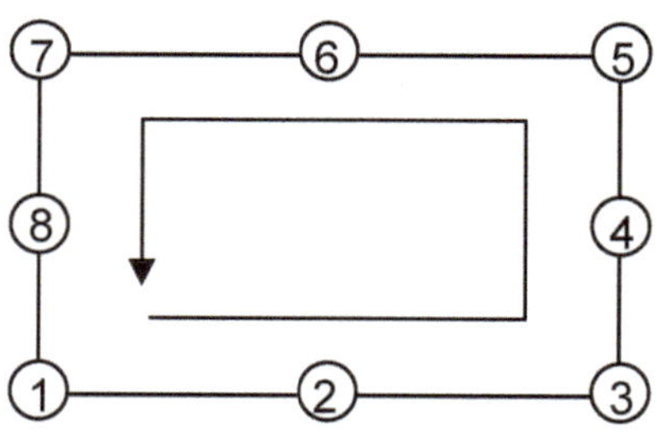

Figure 7.2: Formation of concrete element

Table 7.6: Meshing rules for existing members

Meshing direction	Condition		Meshing rule	Rebar connection rule
Member axis direction (element length)			$l \leq 4b_p$ and $\max(b_{rc1}, b_c, b_{rc2}) \leq l \leq 3\min(b_{rc1}, b_c, b_{rc2})$	
Member thickness direction	① If $b > 4\max(b_{rc1}, b_{rc2})$		A single layer of concrete cover is defined, and the rest of the layer thickness is divided into two or three parts.	Rebars are defined by connecting intermediate nodes or are defined on sides of elements.
	If ① is not applicable		The basic thickness of a single element is 2 times the concrete cover thickness: $b_{rc1} = 2C_1$, $b_{rc2} = 2C_2$	Rebars are defined by connecting intermediate nodes.
			Plain concrete layer b_c is divided as follows: b_c=b-$2C_1$-$2C_2$	
			Criteria for case classification	
		In case of double reinforcement	i) If $b_c \leq \max(C_1, C_2)$ $b_{rc1} = 2C_1+b_c/2$ $b_{rc2} = 2C_2+b_c/2$ * Intermediate nodes to remain at bar locations	RC cross section is divided into four parts.
			ii) If $b_c \geq 1.5\max(b_{rc1}, b_{rc2})$ b_c is divided into two parts.	RC cross section is divided into two parts.
			iii) In cases other than i) and ii) b_c is regarded as a single element. RC cross section is divided into three parts.	
		In case of missing concrete cover	Based on the location of double reinforcement, the concrete cover is removed.	Rebars are defined on sides of elements.
		In case of single reinforcement	The basic rule is to divide the region of interest into two parts by using the reinforcement location as the boundary. If element thickness is too large, it is divided into three parts, and nodes are defined according to the rebar locations.	

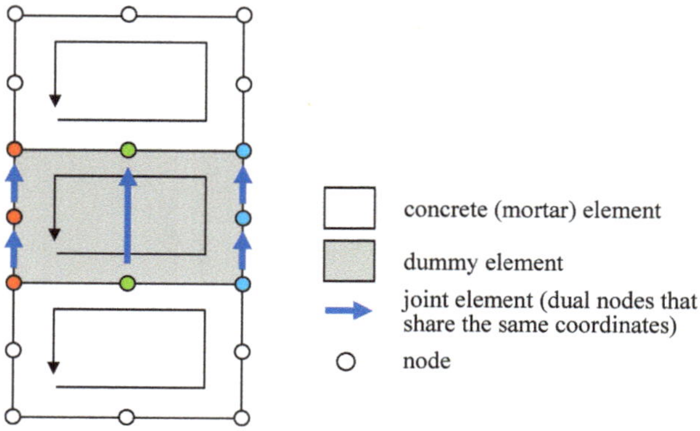

Figure 7.3: Configuration of joint elements and dummy elements (dummy element nodes of the same colour indicate the same coordinates)

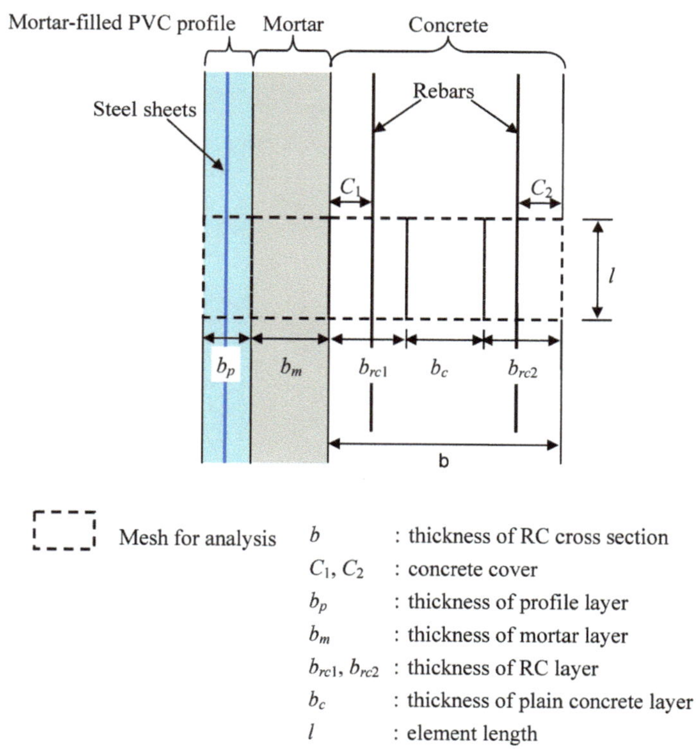

Figure 7.4: Mesh component relationship

Table 7.7: Meshing rules for liner

Meshing direction	Location	Meshing rule	Reinforcement connection rule
Member axis direction (element length)	Profile, mortar	$l \le 4b_p$ and $\max(b_{rc1}, b_c, b_{rc2}) \le l \le 3 \min(b_{rc1}, b_c, br_{c2})$	
	Mortar corners in rectangular, horseshoe-shaped or other sewers	Elements having an aspect ratio of about 1:4 or smaller are placed radially from the centre of radius.	
Member thickness direction (element width)	Profile	Single element for profile thickness	Steel reinforcement element connects middle nodes.
	Mortar	Basically a single layer However, as in the case of a plain concrete layer of an existing member, if $b_m \ge 1.5\max(b_{rc1}, b_{rc2})$, then the layer is divided into two parts.	

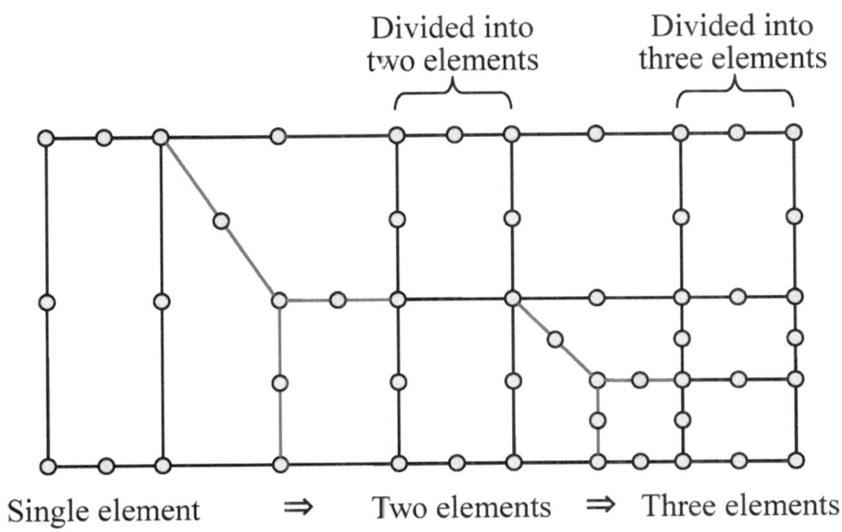

Figure 7.5: Defining a finer mesh in a cross section

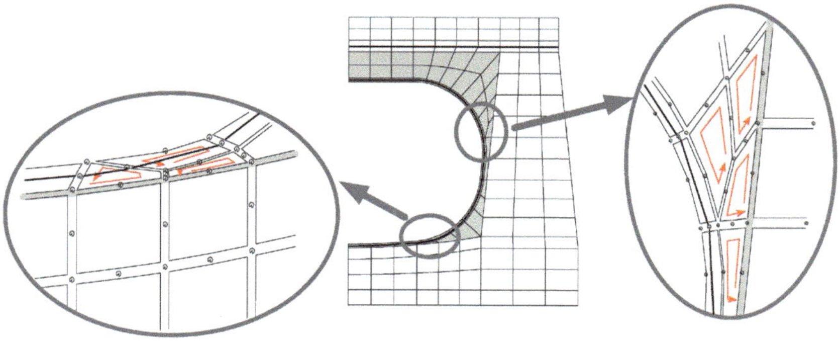

Figure 7.6: Example of meshing

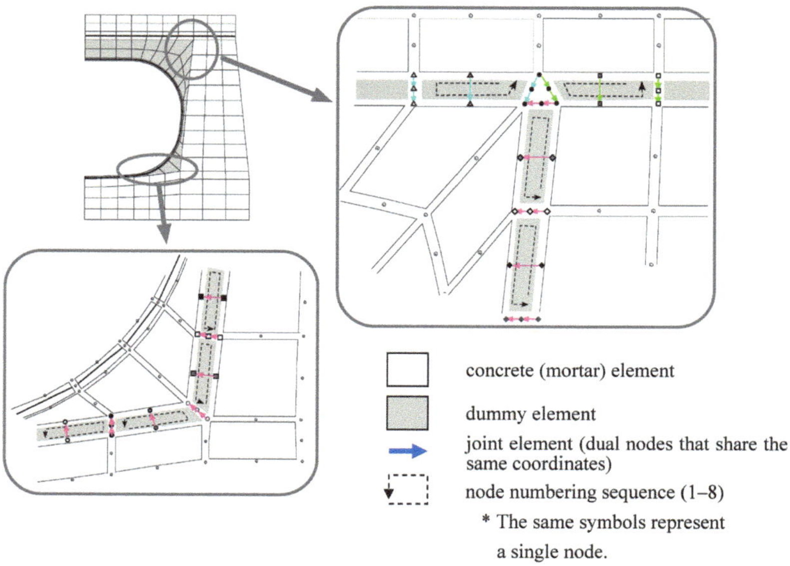

Figure 7.7: Using dummy elements and defining joints

7.7 Normal Loading Analysis

In normal loading analysis, the performance of existing pipes and sewer pipes renovated by the SPR method can be verified on the basis of the limit state design method through nonlinear structural analysis that takes concrete cracking into consideration.

In the analysis, the state of deterioration of the existing pipe can be reflected in the analysis model, and such details can be set separately, if desired, for each member.

7.7.1 Entering analysis conditions

(1) Selecting cross-sectional configuration
The cross-sectional configuration can be chosen from the four types of cross section (rectangular, covered, circular and horseshoe-shaped) shown in Fig. 7.8.

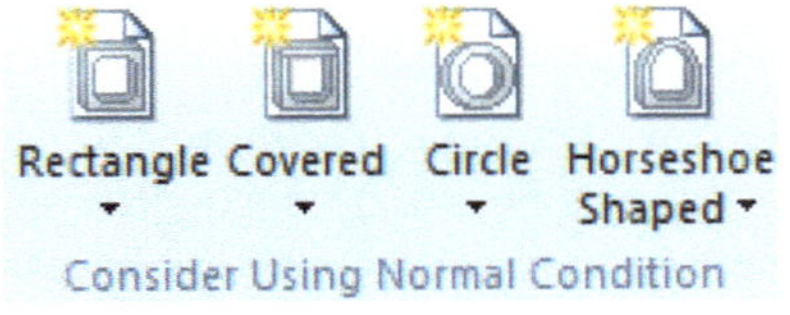

Figure 7.8: Available cross-sectional shapes

(2) Entering verification conditions
Figure 7.9 shows the verification conditions window. There are two verification methods to choose from: the load coefficient method (verification in terms of load coefficient) and the sectional force method (verification in terms of sectional force). It is also possible to perform performance verification only on the existing pipe.

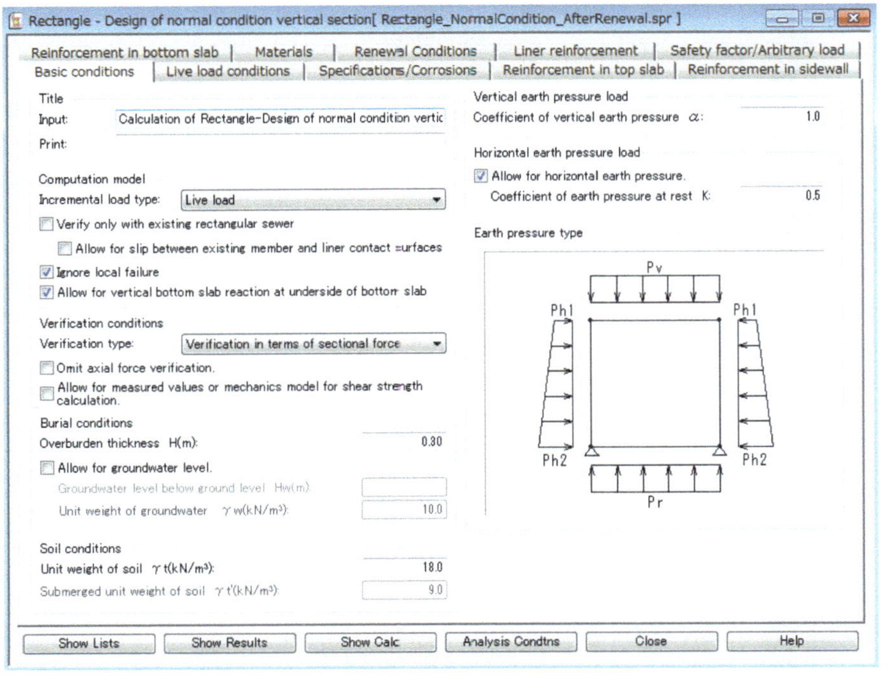

Figure 7.9: Verification conditions window

(3) Entering burial conditions

Figure 7.10 shows the burial conditions window. Vertical earth pressure, horizontal earth pressure and groundwater pressure can be set automatically by entering the overburden thickness over the existing pipe, groundwater level and the unit weight of soil. The coefficient of vertical earth pressure and the coefficient of static earth pressure can be set if desired. Horizontal earth pressure may be ignored.

For a rectangular, covered or horseshoe-shaped sewer, it is possible to allow for bottom slab reaction or rigid foundation (fixed foundation). For a circular sewer, the design support angle can be set in 30-degree steps within the range from 0 to 120 degrees, and the type of vertical earth pressure can be selected from the "Normal earth pressure equation" or the "Loosening earth pressure equation".

(4) Entering live load conditions

Figure 7.11 shows the live load conditions window. Live loads can be set in the vertical and horizontal directions and can be selected from vehicle load, railway load (EA) and overburden load (uniformly distributed load).

For a rectangular, covered or horseshoe-shaped sewer, wheel loads as part of vehicle loads, impact loads and reduction factors can be given if desired. For a circular sewer, vehicle loads can be selected from T-10, T-14, T-20 and T-25.

(5) Entering existing pipe configuration

Figure 7.12 shows the window from which the configuration of the existing pipe can be entered. The configuration of the existing pipe can be freely specified. The configuration of a rectangular or circular sewer can be selected from a list of products registered in advance. For a horseshoe-shaped sewer, "2R horseshoe", "3R horseshoe", "4R horseshoe" or "horseshoe, any shape" can be selected. For a rectangular sewer, the bottom slab configuration can be specified by selecting either "standard type" or "invert type". For a horseshoe-shaped sewer, only "invert type" can be entered.

In cases where corrosion with partial loss of a member has occurred, it can be reflected in the analysis model by entering the location of corrosion and the amount of corrosion for each member. For a horseshoe-shaped sewer, the location of corrosion in the crown can be expressed with an angle. For a circular sewer, the extent of the top, side or bottom zone of the sewer can be expressed with an angle in 15-degree increments between 15 and 90 degrees.

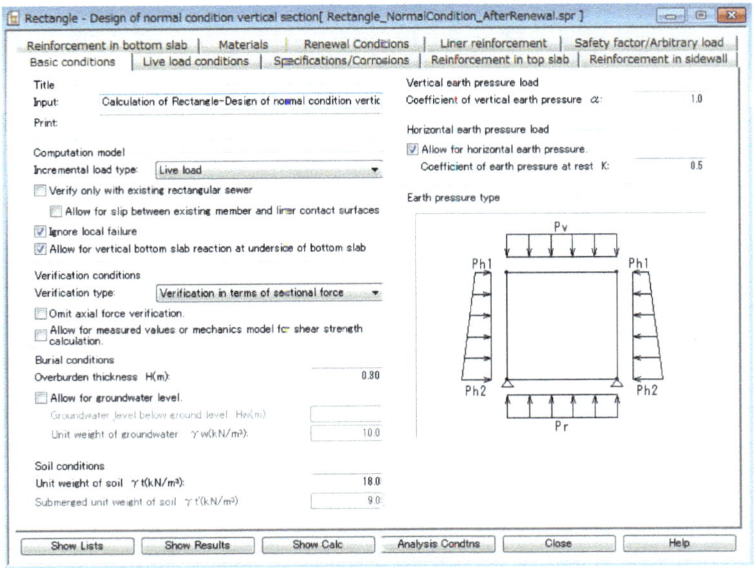

(a)

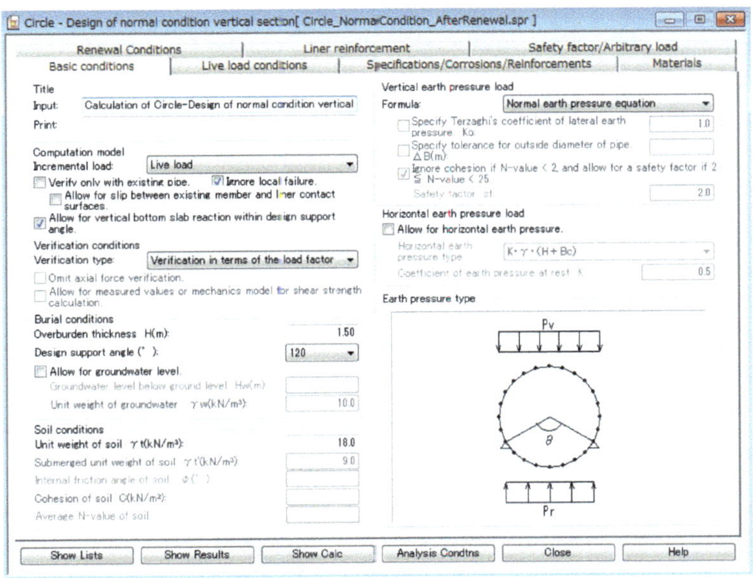

(b)

Figure 7.10: Burial conditions window: (a) rectangular sewer; (b) circular sewer

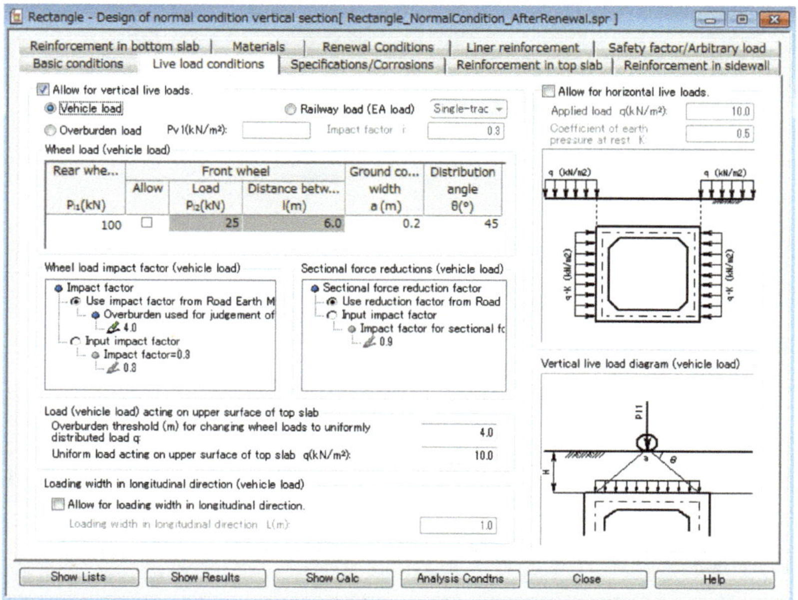

(a)

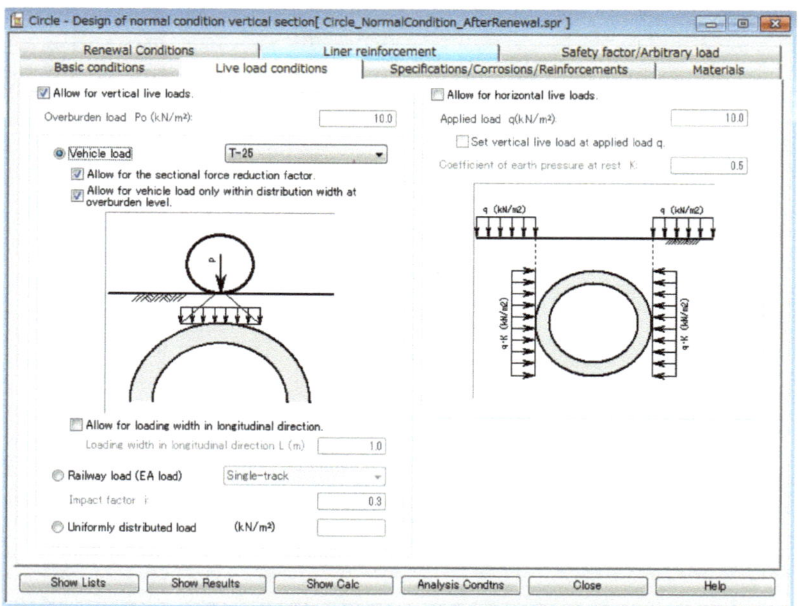

(b)

Figure 7.11: Live load conditions window: (a) rectangular sewer; (b) circular sewer

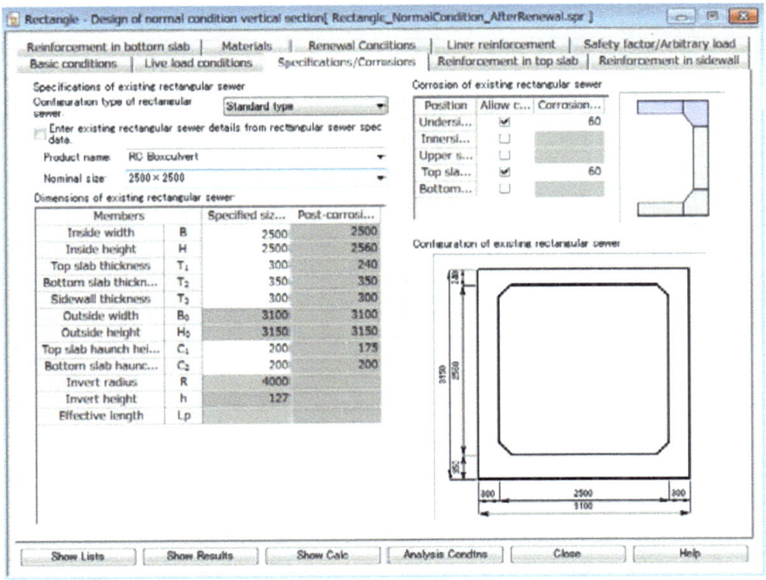

(a)

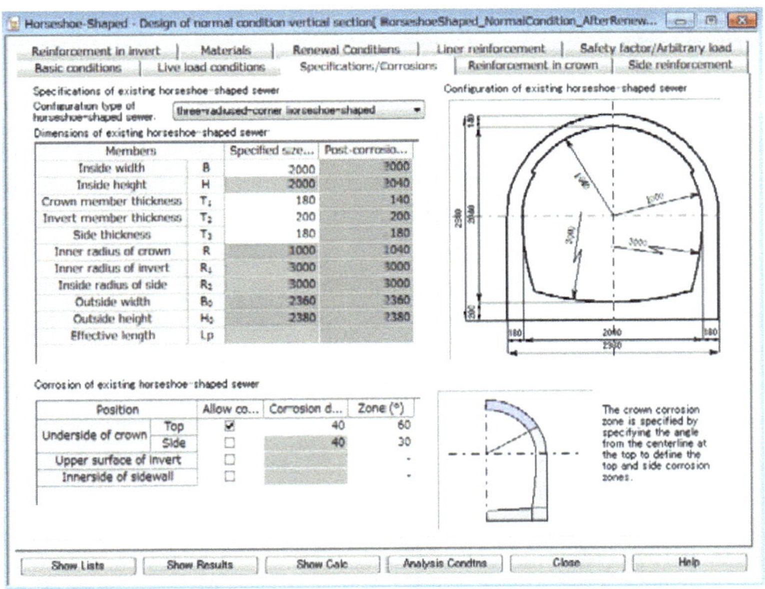

(b)

Figure 7.12: Existing pipe configuration window: (a) rectangular sewer; (b) horseshoe-shaped sewer; (c) circular sewer

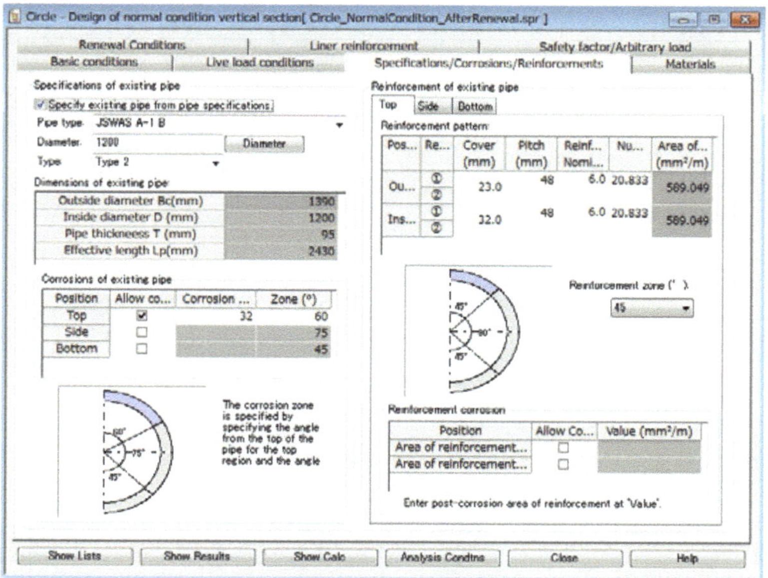

(c)

Figure 7.12: Existing pipe configuration window: (a) rectangular sewer; (b) horseshoe-shaped sewer; (c) circular sewer (continued)

(6) Entering reinforcement conditions

Figure 7.13 shows the reinforcement conditions window. For a rectangular or covered sewer, reinforcement conditions (concrete cover, spacing, nominal bar diameter, number of reinforcing bars) for the top zone (or the cover in the case of a covered sewer), sidewall zones and the bottom slab zone can be specified separately. For a rectangular sewer, similar conditions can be specified separately for the upper and lower haunch zones. For a covered sewer, conditions can be specified for the lower haunches, too.

For a horseshoe-shaped sewer, reinforcement conditions can be specified separately for the crown, side and invert zones. The concrete cover over the internal reinforcement in the sidewall zones, however, must be the same as the concrete cover over the internal reinforcement in the crown zone. For a circular sewer, reinforcement conditions can be specified for the top, side and bottom zones indicated by an angle in 15-degree increments between 15 and 90 degrees.

For a rectangular, covered or horseshoe-shaped sewer, different reinforcement patterns (if any) in each member can be specified in greater detail by specifying the differently-reinforced regions. In cases where corrosion severe enough to cause partial loss of reinforcement has occurred, it can be reflected in the analysis model by entering the location of corrosion and the amount of corrosion.

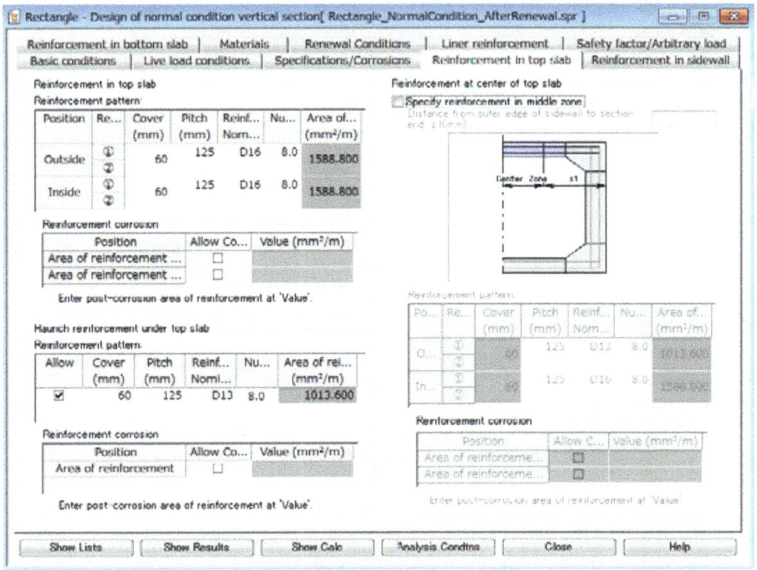

(a)

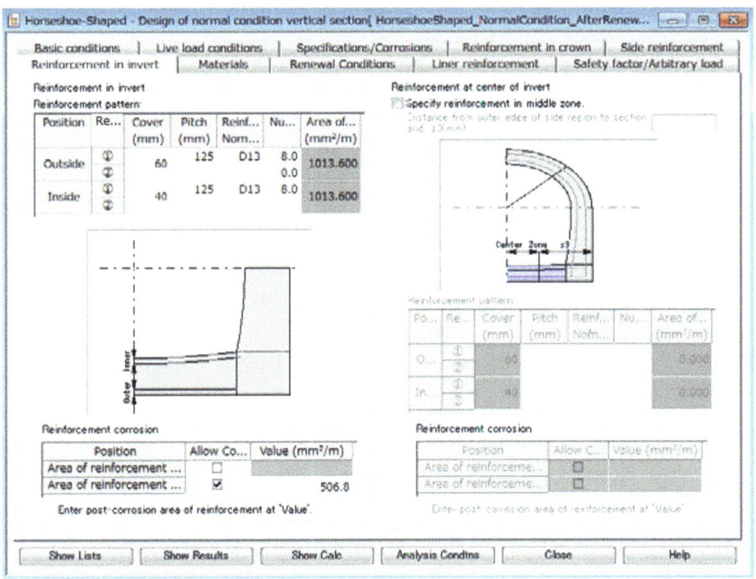

(b)

Figure 7.13: Reinforcement conditions window: (a) rectangular sewer; (b) horseshoe-shaped sewer; (c) covered sewer; (d) circular sewer

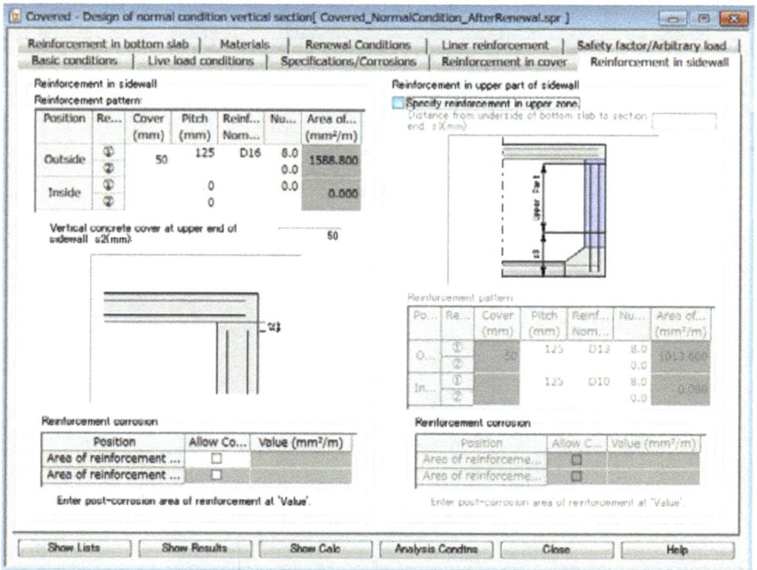

(c)

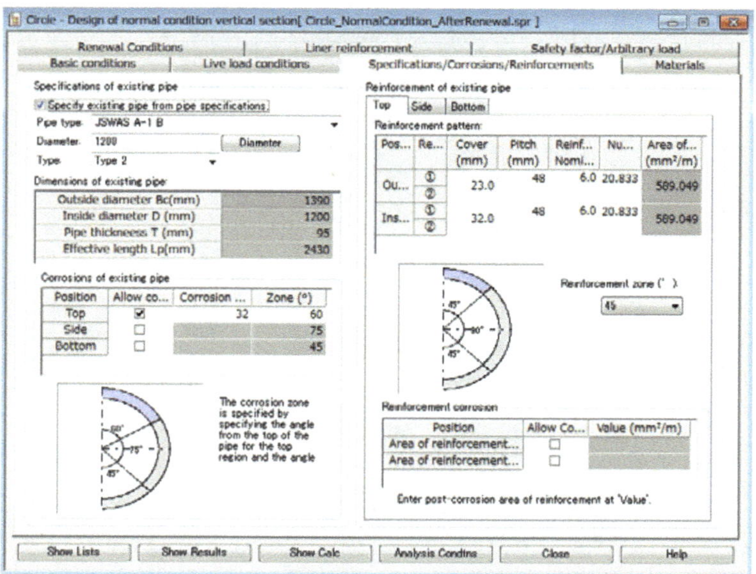

(d)

Figure 7.13: Reinforcement conditions window: (a) rectangular sewer; (b) horseshoe-shaped sewer; (c) covered sewer; (d) circular sewer (continued)

(7) Entering existing pipe material conditions
Figure 7.14 shows the existing pipe material conditions window. The physical properties of concrete and reinforcing bars can be selected from a pre-registered list of materials, but they can also be specified to reflect the state of deterioration of materials.

Material conditions that can be set for concrete are compressive strength, Young's modulus, Poisson's ratio, unit weight and tensile strength. For reinforcing bars, yield strength and Young's modulus can be set if desired. It is also possible to enter only the compressive strength of concrete and automatically input other properties according to the JSCE *Standard Specifications for Concrete Structures*.

For a covered sewer, "materials conditions for sidewall/bottom slab" and "materials conditions for cover" can be set separately.

(8) Entering renovation conditions
Figure 7.15 shows the renovation conditions windows. For renovated sewer configuration, structural types similar to those for the existing pipe can be selected, and any internal space dimensions can be specified as desired. For profile, steel reinforcement and fill mortar, it is also possible to select these materials from a pre-registered list of materials.

The positioning of renovation layers with respect to the existing sewer can be selected from "bottom slab alignment (or invert top alignment or pipe bottom alignment)", "centre-line alignment" and "other (free alignment)". It is also possible to place additional steel reinforcement in the fill mortar of the renovation layer (Fig. 7.16).

(9) Entering safety factor and arbitrary load
Figure 7.17 shows the safety factor and arbitrary load windows. Of the safety factors used in the analysis, the structure factor in the ultimate limit state can be changed as desired. Loads can be specified as desired in the vertical and horizontal directions, and load type can be selected from "dead load" and "live load (incremental load)".

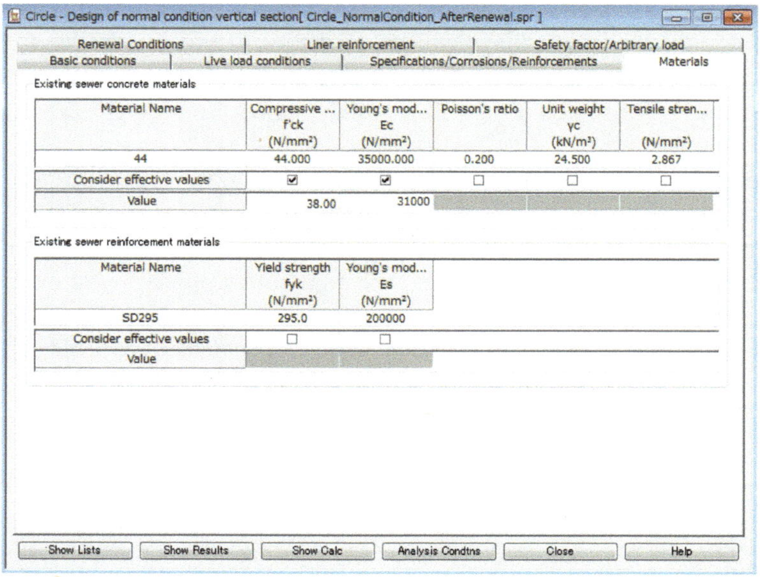

(a)

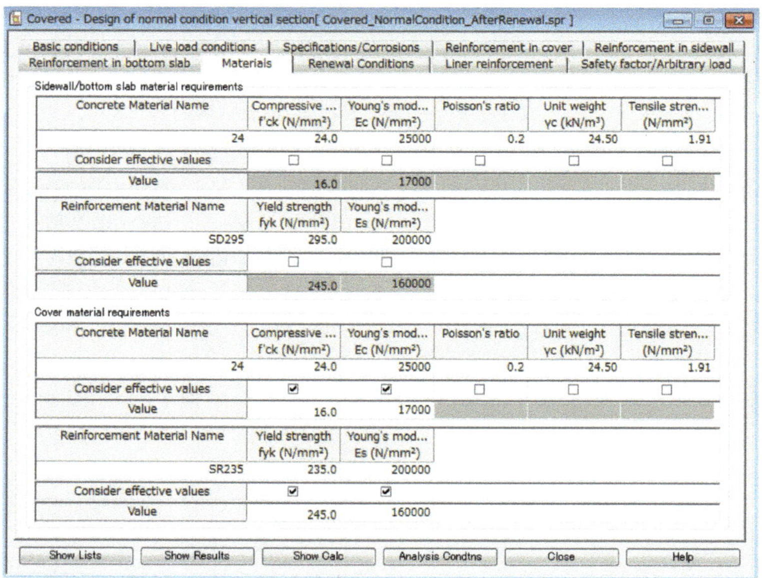

(b)

Figure 7.14: Existing pipe material conditions window: (a) existing pipe (rectangular, horseshoe-shaped, circular); (b) existing pipe (covered sewer)

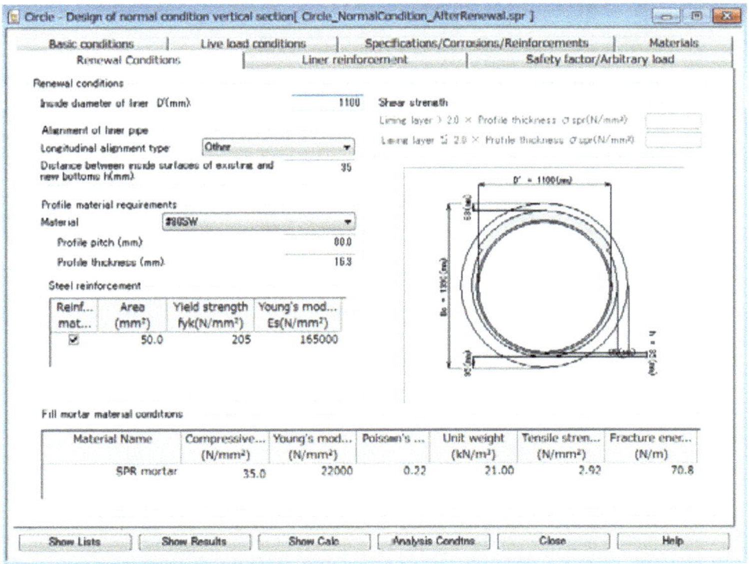

(a)

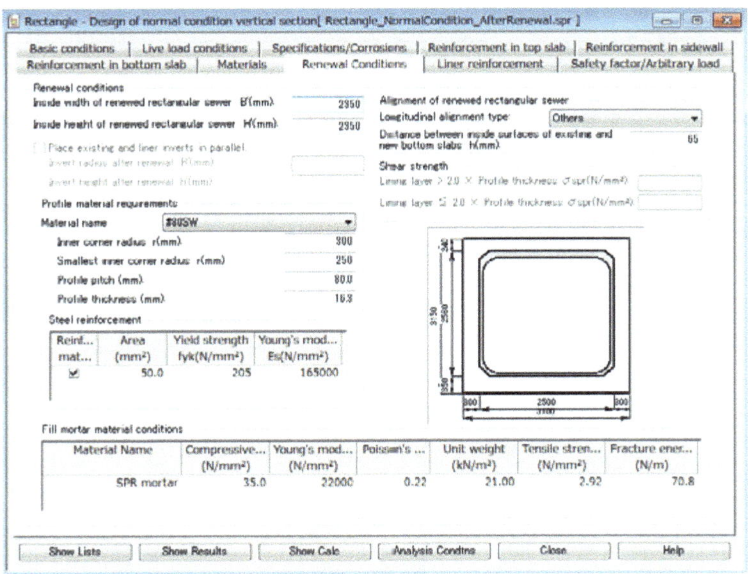

(b)

Figure 7.15: Renovation conditions window: (a) circular sewer; (b) rectangular sewer; (c) horseshoe-shaped sewer

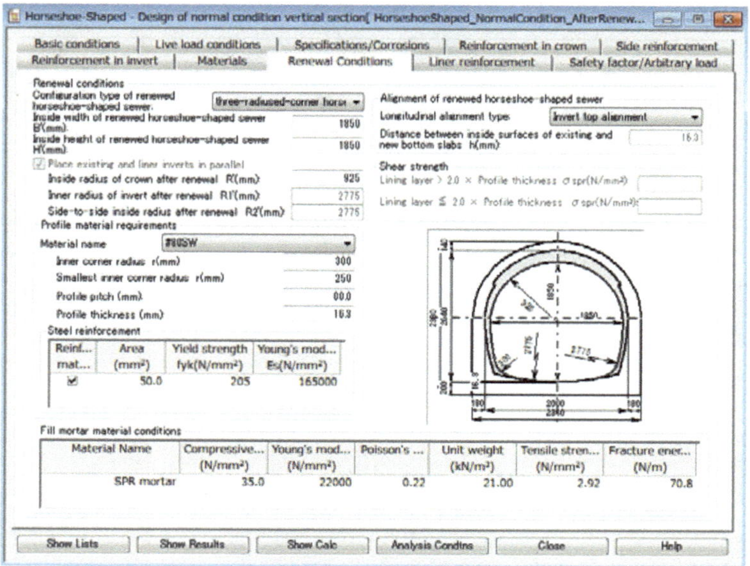

(c)

Figure 7.15: Renovation conditions window: (a) circular sewer; (b) rectangular sewer; (c) horseshoe-shaped sewer (continued)

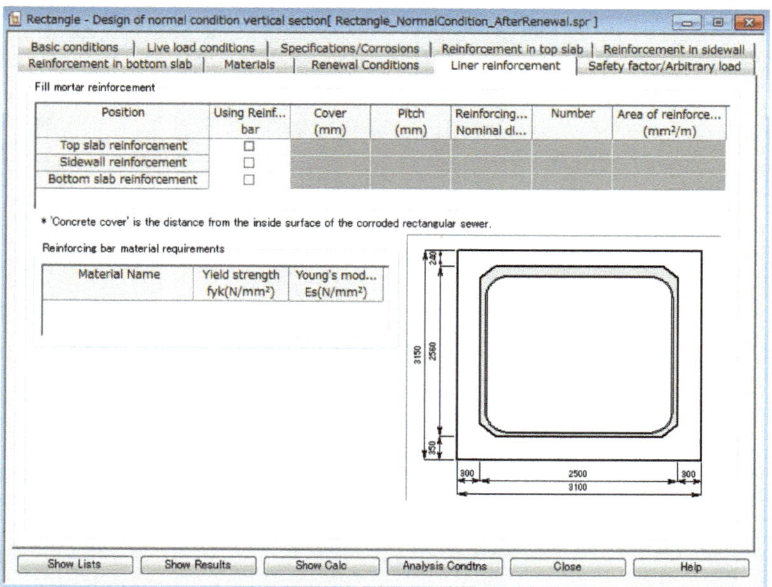

Figure 7.16: Reinforcement conditions window

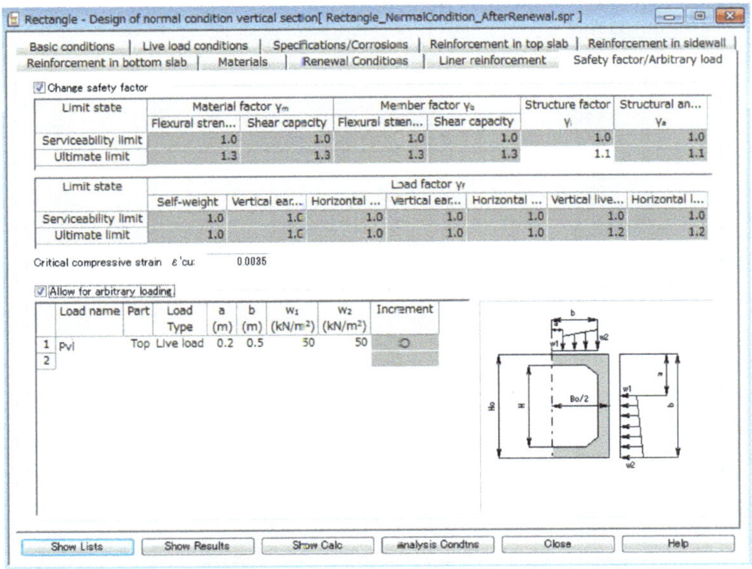

Figure 7.17: Safety factor and arbitrary load input window

7.7.2 Creating an analysis model

Figure 7.18 shows FE meshes of different configurations for different types of sewer created by the automatic meshing function of the SPRana system. Note that these FE meshes cannot be checked on the computer screen or the output sheets of calculation results because the philosophy behind this design-support software is to help the user to get the design results without having to consider the computational details.

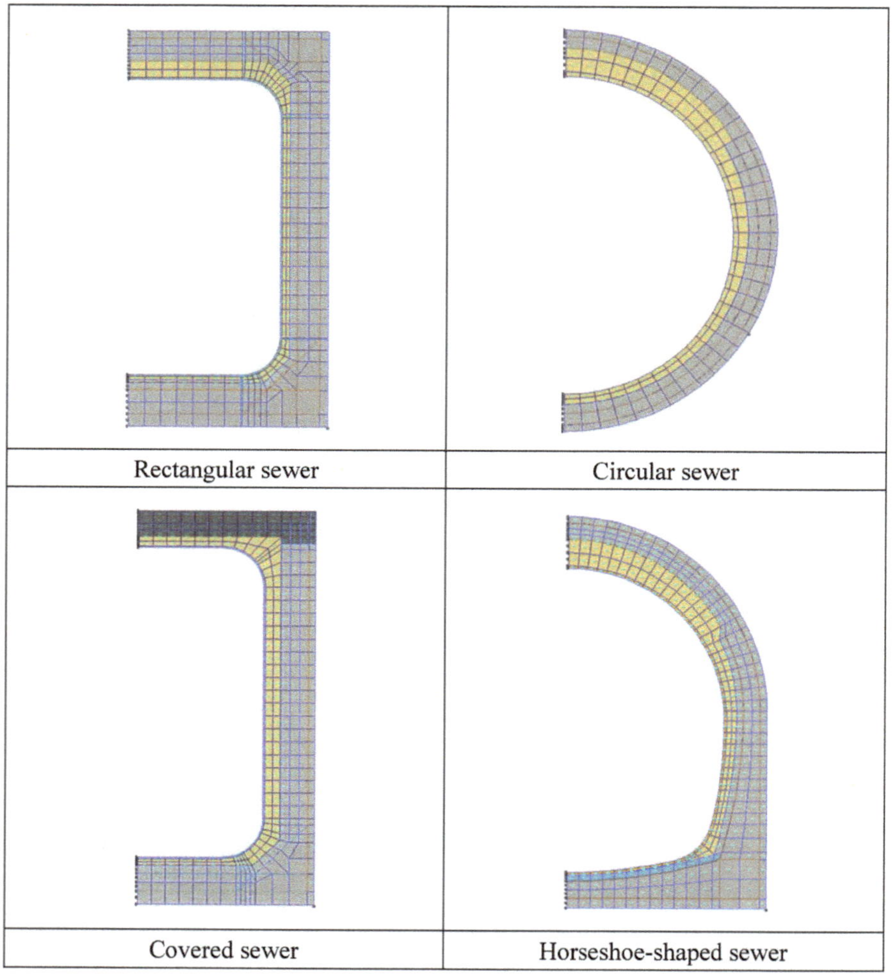

Figure 7.18: Mesh diagrams of various types of sewer

7.7.3 Running a task

After setting up analysis conditions, a task can be executed by pressing the "List", "View result" or "Calculation sheet" button (Fig. 7.19(a)). When a calculation is started, a window appears to show the progress of analysis and the elapsed time (Fig. 7.19(b)). Upon completion of the task, a completion message also showing the execution time appears (Fig. 7.19(c)).

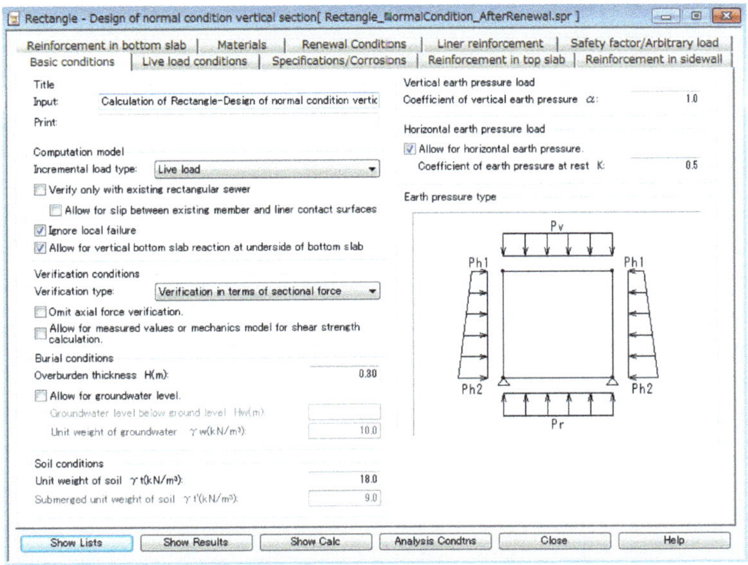

(a)

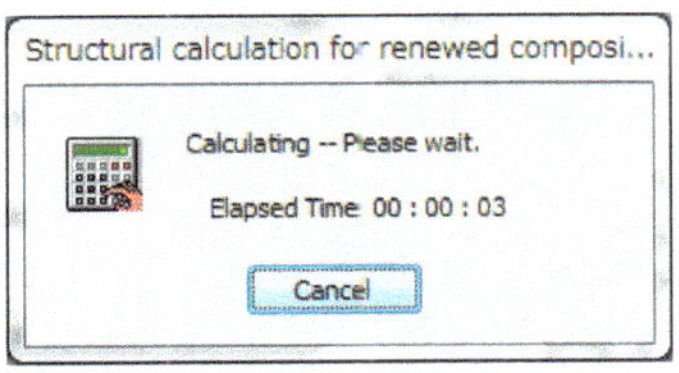

(b)

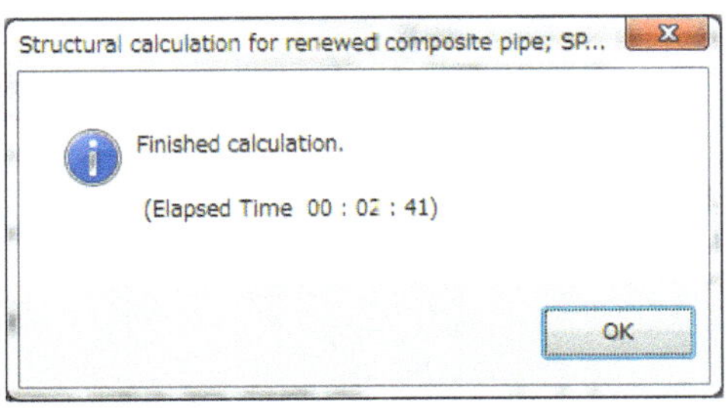

(c)

Figure 7.19: Screen displays related to execution of a task: (a) buttons for running a task; (b) during execution; (c) upon completion of calculation.

7.7.4 Outputting analysis results

Analysis results can be displayed by selecting the “List”, “View result” or “Calculation sheet” button. When the job file is saved, the analysis results are saved together with all the other input data.

By selecting the “List” button, load conditions (Fig. 7.20(a)), sectional forces (Fig. 7.20(b)), cracking (Fig. 7.20(c)) and verification results (Fig. 7.20(d)) can be checked. Sectional forces under the design load and under the maximum load can be checked. Cracking diagrams show cracking behaviours under the design load, at the initiation of cracking and under the maximum load. The user can also check the cracking processes leading up to failure as an animation. Diagrams of sectional forces and cracking can be saved as image data files, and animations can be saved as movie (e.g. avi) files.

By selecting the “Calculation sheet” button, a job report showing analysis conditions and analysis results can be output as a Microsoft Office Word file. The report can be previewed by selecting the “View result” button, and can also be printed and converted to PDF.

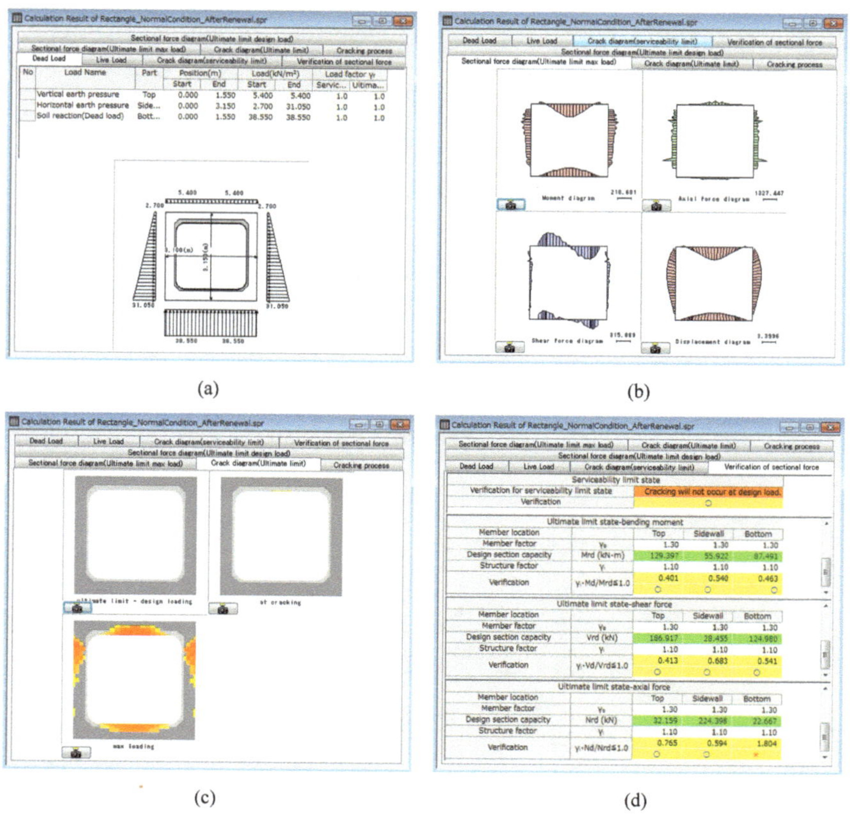

Figure 7.20: On-screen displays of analysis results: (a) dead load; (b) sectional force diagram (at ultimate limit); (c) crack diagram (at ultimate limit); (d) sectional force verification

7.7.5 On-screen warning messages

If any of the messages shown in Table 7.8 appears during or after running the programme, it is necessary to check the input data or the calculation results and determine whether the task has been performed correctly.

Table 7.8: Post-calculation on-screen warning messages

Calculation could not be continued. *Input data may be invalid.*
Pipe failed under a load smaller than the design load. *Calculation may have been terminated.* *Check the following:* • *Material conditions entered* • *Whether the final cracking behaviour can be regarded as causing structural failure*

7.8 Other Functions

In addition to the functions mentioned earlier, the SPRana system has the following functions:

- Earthquake loading analysis based on the response displacement method
- Verification on pipe joint bend angles and pullout displacement (for circular sewers only)
- Verification related to liquefaction and uplift

References

CSD, Co. Ltd. (2015). SPRana <//www.civil.jp/sinsyouhin/kouseihukugou.html>.
JRA (1999). Highway Earthworks: Guidelines for Culverts. Tokyo, Japan Road Association.
JSWA (2003). JSWAS A-1, Reinforced Concrete Sewerage Pipes. Tokyo, Japan Sewage Works Association.
JSWA (2011). Design and Construction Guidelines for Sewer Pipe Rehabilitation. Tokyo, Japan Sewage Works Association.

Toru Kouchi

8 Design Examples of Sewers Renovated by the SPR Method

This chapter introduces four design examples of sewers renovated by the SPR method using the structural analysis theories discussed in the preceding chapters. Note that the safety verification of local buckling of the bottom slab is performed based on the assumption that the invert lining is under direct groundwater pressure, which may occur when the existing cracks in the bottom slab are left untreated. The structural types of the case studies are listed in Table 8.1, which also shows the internal dimensions, overburden thicknesses and types of foundation for the selected sections of four ageing sewers.

Table 8.1: List of design examples on sewer renovation

Design example	Sewer type	Internal dimensions (mm)	Overburden thickness (m)	Foundation conditions
I	Rectangular	4,450 × 2,700	0.26–1.55	Pile foundation
II	Horseshoe-shaped	2,850 × 2,460	2.20–4.99	Pile foundation
III	Circular sewer	ϕ1,210	3.95–5.40	Concrete foundation
IV	Circular sewer	ϕ1,500	0.93–2.35	Sand-filled foundation

8.1 Example I: Rectangular Sewer

8.1.1 Internal investigation of sewer

(1) Investigation of internal cross section and damage condition

The sewer line investigated here is rectangular in cross section, and its internal dimensions and damage condition at each section are shown in Table 8.2. Based on preliminary inspections, it was decided to conduct a structural investigation on Section No. 28, where the structural deterioration was found to be most severe.

Table 8.2: Structural conditions of inspected sewer line (Example I)

Section No.	Results of inspection (measurements)			Sewer conditions
	Width (mm)	Height (mm)	Length (m)	Observed damage and judgement
19	4,410	2,670	70.30	No damage
20	4,430	2,670	60.35	No damage
21	4,400	2,650	47.05	No damage
22	4,400	2,650	63.60	No damage
23	4,430	2,660	57.70	Rebar corrosion
24	4,450	2,650	54.95	No damage
25	4,420	2,650	82.05	No damage
26	4,410	2,640	104.55	No damage
27	4,420	2,670	53.50	No damage
28	4,420	2,640	54.20	Loss of concrete cover, rebar corrosion
29	4,410	2,670	78.55	No damage
30	4,410	2,640	48.15	No damage
31	4,360	2,700	8.05	No damage

(2) Investigation items and results

The items and results of sewer investigation are shown in Table 8.3.

Table 8.3: Investigation items and results

Item			Results
			Investigation results
Cross-sectional shape			Rectangular sewer
Internal dimensions (mm)	Width		4,360–4,450
	Height		2,640–2,700
Overburden thickness (m)			0.26–1.55
Compressive strength of concrete (N/mm^2)			23.4–40.5
Member thickness (mm)	Top slab		235–500
	Sidewall	Left side	242–400
		Right side	250–435
Concrete cover (mm)	Top slab		48.0
	Sidewall	Left side	76.0
		Right side	71.0
Rebar diameter and spacing (mm)	Top slab		ϕ19@135
	Sidewall	Left side	ϕ22@333
		Right side	ϕ22@359
Rebar strength (N/mm^2)	Yield strength		249
	Tensile strength		414
	Based on JIS standard		SR235

8.1.2 Original design documents and determination of cross sections for structural analysis

As the design drawings used for the construction of the target sewer are available, details of the design cross section are determined based on both the results of investigation and data of the original design documents for obtaining critical conditions. Table 8.4 compares the results of structural investigation and data obtained from the original drawings and shows the cross-sectional conditions used for structural analysis.

Table 8.4: Structural details of design cross section

Item			Classification		
			Structural investigation results	As-built drawing	Design cross section
Internal dimensions (mm)			4,450 × 2,700	4,400 × 2,600	4,450 × 2,700
Overburden thickness (m)			0.26–1.55		0.26–1.55
Compressive strength of concrete (N/mm²)			23.4		23.4
Member thickness (mm)	Top slab		235	450	235
	Sidewall		242	400	242
	Bottom slab			500	500
Concrete cover (mm)	Top slab	Inside	48.0	61.0	61.0
		Outside		61.0	61.0
	Sidewall	Inside	76.0	61.0	76.0
		Outside		61.0	76.0
	Haunch			61.0	61.0
	Bottom slab	Inside		61.0	61.0
		Outside		61.0	61.0
Rebar diameter and spacing (mm)	Top slab	Inside	ϕ19@135	ϕ22@320	ϕ19@320
		Outside		ϕ22@320	ϕ22@320
	Sidewall	Inside	ϕ22@359	ϕ22@320	ϕ22@359
		Outside		ϕ22@320	ϕ22@359
	Haunch			ϕ22@320	ϕ22@320
	Bottom slab	Inside		ϕ12@320	ϕ12@320
		Outside		ϕ12@320	ϕ12@320
Rebar strength (N/mm²)	Yield strength		249		235
	Tensile strength		414		380
	Based on JIS standard		SR235		

8.1.3 General conditions for structural analysis

(1) Burial conditions, cross-sectional dimensions and rebar arrangements
Figure 8.1 shows the burial conditions of the existing sewer, its general structural dimensions, rebar arrangements and other structural details. The cross section of a sewer with standard renovation is also shown.

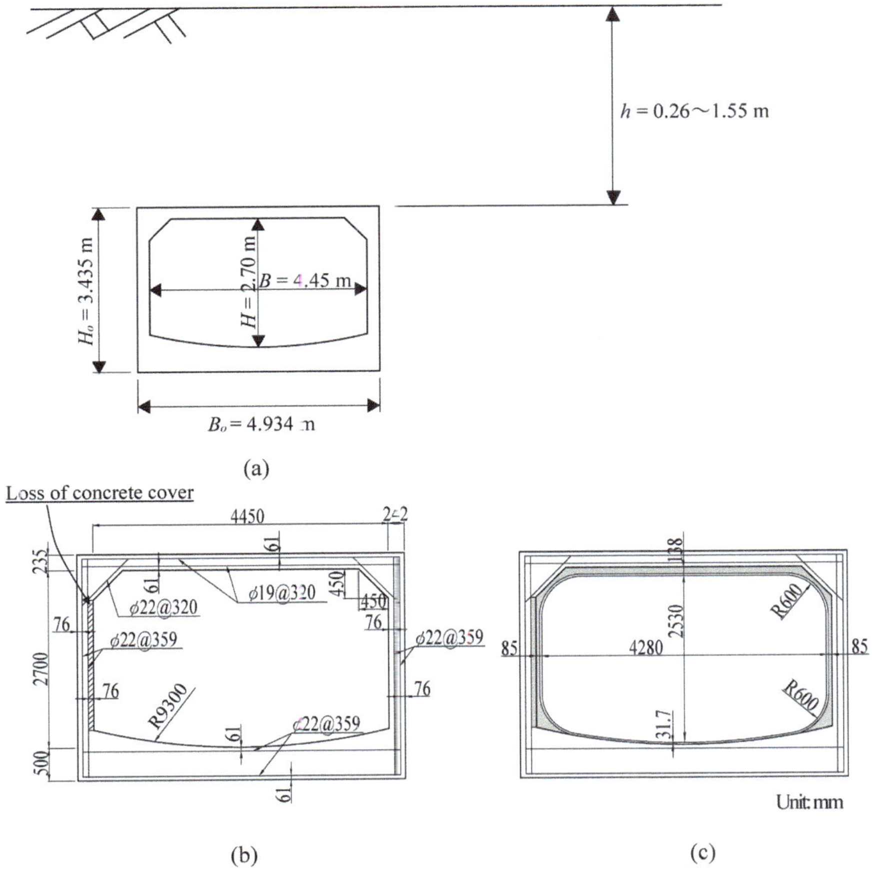

Figure 8.1: Cross section of sewer for structural analysis and renovation design: (a) general structural dimensions and burial conditions; (b) structural details of existing sewer; (c) cross section of sewer with standard renovation

(2) Support conditions
Based on the original design drawing as shown in Fig. 8.2, the foundation is assumed to be a fixed-bottom-slab structure.

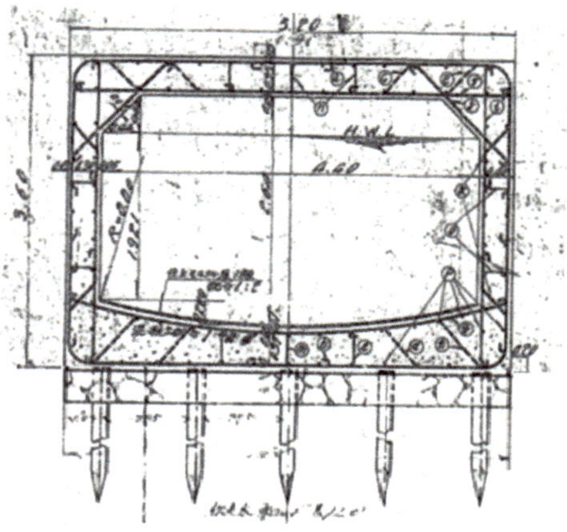

Figure 8.2: As-built drawing (1941)

(3) Material properties

The material properties of the concrete, grouting materials, rebars and profiled steel sheets of the renovation layer are summarised in Table 8.5.

Table 8.5: Material properties

Material	Item	Specification	Unit	Remarks
Concrete members	Compressive strength	23.4	N/mm^2	Minimum test value
	Tensile strength	1.88	N/mm^2	JSCE (2013)
	Modulus of elasticity	24.70	kN/mm^2	〃
	Poisson's ratio	0.2		〃
	Fracture energy	83.6	N/m	〃
Rebar	Yield strength	235	N/mm^2	Design value
	Modulus of elasticity	200	kN/mm^2	JSCE (2013)
Grout (Mortar No. 2)	Compressive strength	21.0	N/mm^2	Design value
	Tensile strength	1.83	N/mm^2	〃
	Modulus of elasticity	6.60	kN/mm^2	〃
	Poisson's ratio	0.25		〃
	Fracture energy	17.6	N/m	〃
Grout (Mortar No. 3)	Compressive strength	35.0	N/mm^2	Design value
	Tensile strength	2.92	N/mm^2	〃
	Modulus of elasticity	22.0	kN/mm^2	〃
	Poisson's ratio	0.22		〃
	Fracture energy	70.8	N/m	〃
Profiled steel sheet #792SU	Yield point	210	N/mm^2	Design value
	Modulus of elasticity	170	kN/mm^2	〃
	Cross-sectional area	1571	mm^2/m	〃

(4) Soil conditions

In view of the boring data obtained from a nearby location, soil conditions are assumed as shown in Table 8.6.

Table 8.6: Soil conditions

Boring data	Soil No.	Soil type	Depth (m)	Thickness, H_i (m)	Average N-value	Average shear wave velocity, V_{si} (m/s)	H_i/V_{si} (s)
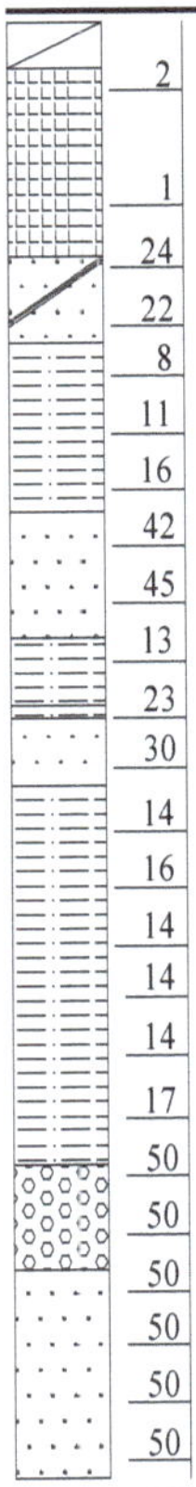	1	Fill	4.22	4.22	2	119.72	0.035
	2	Silty sand	5.60	1.39	23	288.59	0.005
	3	Silt	6.70	1.10	8	208.44	0.005
	4	Silt	8.50	1.80	14	260.74	0.007
	5	Sand	10.80	2.30	44	363.55	0.006
	6	Silt	12.40	1.60	18	288.31	0.006
	7	Sand	13.40	1.00	30	311.91	0.003
	8	Silt	20.00	6.60	15	268.03	0.025
	Basement layer	Sand and gravel			> 50	400.0	

Evaluation of ground property: $T_G = 4\sum_{i=1}^{n}\frac{H_i}{V_{si}} = 0.368$ (s), Class II

Classification of ground: $T_G < 0.2$, Class I; $0.2 \le T_G < 0.6$, Class II; $0.6 \le T_G$, Class III

(5) Load conditions

Since the overburden thickness along this sewer line is less than 4 m, the minimum overburden thickness (0.26 m) is used to define load conditions. Parallel live loads (T-25) are assumed because there is a road running along this sewer line. Load conditions are calculated based on relevant codes (JRA, 2013a-b).

1) Static earth pressure (Fig. 8.3)

$$P_{vd} = \alpha \cdot \gamma \cdot h \text{ (kN/m}^2\text{)} \tag{8.1}$$

where P_{vd}: vertical earth pressure; α: coefficient of vertical earth pressure (= 1.0); γ: unit weight of soil (= 18.0 kN/m³); h: overburden thickness (= 0.260 m).

$$P_{hd1} = K_0 \cdot \gamma \cdot h \text{ (kN/m}^2\text{)} \tag{8.2}$$

$$P_{hd2} = K_0 \cdot \gamma \cdot (h + H_o) \text{ (kN/m}^2\text{)} \tag{8.3}$$

where P_{hd1}, P_{hd2}: horizontal earth pressure; K_0: coefficient of horizontal earth pressure (= 0.5); H_o: outside diameter (height) of pipe (= 3.435 m).

2) Live loads (Fig. 8.4)

$$P_{vl} = \frac{Pl/2 \times (1+i)}{(2 \cdot h \cdot \tan\theta + b) \cdot (2 \cdot h \cdot \tan\theta + a)} \times v / b_e \text{ (kN/m}^2\text{)} \tag{8.4}$$

where P_{vl}: vertical live load; Pl: rear wheel load (= 100 kN); i: impact factor (= 0.3); a: ground contact width in the longitudinal direction (= 0.2 m); b: ground contact width in the transverse direction (= 0.5 m); θ: load distribution angle (= 45°); v: load distribution width in the longitudinal direction (= 0.720 m); b_e: effective width (= 1.010 m).

$$b_e = v + ((c - a)/2 - h \times \tan\theta) \tag{8.5}$$

where c: distance between rear wheels (= 1.300 m).

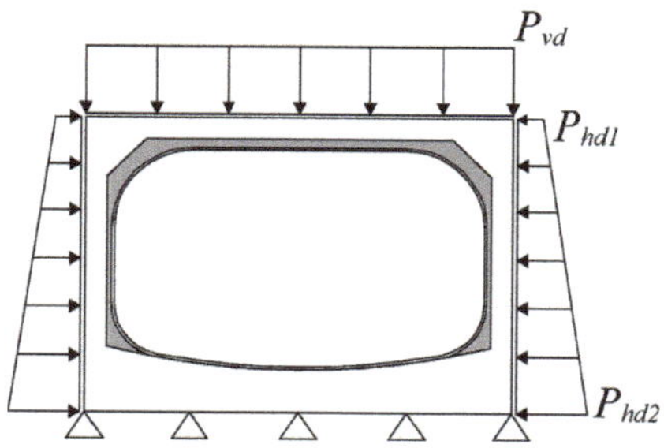

Static earth pressure (kN/m²)		
Vertical load	Horizontal load	
P_{vd}	P_{hd1}	P_{hd2}
4.68	2.34	33.25

Figure 8.3: Static earth pressure

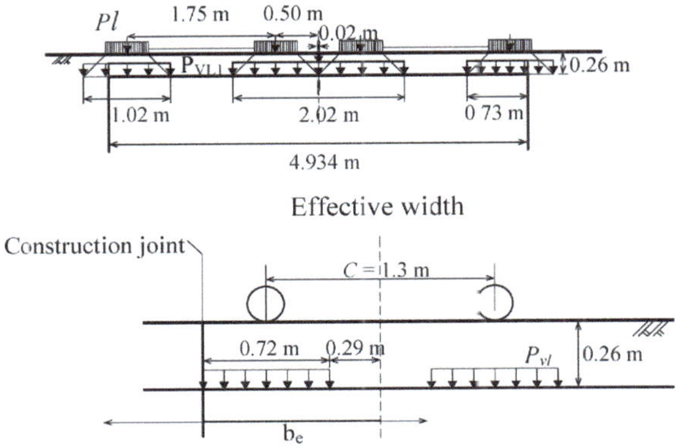

Live load (kN/m^2)
P_{vl}
63.11

Figure 8.4: Live loads

(6) Analysis model

Figures 8.5 and 8.6 show FE models for structural analysis under normal load conditions and under seismic load conditions, respectively. In view of the asymmetry of the sidewall members, a full cross-sectional model is used for structural analysis under normal loads. In the seismic performance evaluation, the ground is modelled with finite elements having sufficient coverage for pipe–soil coupled analysis.

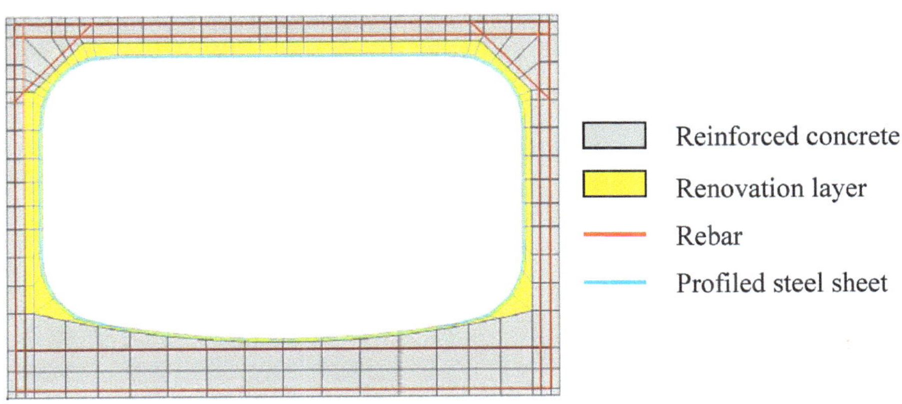

Figure 8.5: FE model for structural analysis under normal loads

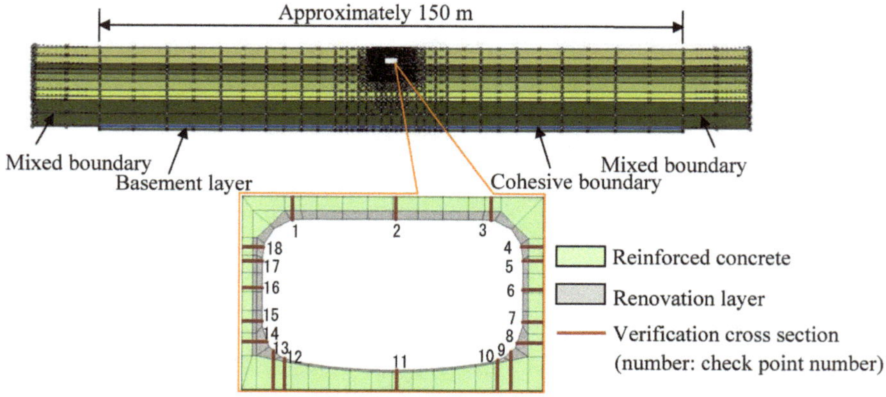

Figure 8.6: FE model for seismic performance analysis

8.1.4 Numerical results under normal load conditions

(1) Existing cross section

Figures 8.7 and 8.8 show the cracking patterns, sectional forces and structural deformations based on the FE model of the existing cross section.

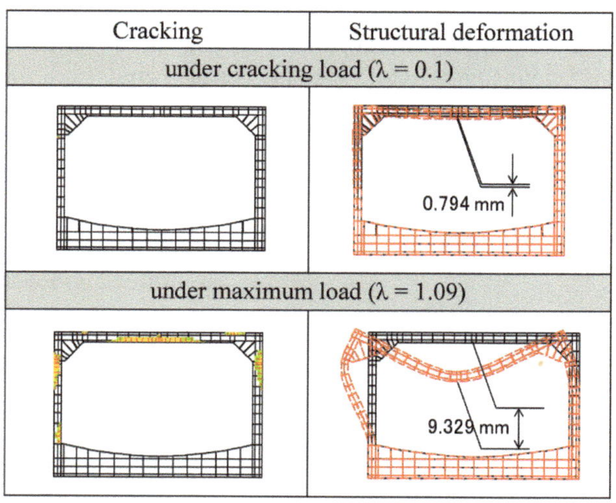

Figure 8.7: Cracking pattern and structural deformation (existing cross section)

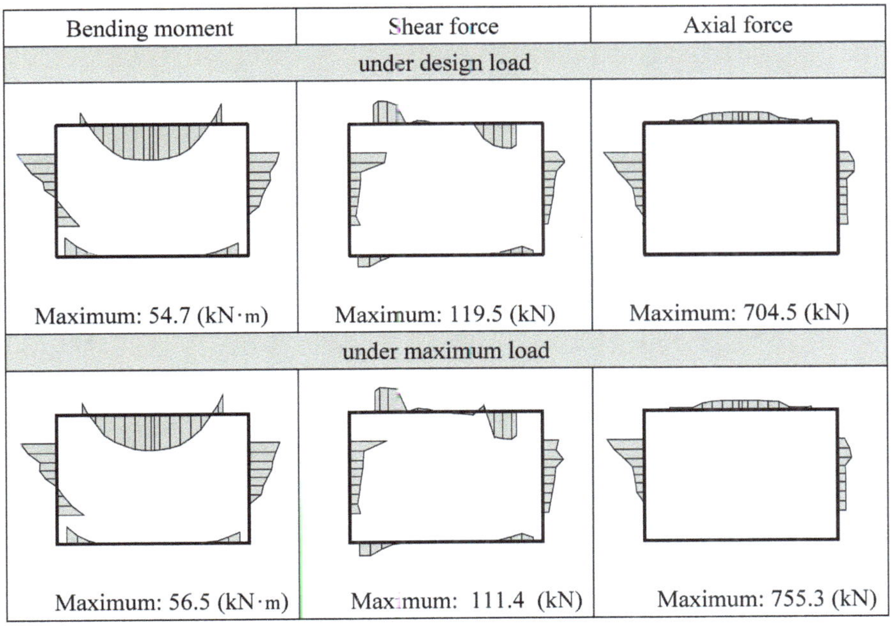

Figure: 8.8: Sectional forces (existing cross section)

(2) Standard renovation cross section with SPR Mortar No. 2

Figures 8.9 and 8.10 show the cracking patterns, sectional forces and structural deformations based on the FE model of the standard renovation cross section with SPR Mortar No. 2.

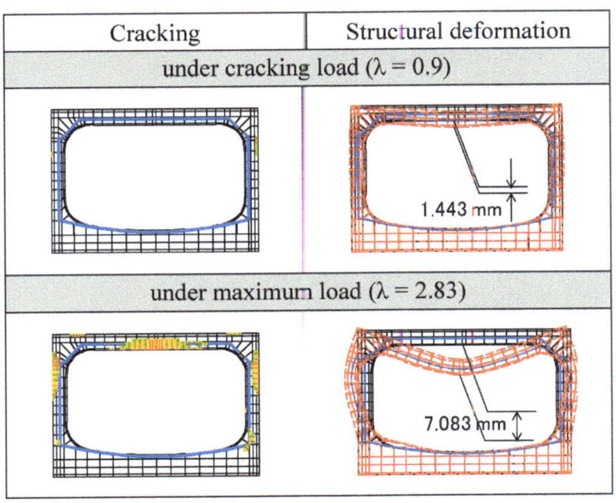

Figure 8.9: Cracking pattern and structural deformation (standard renovation cross section, Mortar No. 2)

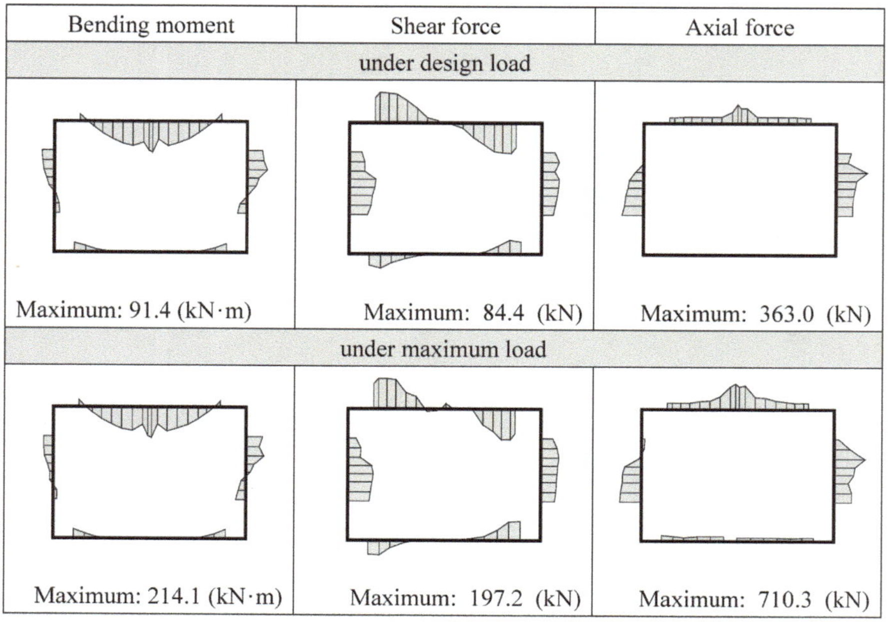

Figure 8.10: Sectional forces (standard renovation cross section, Mortar No. 2)

(3) New cross section 1 of renovated sewer with SPR Mortar No. 2
Figures 8.11 and 8.12 show the cracking patterns, sectional forces and structural deformations based on the FE model of the new cross section 1 of renovated sewer (increasing the thickness of sidewalls by 60 mm) with SPR Mortar No. 2.

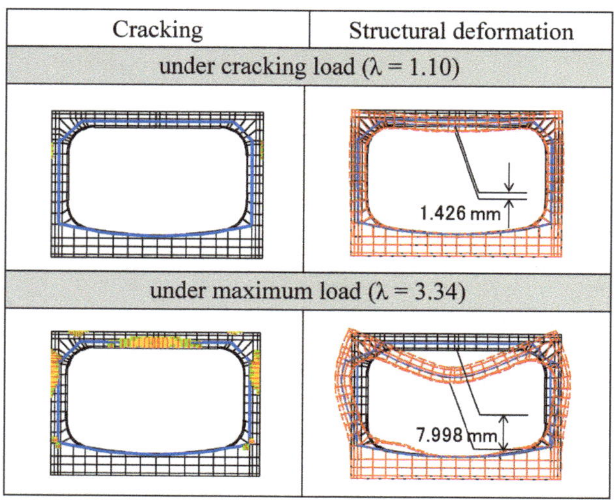

Figure 8.11: Cracking pattern and structural deformation (new cross section 1, Mortar No. 2)

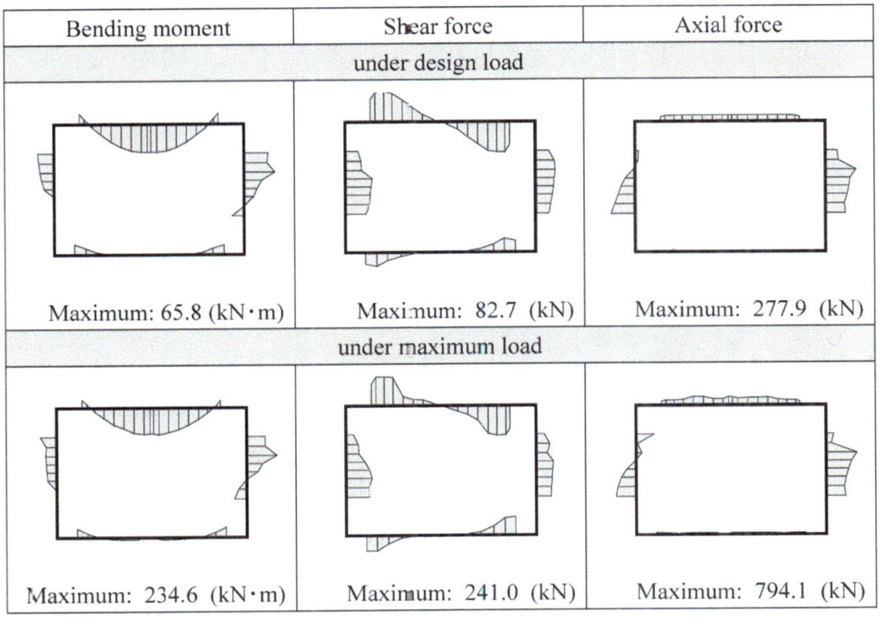

Figure 8.12: Sectional forces (new cross section 1, Mortar No. 2)

(4) Standard cross section of renovated sewer with SPR Mortar No. 3
Figures 8.13 and 8.14 show the cracking patterns, sectional forces and structural deformations based on the FE model of the standard cross section of renovated sewer with SPR Mortar No. 3.

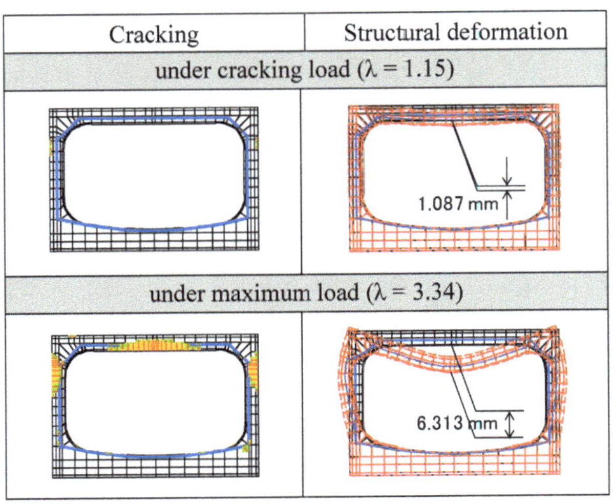

Figure 8.13: Cracking pattern and structural deformation (standard renovation, Mortar No. 3)

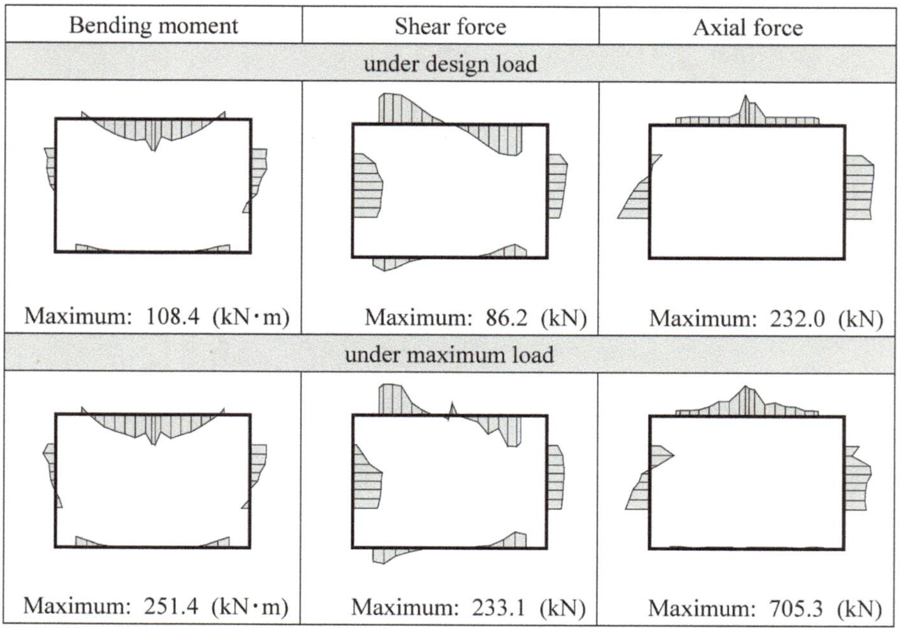

Figure 8.14: Sectional forces (standard renovation, Mortar No. 3)

8.1.5 Verification of safety under normal load conditions

The results of safety verification on the existing sewer and the other three renovation conditions are shown in Table 8.7. Based on these results, the following comments are made:

- In the serviceability limit state, neither the existing sewer nor the standard renovation cross section with Mortar No. 2 will meet the performance requirements because the load coefficient for cracking is less than 1.0. The standard renovation cross section with Mortar No. 3 will satisfy the performance requirements because the load coefficient for cracking is greater than 1.0.
- In the ultimate limit state, the existing sewer will not meet the performance requirements because the maximum load coefficient is less than 2.5. The standard renovation cross section with Mortar No. 2 or No. 3 will satisfy the performance requirements because the maximum load coefficient is greater than 2.5.
- If the liner thickness of the sidewalls is increased by 60 mm from the standard renovation cross section using Mortar No. 2, a renovated sewer with a new cross section 1 (internal dimensions 4,160 × 2,530 mm) will meet the performance requirements in both the serviceability limit state and the ultimate limit state, with the obtained load coefficient in each case exceeding the minimum required load coefficient.

Table 8.7: Results of safety verification based on load coefficients

Renovation condition		Parallel live load		Judgement	Remarks
Cross section	**Mortar**	**Cracking load coefficients**	**Maximum load coefficients**		
Existing cross section		0.10	1.09	NG	
Standard renovation	No. 2	0.90	2.83	NG	
	No. 3	1.15	3.34	OK	
New cross section 1	No. 2	1.10	3.34	OK	Increasing the thickness of sidewalls by 60 mm
Required load coefficients		> 1.0	≥ 2.5		

8.1.6 Results of seismic performance analysis

(1) Existing cross section

1) Level 1 earthquake ground motion

Figure 8.15 shows the cracking and rebar yielding of the existing cross section under Level 1 earthquake ground motion.

Rebar yielding	None
Judgement	OK

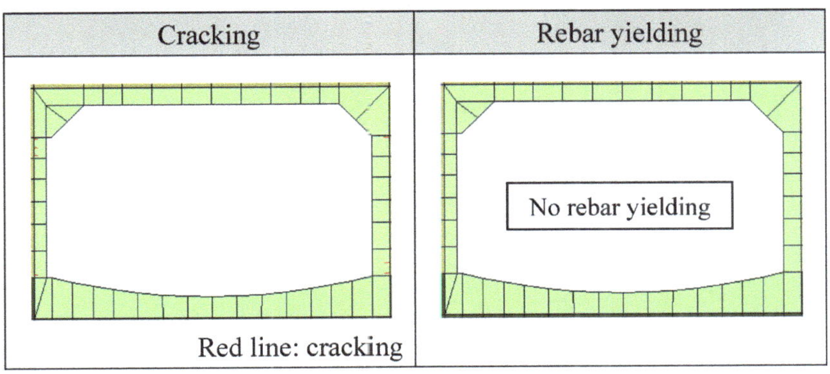

Figure 8.15: Cracking and rebar yielding in existing cross section (Level 1 earthquake ground motion)

2) Level 2 earthquake ground motion

Figure 8.16 shows the cracking and rebar yielding of the existing cross section under Level 2 earthquake ground motion. Table 8.8 shows the verification results under Level 2 ground motion.

Table 8.8: Verification results for existing cross section under Level 2 earthquake ground motion

Cross section			Existing cross section (4,450 × 2,680)							
Input ground motion			Level 2 earthquake ground motion (Inland Type 1)				Level 2 earthquake ground motion (Ocean Type 2)			
Verification item			Response value	Response value/ limit value	Judge- ment	Limit value	Response value	Response value/ limit value	Judge- ment	Limit value
Maximum compressive strain			0.0007		OK	0.0031	0.0004		OK	0.0031
Shear force at check point (kN)	Top slab	1	79.01	1.04	NG	76.03	54.13	0.59	OK	91.24
		2	24.79	0.27	OK	91.24	23.02	0.25	OK	91.24
		3	87.59	1.15	NG	76.03	79.48	1.05	NG	76.03
	Sidewall	4	72.91	0.99	OK		78.56	1.06	NG	
		5	67.38	0.91	OK		74.23	1.01	NG	
		6	73.02	0.99	OK	73.82	66.54	0.90	OK	73.82
		7	130.5	1.77	NG		63.2	0.86	OK	
		8	165.02	2.24	NG		80.69	1.09	NG	
	Bottom slab	9	163.58	1.51	NG	108.43	115.9	1.07	NG	108.43
		10	145.86	1.42	NG	102.45	97.33	0.95	OK	102.45
		11	50.77	0.77	OK	65.94	22.52	0.28	OK	79.13
		12	93.18	0.91	OK	102.45	87.31	0.85	OK	102.45
		13	110.28	1.02	NG	108.43	102.99	0.95	OK	108.43
	Sidewall	14	83.58	1.91	NG		78.56	1.79	NG	
		15	51.72	1.18	NG		56.36	1.29	NG	
		16	43.66	1.00	NG	43.85	27.63	0.63	OK	43.85
		17	62.64	1.43	NG		40.45	0.92	OK	
		18	144.14	3.29	NG		50.29	1.15	NG	
Final judgement			NG				NG			

(Shaded) Member with rebar yielding (structural member coefficient $\gamma_b = 1.56$)

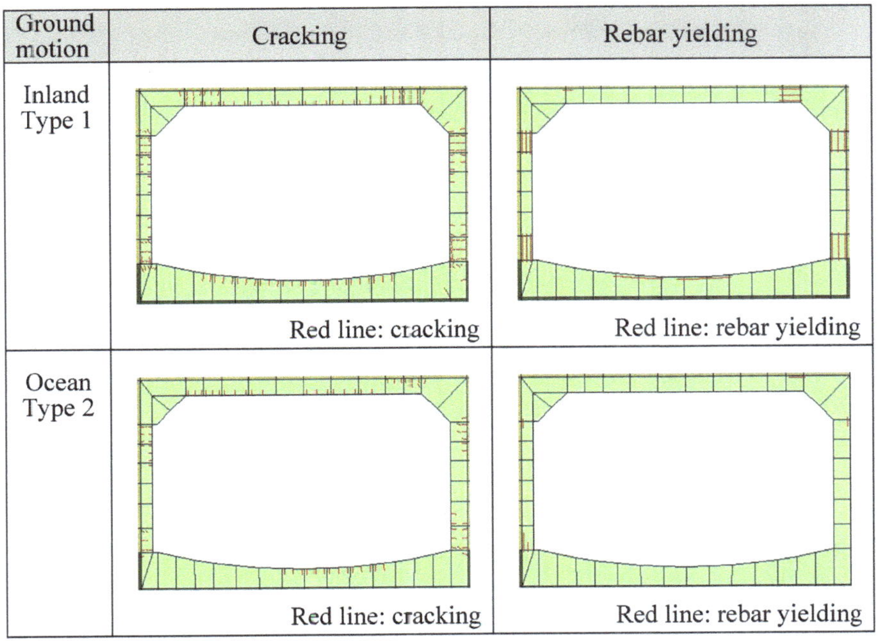

Figure 8.16: Cracking and rebar yielding in existing cross section (Level 2 earthquake ground motion)

(2) Renovation with Mortar No. 2

1) New cross section 1

Numerical analyses under normal load conditions indicated that the standard renovation would not meet the performance requirements while new cross section 1 formed by increasing the sidewall thickness by 60 mm would satisfy these requirements. In the seismic performance analysis, therefore, a safety evaluation was conducted for new cross section 1. Figure 8.17 shows the cracking and rebar yielding under Level 2 earthquake ground motion, and Table 8.9 shows the verification results.

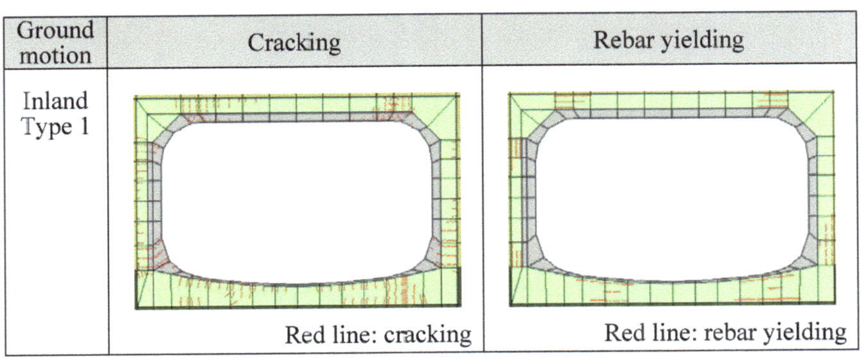

Figure 8.17: Cracking and rebar yielding in new cross section 1 with Mortar No. 2 (Level 2 earthquake ground motion)

Table 8.9: Verification results for new cross section 1 with Mortar No. 2 under Level 2 earthquake ground motion

Cross section			**New cross section 1 (4,160 × 2,530)**			
Renovation condition			**Increasing sidewall thickness by 60 mm**			
Mortar			**Mortar No. 2**			
Input ground motion			Level 2 earthquake ground motion (Inland Type 1)			
Verification item			Response value	Response value/ limit value	Judgement	Limit value
Maximum compressive strain			0.0011		OK	0.0031
Shear force at check point (kN)	Top slab	1	126.10	0.33	OK	384.96
		2	55.74	0.14	OK	
		3	115.61	0.30	OK	
	Sidewall	4	92.06	0.23	OK	397.71
		5	105.79	0.27	OK	
		6	143.49	0.36	OK	
		7	210.02	0.53	OK	
		8	224.79	0.57	OK	
	Bottom slab	9	213.91	0.32	OK	663.29
		10	203.93	1.63	NG	125.10
		11	79.78	0.60	OK	132.40
		12	152.48	1.22	NG	125.10
		13	165.77	0.25	OK	663.29
	Sidewall	14	148.58	0.40	OK	367.74
		15	148.32	0.40	OK	
		16	153.59	0.42	OK	
		17	150.09	0.41	OK	
		18	146.42	0.40	OK	
Final judgement			NG			

(shaded) Member with rebar yielding (structural member coefficient $\gamma_b = 1.56$)

2) New cross section 2

It was found that new cross section 1 would not meet the earthquake resistance requirements under Level 2 earthquake ground motion. New cross section 2, therefore, was defined by increasing the liner thickness of the bottom slab in new cross section 1 by 20 mm (internal dimensions 4,160 × 2,510 mm). The mortar used is Mortar No. 2. Figure 8.18 shows the cracking and rebar yielding under Level 2 earthquake ground motion in new cross section 2, and Table 8.10 shows the verification results.

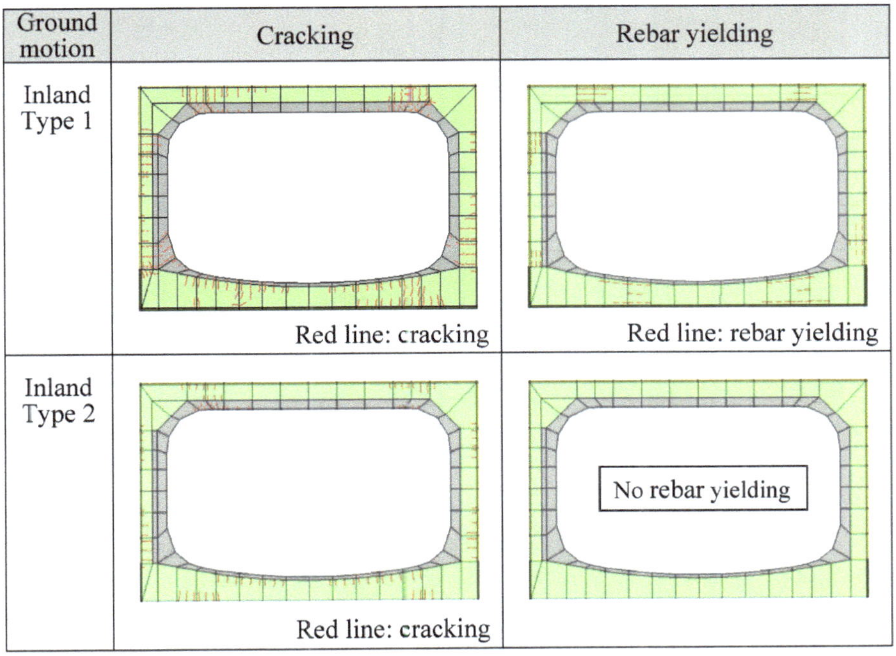

Figure 8.18: Cracking and rebar yielding in new cross section 2 with Mortar No. 2 (Level 2 earthquake ground motion)

(3) Renovation with Mortar No. 3

1) Standard renovation cross section

Figure 8.19 shows the cracking and rebar yielding under Level 2 earthquake ground motion of a standard renovation cross section, and Table 8.11 shows the verification results.

Table 8.10: Verification results for new cross section 2 with Mortar No. 2 under Level 2 earthquake ground motion

Cross section			**New cross section 2 (4,160 × 2,510)**							
Renovation condition			**Increasing sidewall thickness by 60 mm and bottom slab thickness by 20 mm**							
Mortar			**Mortar No. 2**							
Input ground motion			Level 2 earthquake ground motion (Inland Type 1)				Level 2 earthquake ground motion (Inland Type 2)			
Verification item			Response value	Response value/ limit value	Judge-ment	Limit value	Response value	Response value/ limit value	Judge-ment	Limit value
Maximum compressive strain			0.0010		OK	0.0031	less than 0.0031		OK	0.0031
Shear force at check point (kN)	Top slab	1	128.12	0.33	OK	384.96	96.98	0.21	OK	461.95
		2	56.41	0.15	OK	381.54	54.58	0.14	OK	381.54
		3	140.14	0.36	OK	384.96	108.13	0.23	OK	461.95
	Sidewall	4	89.30	0.22	OK		111.31	0.23	OK	
		5	108.22	0.27	OK		113.34	0.24	OK	
		6	145.91	0.37	OK	397.71	112.85	0.24	OK	477.25
		7	206.82	0.52	OK		133.45	0.28	OK	
		8	227.14	0.57	OK		148.58	0.31	OK	
	Bottom slab	9	211.19	0.29	OK	716.90	159.89	0.22	OK	716.90
		10	201.57	0.92	OK	218.50	149.49	0.57	OK	262.20
		11	83.60	0.38	OK	217.71	69.09	0.32	OK	217.71
		12	152.71	0.58	OK	262.20	157.94	0.60	OK	262.20
		13	165.48	0.23	OK	716.90	163.74	0.23	OK	716.90
	Sidewall	14	154.39	0.42	OK		168.48	0.38	OK	
		15	149.39	0.41	OK		153.21	0.35	OK	
		16	154.42	0.42	OK	367.74	116.31	0.26	OK	441.29
		17	150.65	0.41	OK		93.91	0.21	OK	
		18	146.82	0.40	OK		94.01	0.21	OK	
Final judgement			OK				OK			

Member with rebar yielding (structural member coefficient $\gamma_b = 1.56$)

Table 8.11: Verification results for standard renovation cross section with Mortar No. 3 under Level 2 earthquake ground motion

Cross section			**Standard renovation cross section (4,280 × 2,530)**			
Mortar			**Mortar No. 3**			
Input ground motion			Level 2 earthquake ground motion (Inland Type 1)			
Verification item			Response value	Response value/ limit value	Judgement	Limit value
Maximum compressive strain			0.0012		OK	0.0031
Shear force at check point (kN)	Top slab	1	122.63	0.28	OK	438.15
		2	63.35	0.12	OK	525.78
		3	113.14	0.26	OK	438.15
	Sidewall	4	87.73	0.30	OK	296.38
		5	101.54	0.34	OK	
		6	134.47	0.45	OK	
		7	191.57	0.65	OK	
		8	221.45	0.75	OK	
	Bottom slab	9	190.30	0.25	OK	758.83
		10	180.37	1.23	NG	146.80
		11	83.11	0.60	OK	138.14
		12	150.45	0.85	OK	176.17
		13	164.64	0.22	OK	758.83
	Sidewall	14	144.89	0.49	OK	296.38
		15	153.94	0.52	OK	
		16	158.97	0.54	OK	
		17	157.24	0.53	OK	
		18	155.37	0.52	OK	
Final judgement			NG			

Member with rebar yielding (structural member coefficient $\gamma_b = 1.56$)

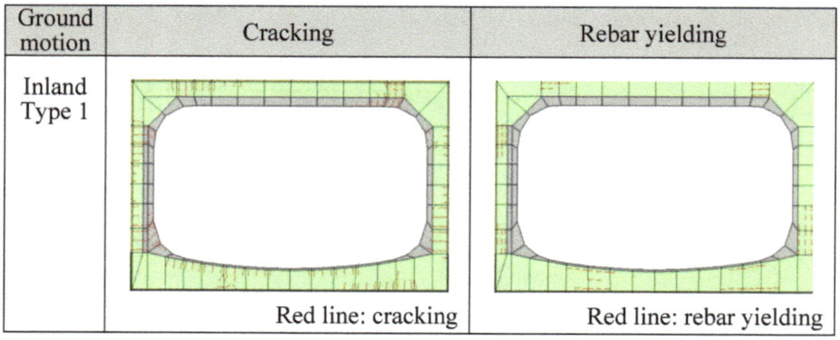

Figure 8.19: Cracking and rebar yielding in standard renovation cross section with Mortar No. 3 (Level 2 earthquake ground motion)

2) New cross section 3

It was found that a standard renovation cross section would not meet the earthquake resistance requirements under Level 2 earthquake ground motion. New cross section 3, therefore, was defined by raising the centre of the liner by 10 mm, i.e., increasing the liner thickness of the bottom slab by 10 mm and reducing the liner thickness of the top plate by 10 mm (internal dimensions 4,280 × 2,520 mm). The mortar used is Mortar No. 3. Figure 8.20 shows the cracking and rebar yielding under Level 2 earthquake ground motion of new cross section 3, and Table 8.12 shows the verification results.

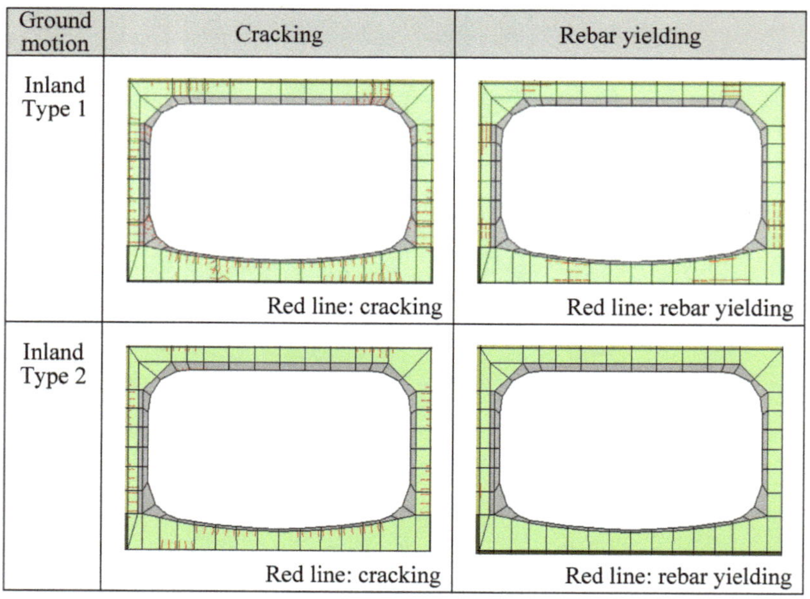

Figure 8.20: Cracking and rebar yielding in new cross section 3 with Mortar No. 3 (Level 2 earthquake ground motion)

Table 8.12: Verification results for new cross section 3 with Mortar No. 3 under Level 2 earthquake ground motion

Cross section			**New cross section 3 (4,280 × 2,520)**							
Renovation condition			**Increasing bottom slab thickness by 10 mm, and reducing top plate thickness by 10 mm**							
Mortar			**Mortar No. 3**							
Input ground motion			Level 2 earthquake ground motion (Inland Type 1)				Level 2 earthquake ground motion (Inland Type 2)			
Verification item			Response value	Response value/ limit value	Judgement	Limit value	Response value	Response value/ limit value	Judgement	Limit value
Maximum compressive strain			0.0001		OK	0.0031	0.0001		OK	0.0031
Shear force at check point (kN)	Top slab	1	122.47	0.30	OK	411.97	99.50	0.20	OK	494.36
		2	61.58	0.12	OK	494.36	56.81	0.11	OK	494.36
		3	160.66	0.39	OK	411.97	111.08	0.22	OK	494.36
	Sidewall	4	86.40	0.29	OK	296.38	113.47	0.32	OK	355.66
		5	100.02	0.34	OK		114.81	0.32	OK	
		6	133.12	0.45	OK		113.71	0.32	OK	
		7	191.54	0.65	OK		132.19	0.37	OK	
		8	216.38	0.73	OK		147.94	0.42	OK	
	Bottom slab	9	189.82	0.24	OK	790.25	157.13	0.20	OK	790.25
		10	179.30	0.83	OK	215.24	146.42	0.57	OK	258.29
		11	86.08	0.41	OK	210.15	68.76	0.33	OK	210.15
		12	151.17	0.59	OK	258.29	154.10	0.60	OK	258.29
		13	165.76	0.21	OK	790.25	160.63	0.20	OK	790.25
	Sidewall	14	146.86	0.50	OK	296.38	168.17	0.57	OK	296.38
		15	152.90	0.52	OK		152.46	0.51	OK	
		16	158.75	0.54	OK		114.74	0.39	OK	
		17	156.02	0.53	OK		95.23	0.32	OK	
		18	154.20	0.52	OK		95.87	0.32	OK	
Final judgement			OK				OK			

(shaded) Member with rebar yielding (structural member coefficient $\gamma_b = 1.56$)

8.1.7 Verification of safety under earthquake loading

The results of seismic safety verifications on the existing sewer and the other four types of renovation conditions are shown in Table 8.13. Based on these results, the following comments are made:

- The verification results indicate that under Level 1 earthquake ground motion, the existing sewer meets the earthquake resistance requirements.
- The verification results indicate that under Level 2 earthquake ground motion, the existing sewer does not meet the earthquake resistance requirements, but new cross section 2 with Mortar No. 2 and new cross section 3 with Mortar No. 3 meet the earthquake resistance requirements.
- With new cross section 3 and Mortar No. 3, the load coefficient for cracking is 1.12 and the maximum load coefficient is 3.30, satisfying the performance requirements under normal load conditions.

Table 8.13: Results of safety verification under earthquake loading

Renovation condition		Level 1 earthquake ground motion	Level 2 earthquake ground motion			Remarks
Cross section	Mortar		Ultimate displacement	Shear force	Judgement	
Existing cross section		OK	OK	NG	NG	
Standard renovation	No. 3		OK	NG	NG	
New cross section 1	No. 2		OK	NG	NG	Increasing both sidewall thicknesses by 60 mm
New cross section 2	No. 2		OK	OK	OK	Increasing both sidewall thicknesses by 60 mm, and bottom slab thickness by 20 mm
New cross section 3	No. 3		OK	OK	OK	Increasing bottom slab thickness by 10 mm, and reducing top plate thickness by 10 mm

Note: Ultimate displacement under Level 2 earthquake ground motion is verified by checking the maximum compressive strain.

8.1.8 Safety verification of local buckling of the bottom slab

The possibility of local buckling of the renovation layer at the bottom slab under groundwater pressure is checked for three renovation conditions: the standard renovation cross section with Mortar No. 2, new cross section 2 with Mortar No. 2 and new cross section 3 with Mortar No. 3. The buckling equation employed is

$$P_{cr} = \frac{1.28E}{(1-\nu^2)}\left(\frac{I}{R^3}\right)^{3/5}\left(\frac{\pi A}{L}\right)^{2/5} \quad (\text{N/mm}^2) \tag{8.6}$$

where P_{cr} = buckling strength; I = moment of inertia of area; R = invert radius of renovation layer; L = invert length of renovation layer; A = cross sectional area of renovation layer; E = modulus of elasticity; ν = Poisson's ratio. Note that Eq. (8.6) was derived in Section 5.4.

Buckling analyses on the three renovation conditions show that, while the possibility of local buckling exists for the case of standard renovation with Mortar No. 2, the other two renovation conditions have sufficient buckling strength against the maximum groundwater pressure. Details of the verification study on local buckling are presented in Table 8.14.

Table 8.14: Safety verification of local buckling of bottom slab

Item	Variable	Unit	Renovation condition		
			Standard renovation cross section	New cross section 2	New cross section 3
			Mortar No. 2	Mortar No. 2	Mortar No. 3
Maximum overburden thickness	h	m	1.550		
Height of sewer	H_o	m	3.435		
Poisson's ratio of liner	ν		0.30		
Liner thickness at bottom slab	t	mm	31.7	51.7	41.7
Invert radius	R	mm	9,300		
Invert length of liner	L	mm	3,300	3,180	3,300
Moment of inertia of area	I	mm³/mm	2.655	11.516	6.043
Cross-sectional area of liner	A	mm²/mm	31.7	51.7	41.7
Equivalent stiffness of liner	E	N/mm²	12.71×10^3	10.35×10^3	21.25×10^3
Buckling strength	P_{cr}	N/mm²	0.036	0.087	0.110
Maximum groundwater pressure	P_w	N/mm²	0.049		
Safety factor	γ		1.5		
Safety verification			$P_{cr} < \gamma P_w$	$P_{cr} > \gamma P_w$	$P_{cr} > \gamma P_w$
Judgement			NG	OK	OK

8.1.9 Determination of renovation methods

(1) Structural strength

Though the existing sewer can withstand a Level 1 earthquake force, it is not strong enough to withstand a Level 2 earthquake ground motion, and it fails to meet the performance requirements under normal load conditions. Therefore, structural renovation has to be performed on this sewer line. In cases where Mortar No. 2 is used, the required structural strength is achieved by increasing both the sidewall thickness by 60 mm and the bottom slab thickness by 20 mm from the standard renovation cross section. In cases where Mortar No. 3 is used, the required structural strength is achieved by reducing the top plate thickness by 10 mm and increasing the bottom slab thickness by 10 mm from the standard renovation cross section.

(2) Achieving the required discharge capacity

The required discharge capacity of the renovated sewer is studied by calculating the flow ratio based on the following equation:

$$\text{Flow ratio} = \text{Discharge capacity } (m^3/s)/\text{Total flow } (m^3/s) \tag{8.7}$$

where discharge capacity = $A{\cdot}v$; A = cross-sectional area (m^2); v = flow velocity (m/s). The flow velocity is defined as $v = 1/n \cdot R^{2/3} \cdot I^{1/2}$, where n = roughness coefficient; R = wetted perimeter (m); I = slope. Note that for the existing sewer of reinforced concrete n = 0.013, and for the renovated sewer with the PVC liner n = 0.010. The lower roughness coefficient is partially or totally compensating for the reduction in the cross-sectional area of a renovated sewer. The discharge capacity at each section of the renovated sewer is calculated and shown in Table 8.15.

(3) Conclusions

Based on the various analysis results discussed above, sewer renovation will be carried out for this sewer line to strengthen its earthquake resistance, and to mitigate flooding damage by keeping its discharge capacity adequate. Since this sewer line contains a section where the flow ratio of the existing sewer is lower than 1.0, it is necessary to maximise the cross section of the sewer. Therefore, the renovation methods are determined based on the structural details of new cross section 3 with Mortar No. 3. The renovation plan for each section of this sewer line is listed in Table 8.15.

Table 8.15: List of renovation plans (Example I)

Section No.		**19**	**20**	**21**	**22**	**23**	**24**	**25**
Sewer length (m)		70.30	50.35	47.05	63.60	57.70	54.95	82.05
Dimensions	Existing sewer	4,410 × 2,670	4,430 × 2,670	4,400 × 2,650	4,400 × 2,650	4,430 × 2,660	4,450 × 2,650	4,420 × 2,650
	After renovation	4,200 × 2,420	4,200 × 2,420	4,200 × 2,420	4,200 × 2,420	4,200 × 2,420	4,200 × 2,420	4,200 × 2,420
Slope (‰)	Existing sewer	1.7	1.8	0.9	2.4	2.2	1.0	1.7
	After renovation	1.7	1.4	1.4	1.9	1.9	1.9	1.9
Flow ratio	Existing sewer	1.16	1.22	0.84	1.34	1.33	0.84	1.07
	After renovation	1.23	1.12	1.11	1.30	1.30	1.21	1.11
Section No.		**26**	**27**	**28**	**29**	**30**	**31**	
Sewer length (m)		104.55	53.50	94.20	78.55	48.15	8.05	
Dimensions	Existing sewer	4,410 × 2,640	4,420 × 2,670	4,420 × 2,640	4,410 × 2,670	4,410 × 2,640	4,360 × 2,700	
	After renovation	4,200 × 2,420	4,200 × 2,420	4,230 × 2,420	4,210 × 2,440	4,190 × 2,400	4,190 × 2,400	
Slope (‰)	Existing sewer	1.6	1.7	1.6	1.9	1.2	3.5	
	After renovation	1.6	1.6	1.6	1.6	1.6	3.5	
Flow ratio	Existing sewer	1.04	1.07	0.99	1.06	0.84	1.77	
	After renovation	1.12	1.09	1.09	1.07	1.04	1.87	

8.2 Example II: Horseshoe-shaped Sewer

8.2.1 Internal investigation of sewer

(1) Investigation of internal cross section and damage condition
This sewer is horseshoe-shaped in cross section, and its internal dimensions and damage condition at each section are shown in Table 8.16. Based on preliminary inspections, it was decided to conduct a structural investigation of Section No. 16, where the structural deterioration was found to be most severe.

Table 8.16: Structural conditions of inspected sewer line (Example II)

Section No.	Results of inspection (measurements)			Sewer conditions
	Width (mm)	Height (mm)	Length (m)	Observed damage and judgement
10	2,820	2,140	98.90	Cracks, groundwater infiltration, fracture
11	2,850	2,160	61.30	Cracks, groundwater infiltration
12	2,800	2,170	104.70	Cracks, groundwater infiltration
13	2,730	2,460	21.90	No damage
14	2,720	2,450	64.90	Cracks, groundwater infiltration
15	2,720	2,450	59.90	Groundwater infiltration, deposits
16	2,730	2,440	102.35	Cracks, groundwater infiltration, surface deterioration, rebar corrosion
17	2,730	2,470	18.50	No damage

(2) Investigation items and results

The items and results of sewer investigation are shown in Table 8.17.

Table 8.17: Investigation items and results

Item			Results
			Investigation results
Cross-sectional shape			Horseshoe-shaped sewer
Internal dimensions (mm)	Width		2,720–2,850
	Height		2,140–2,460
Overburden thickness (m)			2.20–4.99
Compressive strength of concrete (N/mm²)			24.1–30.6
Member thickness (mm)	Top slab		244
	Sidewall	Left side	400
		Right side	435
Concrete cover (mm)	Top slab		21.0
	Sidewall	Left side	21.0
		Right side	22.0
Rebar diameter and spacing (mm)	Top slab		ϕ11.8@168
	Sidewall	Left side	ϕ13.2@160
		Right side	ϕ13.2@178
Rebar strength (N/mm²)	Yield strength		262
	Tensile strength		354
	Based on JIS standard		None

8.2.2 Original design documents and determination of cross sections for structural analysis

As the design drawings used for the construction of the target sewer are available, details of the design cross section are determined based on both the results of investigation and data of the original design documents for obtaining the critical conditions. Table 8.18 compares the results of structural investigation and data obtained from the original drawings and shows the cross-sectional conditions used for structural analysis.

Table 8.18: Structural details of design cross section

Item			Classification		
			Structural investigation results	**As-built drawing**	**Design cross section**
Internal dimensions (mm)			2,850 × 2,460	2,730 × 2,450	2,850 × 2,460
Overburden thickness (m)			2.20–4.99		2.20–4.99
Compressive strength of concrete (N/mm²)			24.1		24.1
Member thickness (mm)	Top slab		244	242	242
	Sidewall		400	530	400
	Bottom slab			364	364
Concrete cover (mm)	Top slab	Inside	21.0	46.0	46.0
		Outside		46.0	46.0
	Sidewall	Inside	21.0	46.0	46.0
		Outside		46.0	46.0
	Bottom slab	Inside		46.0	46.0
		Outside		55.0	55.0
Rebar diameter and spacing (mm)	Top slab	Inside	ϕ11.8@168	ϕ12.7@171	ϕ11.8@178
		Outside		ϕ12.7@171	ϕ12.7@178
	Sidewall	Inside	ϕ13.2@178	ϕ12.7@171	ϕ11.8@178
		Outside		ϕ12.7@171	ϕ12.7@178
	Bottom slab	Inside		ϕ12.7@171	ϕ12.7@171
		Outside		ϕ12.7@171	ϕ12.7@171
Rebar strength (N/mm²)	Yield strength		262		262
	Tensile strength		354		354
	Based on JIS standard		None		None

8.2.3 General conditions for structural analysis

(1) Burial conditions, cross-sectional dimensions and rebar arrangements

Figure 8.21 shows the burial conditions of the existing sewer, its general structural dimensions, rebar arrangements and other structural details. The cross section of a sewer with standard renovation is also shown.

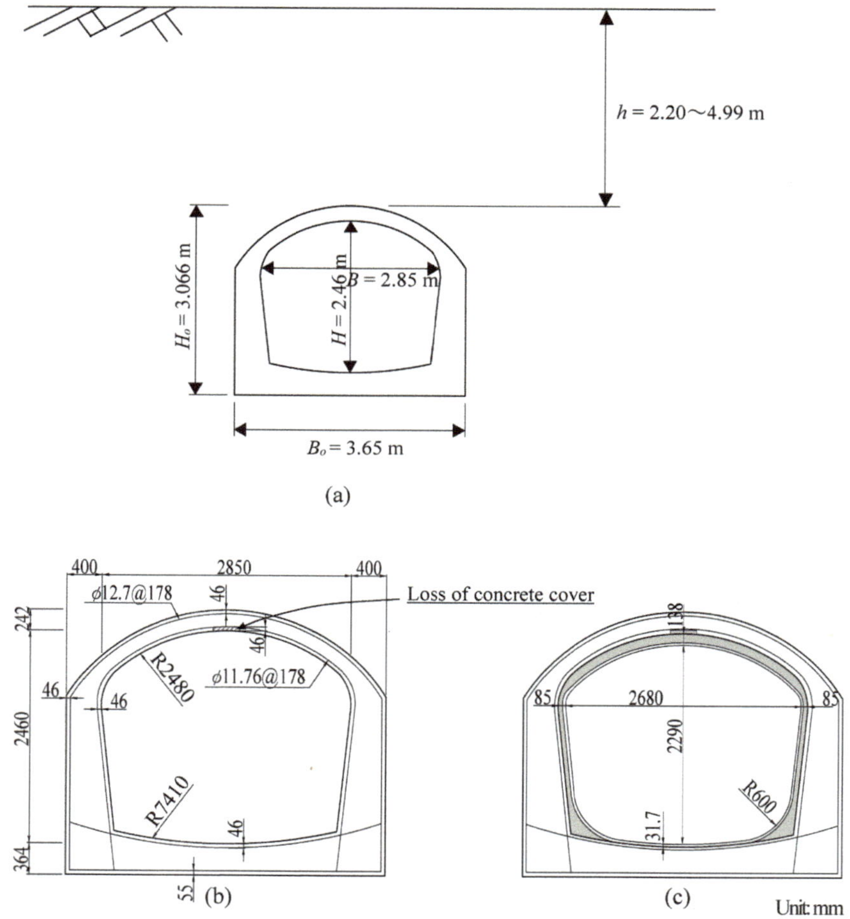

Figure 8.21: Cross section of sewer for structural analysis and renovation design: (a) general structural dimensions and burial conditions; (b) structural details of existing sewer; (c) cross section of sewer with standard renovation

(2) Support conditions

Based on the original design drawing as shown in Fig. 8.22, the foundation is assumed to be a fixed-bottom-slab structure.

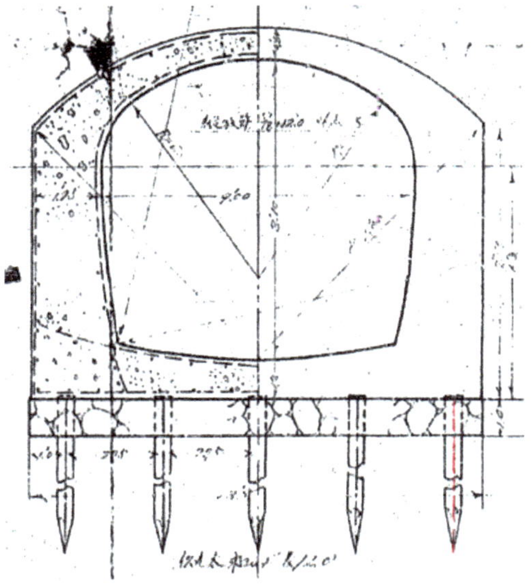

Figure 8.22: As-built drawing (1926)

(3) Material properties

Material properties of concrete, grouting materials, rebars and profiled steel sheets of the renovation layer are summarised in Table 8.19.

Table 8.19: Materials properties

Material	Item	Specification	Unit	Remarks
Concrete members	Compressive strength	24.1	N/mm²	Minimum test value
	Tensile strength	1.92	N/mm²	JSCE (2013)
	Modulus of elasticity	25.1	kN/mm²	″
	Poisson's ratio	0.2		″
	Fracture energy	84.5	N/m	″
Rebar	Yield strength	262	N/mm²	Test value
	Modulus of elasticity	200	kN/mm²	JSCE (2013)
Grout (Mortar No. 2)	Compressive strength	21.0	N/mm²	Design value
	Tensile strength	1.83	N/mm²	″
	Modulus of elasticity	6.60	kN/mm²	″
	Poisson's ratio	0.25		″
	Fracture energy	17.6	N/m	″
Grout (Mortar No. 3)	Compressive strength	35.0	N/mm²	Design value
	Tensile strength	2.92	N/mm²	″
	Modulus of elasticity	22.0	kN/mm²	″
	Poisson's ratio	0.22		″
	Fracture energy	70.8	N/m	″
Profiled steel sheet #792SU	Yield strength	210	N/mm²	Design value
	Modulus of elasticity	170	kN/mm²	″
	Cross-sectional area	1571	mm²/m	″

(4) Soil conditions

In view of the boring data obtained from a nearby location, soil conditions are assumed as shown in Table 8.20.

Table 8.20: Soil conditions

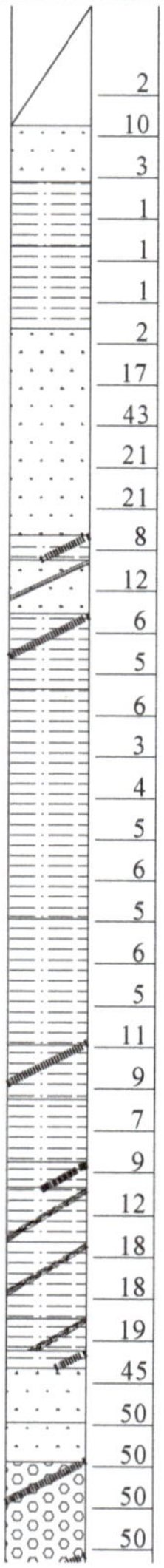

Boring data	Soil No.	Soil type	Depth (m)	Thickness, H_i (m)	Average N-value	Average shear wave velocity, V_{si} (m/s)	H_i/V_{si} (s)
	1	Surface soil	2.90	2.90	2	108.64	0.027
	2	Sand	4.10	1.20	10	200.99	0.006
	3	Silt	7.90	3.80	2	119.72	0.032
	4	Sand	8.60	0.70	2	105.58	0.007
	5	Sand	9.60	1.00	17	248.52	0.004
	6	Sand	10.70	1.10	43	360.22	0.003
	7	Sand	12.80	2.10	21	270.44	0.008
	8	Sandy silt	13.30	0.50	8	208.44	0.002
	9	Sandy silt	14.80	1.50	12	245.15	0.006
	10	Silt	25.10	10.30	5	172.72	0.060
	11	Sandy silt	28.50	3.40	9	218.50	0.016
	12	Loamy clay	32.90	4.40	17	281.79	0.016
	13	Sand	34.20	1.30	45	348.01	0.004
	Basement layer	Sand and gravel			> 50	400.00	

Evaluation of ground property: $T_G = 4\sum_{i=1}^{n} \frac{H_i}{V_{si}} = 0.300$ (s), Class II

Classification of ground: $T_G < 0.2$, Class I; $0.2 \le T_G < 0.6$, Class II; $0.6 \le T_G$, Class III

(5) Load conditions

Since the overburden thickness along this sewer line exceeds 4 m in some sections, the maximum overburden thickness (4.99 m) is used to define load conditions. As the overburden thickness is greater than 4 m, a live load of 10 kN/m² is applied uniformly.

1) Static earth pressure

Applied static earth pressures are shown in Fig. 8.23. For details of earth pressure computation, refer to Example I. Note that the coefficient of horizontal earth pressure, K_0, is 0.3 for the present case.

2) Live loads

Applied live loads are shown in Fig. 8.24. Note that the coefficient of horizontal earth pressure, K_0, is 0.3 for the present case.

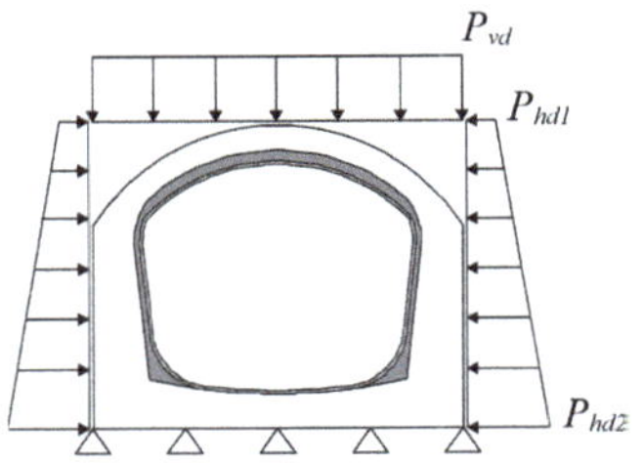

Static earth pressure (kN/m²)		
Vertical load	Horizontal load	
P_{vd}	P_{hd1}	P_{hd2}
89.82	26.95	43.50

Figure 8.23: Static earth pressure

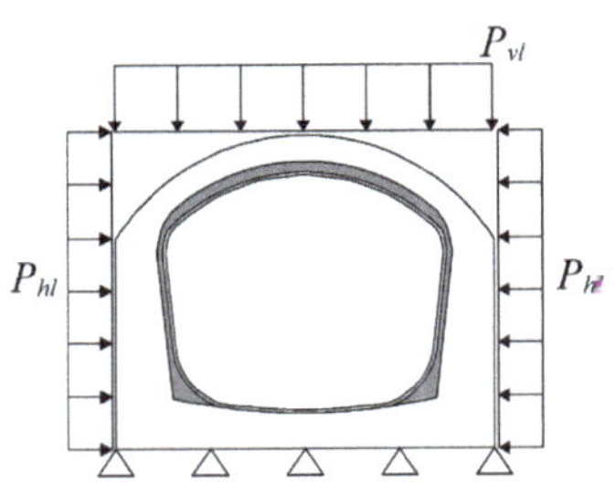

Live load (kN/m²)	
Vertical load	Horizontal load
P_{vl}	P_{hl}
10.00	3.00

Figure 8.24: Live loads

(6) Analysis model

Figures 8.25 and 8.26 show FE models for structural analysis under normal load conditions and under seismic load conditions, respectively. In view of the structural symmetry of the cross section, a half cross-sectional model is used for structural analysis under normal loads. In the seismic performance evaluation, the ground is modelled with finite elements having sufficient coverage for pipe–soil coupled analysis.

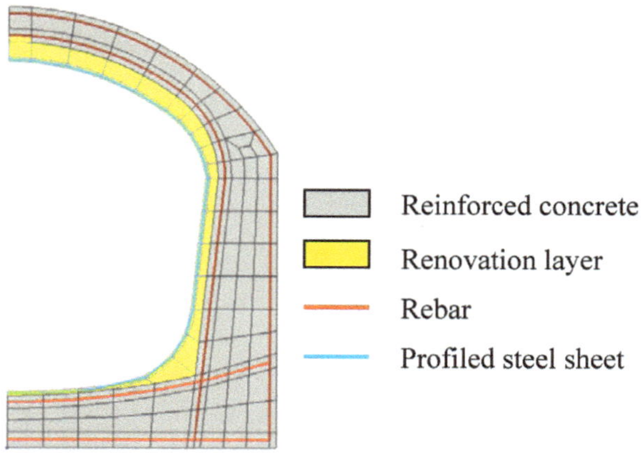

Figure 8.25: FE model for structural analysis under normal loads

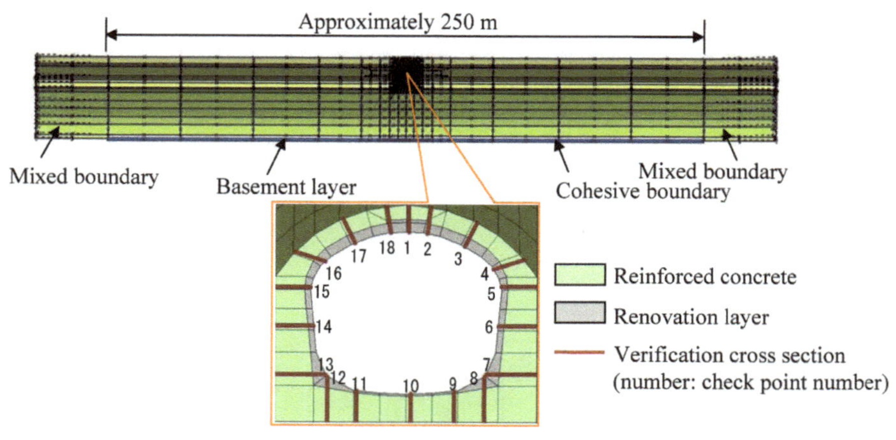

Figure 8.26: FE model for seismic performance analysis

8.2.4 Numerical results under normal load conditions

(1) Existing cross section

Figures 8.27 and 8.28 show the cracking patterns, sectional forces and structural deformations based on the FE model of the existing cross section.

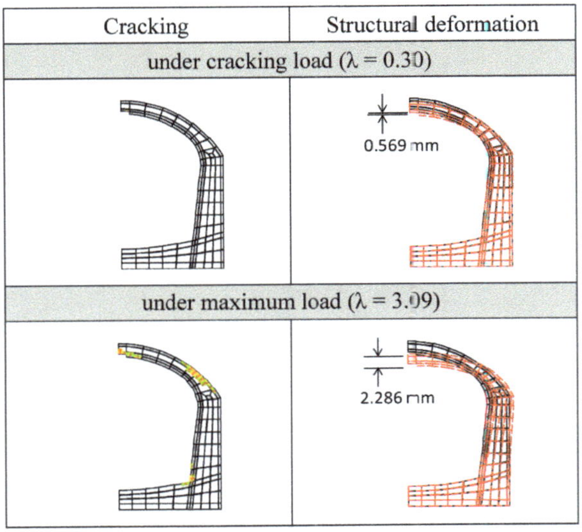

Figure 8.27: Cracking pattern and structural deformation (existing cross section)

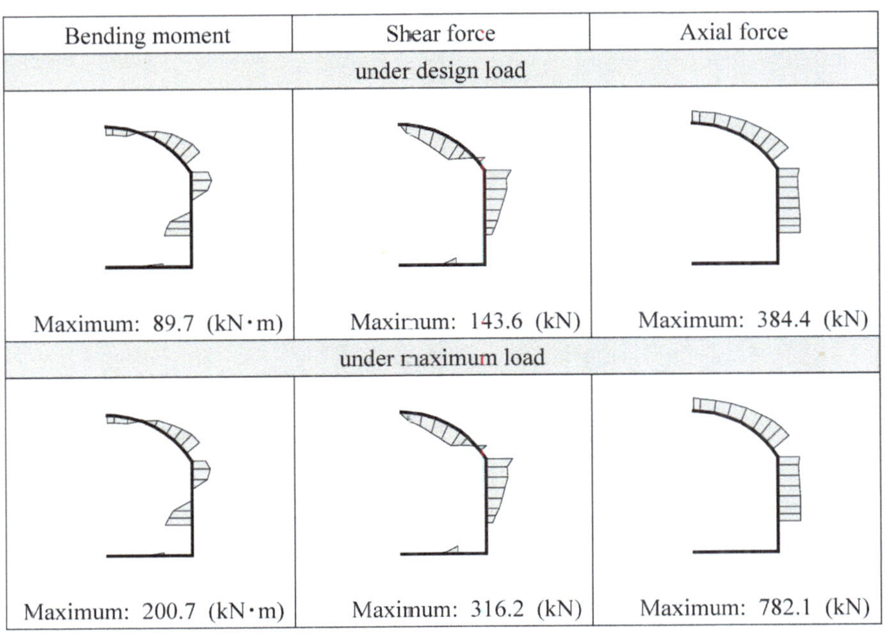

Figure 8.28: Sectional forces (existing cross section)

(2) Standard renovation cross section with Mortar No. 2

Figures 8.29 and 8.30 show the cracking patterns, sectional forces and structural deformations based on the FE model of the standard renovation cross section with Mortar No. 2.

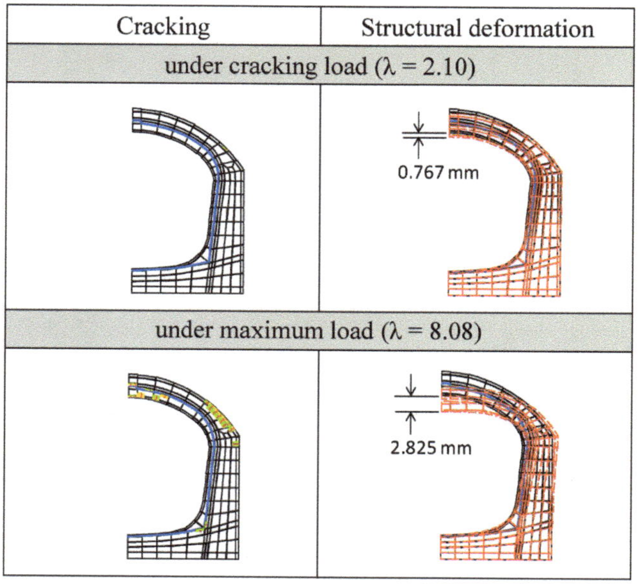

Figure 8.29: Cracking pattern and structural deformation (standard renovation cross section, Mortar No. 2)

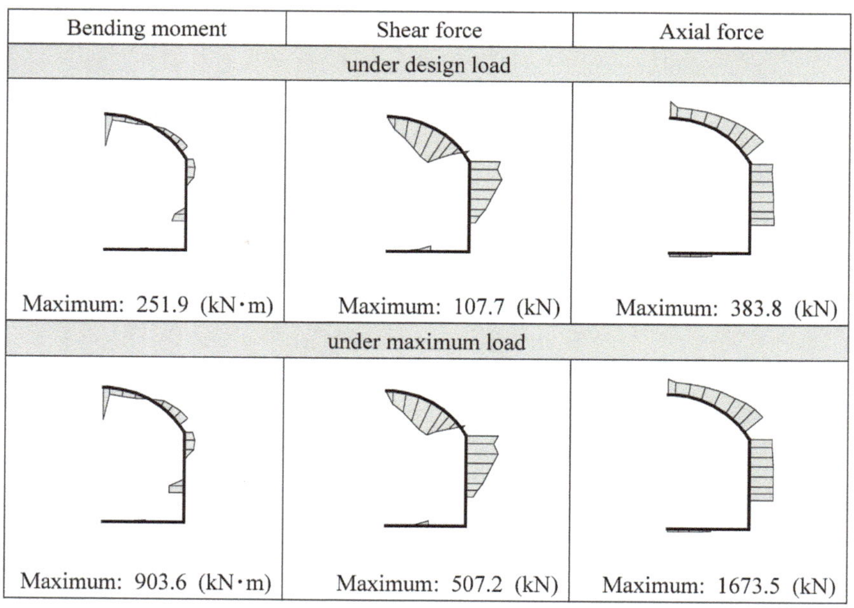

Figure 8.30: Sectional forces (standard renovation cross section, Mortar No. 2)

(3) Standard renovation cross section with Mortar No. 3

Figures 8.31 and 8.32 show the cracking patterns, sectional forces and structural deformations based on the FE model of the standard renovation cross section with Mortar No. 3.

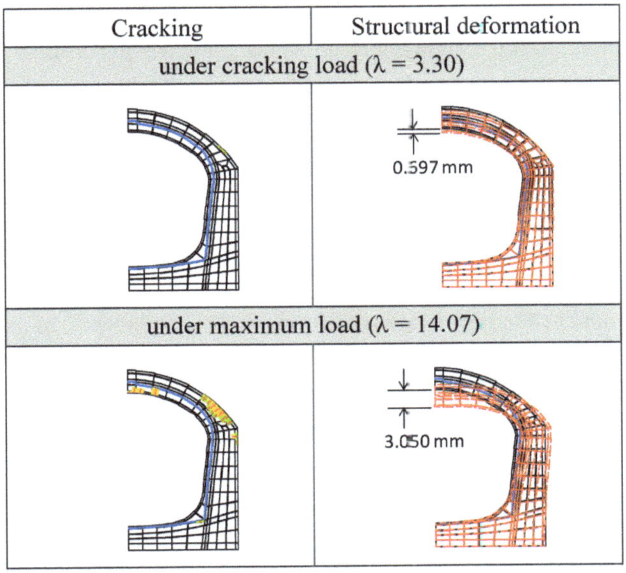

Figure 8.31: Cracking pattern and structural deformation (standard renovation cross section, Mortar No. 3)

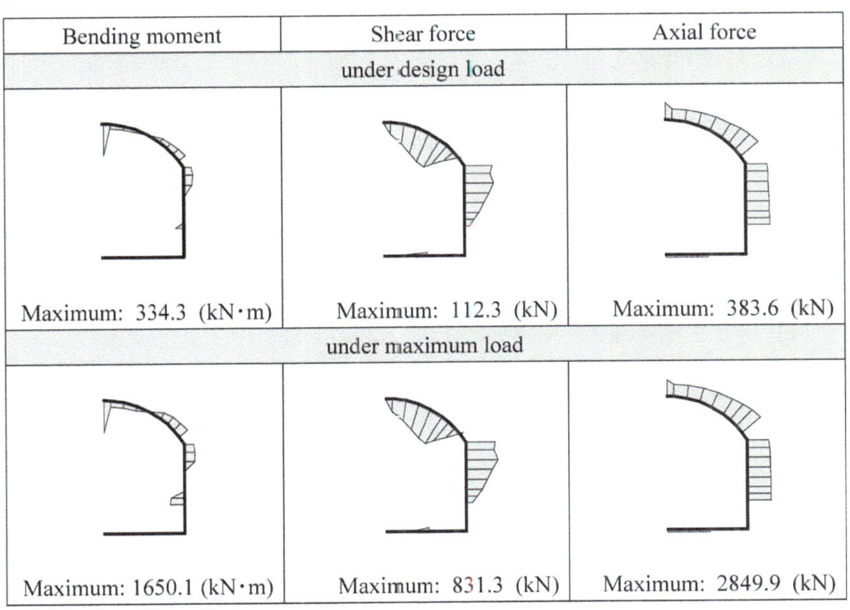

Figure 8.32: Sectional forces (standard renovation cross section, Mortar No. 3)

8.2.5 Verification of safety under normal load conditions

The results of safety verification on the existing sewer and the other two renovation conditions are shown in Table 8.21. Based on these results, the following comments are made:

- In the serviceability limit state, the existing sewer will not meet the performance requirements because the load coefficient for cracking is less than 1.0. The standard renovation cross section with Mortar No. 2 or No. 3 will satisfy the performance requirements because the load coefficient for cracking is greater than 1.0.
- In the ultimate limit state, the existing sewer will meet the performance requirements because its maximum load coefficient is greater than 2.5. The standard renovation cross section with Mortar No. 2 or No. 3 will also satisfy the performance requirements because the maximum load coefficient exceeds 2.5.

Table 8.21: Results of safety verification based on load coefficients

Renovation condition		**Parallel live load ($10\ kN/m^2$)**		**Judgement**	**Remarks**
Cross section	**Mortar**	**Cracking load coefficients**	**Maximum load coefficients**		
Existing cross section		0.30	3.09	NG	
Standard renovation	No. 2	2.10	8.08	OK	
	No. 3	3.30	14.07	OK	
Required load coefficients		> 1.0	≥ 2.5		

8.2.6 Results of seismic performance analysis

(1) Existing cross section

1) Level 1 earthquake ground motion

Figure 8.33 shows the cracking and rebar yielding of the existing cross section under Level 1 earthquake ground motion.

Rebar yielding	None
Judgement	OK

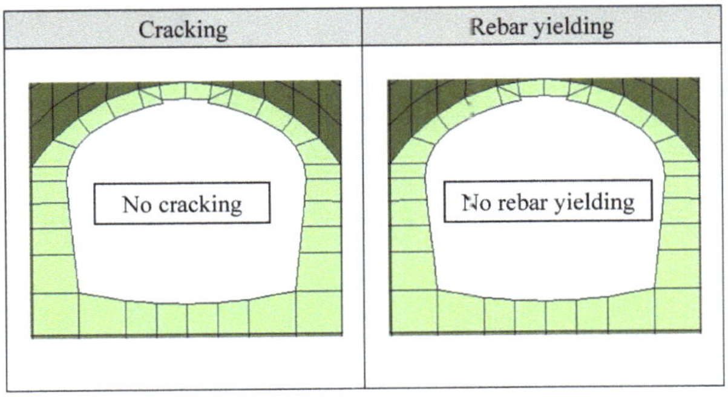

Figure 8.33: Cracking and rebar yielding in existing cross section (Level 1 earthquake ground motion)

2) Level 2 earthquake ground motion
Figure 8.34 shows the cracking and rebar yielding of the existing cross section under Level 2 earthquake ground motion. Table 8.22 shows the verification results under Level 2 ground motion.

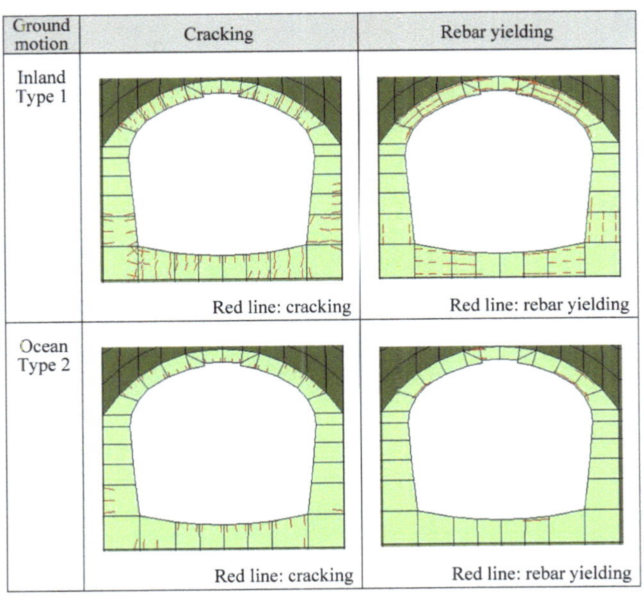

Figure 8.34: Cracking and rebar yielding in existing cross section (Level 2 earthquake ground motion)

Table 8.22: Verification results for existing cross section under Level 2 earthquake ground motion

Cross section			Existing cross section (2,850 × 2,460)							
Input ground motion			Level 2 earthquake ground motion (Inland Type 1)				Level 2 earthquake ground motion (Ocean Type 2)			
Verification item			Response value	Response value/ limit value	Judge- ment	Limit value	Response value	Response value/ limit value	Judge- ment	Limit value
Maximum compressive strain			0.0005		OK	0.0031	Less than 0.0031		OK	0.0031
Shear force at check point (kN)	Top slab	1	78.49	1.32	NG	59.46	54.46	0.76	OK	71.36
		2	78.14	1.31	NG		62.35	0.87	OK	
	Right arch	3	67.48	0.97	OK	69.47	62.08	0.89	OK	69.47
		4	93.21	1.34	NG		81.03	1.17	NG	
	Right sidewall	5	140.89	1.59	NG	88.65	122.94	1.16	NG	106.38
		6	170.03	1.92	NG		81.50	0.77	OK	
		7	241.97	2.73	NG		152.28	1.43	NG	
	Bottom slab	8	290.67	2.80	NG	103.74	214.04	1.72	NG	124.49
		9	244.61	2.62	NG	93.53	162.93	1.74	NG	93.53
		10	183.19	2.08	NG	88.17	106.93	1.01	NG	105.80
		11	204.17	2.18	NG	93.53	186.02	1.66	NG	112.24
		12	249.49	2.40	NG	103.74	232.34	1.87	NG	124.49
	Left sidewall	13	197.6	2.23	NG	88.65	172.18	1.62	NG	106.38
		14	151.00	1.70	NG		101.89	0.96	OK	
		15	194.36	2.19	NG		103.10	0.97	OK	
	Right arch	16	149.01	2.14	NG	69.47	67.43	0.97	OK	69.47
		17	64.68	0.93	OK		49.26	0.71	OK	
	Top slab	18	84.60	1.42	NG	59.46	45.17	0.76	OK	59.46
Final judgement			NG				NG			

▭ Member with rebar yielding (structural member coefficient $\gamma_b = 1.56$)

(2) Renovation with Mortar No. 2

1) Standard renovation cross section

Figure 8.35 shows the cracking and rebar yielding of the standard renovation cross section with Mortar No. 2 under Level 2 earthquake ground motion. Table 8.23 shows the verification results.

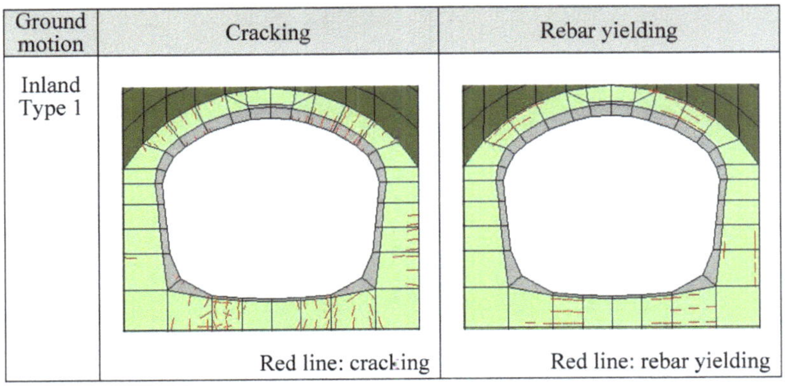

Figure 8.35: Cracking and rebar yielding in standard renovation cross section with Mortar No. 2 (Level 2 earthquake ground motion)

2) New cross section 1

It was found that the standard renovation cross section with Mortar No. 2 would not meet the earthquake resistance requirements under Level 2 earthquake ground motion. New cross section 1, therefore, was defined by keeping the inner diameter of the standard renovation cross section and raising the centre of the renovated sewer by 70 mm, thus increasing the liner thickness of the bottom slab by 70 mm and reducing the liner thickness of the top plate by 70 mm (internal dimensions 2,680 × 2,290 mm). Figure 8.36 shows the cracking and rebar yielding of new cross section 1 under Level 2 earthquake ground motion. Table 8.24 shows the verification results.

Table 8.23: Verification results for standard renovation cross section with Mortar No. 2 under Level 2 earthquake ground motion

Cross section			**Standard renovation cross section (2,680 × 2,290)**			
Mortar			**Mortar No. 2**			
Input ground motion			Level 2 earthquake ground motion (Inland Type 1)			
Verification item			Response value	Response value/ limit value	Judgement	Limit value
Maximum compressive strain			0.0003		OK	0.0031
Shear force at check point (kN)	Top slab	1	115.88	0.26	OK	442.07
		2	98.58	0.22	OK	
	Right arch	3	79.14	0.21	OK	377.73
		4	122.37	0.32	OK	
	Right sidewall	5	153.12	0.55	OK	278.52
		6	205.54	0.72	OK	284.07
		7	275.00	0.57	OK	482.21
	Bottom slab	8	313.06	0.74	OK	424.73
		9	299.74	2.17	NG	137.92
		10	235.17	1.48	NG	159.08
		11	238.62	1.73	NG	137.92
		12	257.29	0.61	OK	424.73
	Left sidewall	13	214.12	0.37	OK	578.62
		14	199.81	0.59	OK	340.89
		15	233.07	0.70	OK	334.22
	Left arch	16	149.01	0.39	OK	377.73
		17	95.95	0.25	OK	
	Top slab	18	123.82	0.28	OK	442.07
Final judgement			NG			

▒ Member with rebar yielding (structural member coefficient $\gamma_b = 1.56$)

Table 8.24: Verification results for new cross section 1 with Mortar No. 2 under Level 2 earthquake ground motion

Cross section			**New cross section 1 (2,680 × 2,290)**							
Renovation conditions			**Increasing bottom slab thickness by 70 mm, and reducing top plate thickness by 70 mm**							
Mortar			**Mortar No. 2**							
Input ground motion			Level 2 earthquake ground motion (Inland Type 1)				Level 2 earthquake ground motion (Inland Type 2)			
Verification item			Response value	Response value/ limit value	Judge-ment	Limit value	Response value	Response value/ limit value	Judge-ment	Limit value
Maximum compressive strain			0.0004		OK	0.0031	0.0003		OK	0.0031
Shear force at check point (kN)	Top slab	1	95.71	0.38	OK	254.43	105.33	0.41	OK	254.43
		2	84.49	0.33	OK		108.54	0.43	OK	
	Right arch	3	66.48	0.30	OK	221.36	73.37	0.33	OK	221.36
		4	112.50	0.51	OK		128.43	0.58	OK	
	Sidewall	5	155.13	0.56	OK	278.52	214.13	0.64	OK	334.22
		6	211.45	0.74	OK	284.07	176.19	0.53	OK	340.89
		7	280.18	0.57	OK	489.43	247.16	0.74	OK	587.32
	Bottom slab	8	324.16	0.55	OK	581.09	302.04	0.43	OK	697.31
		9	315.86	0.98	OK	320.70	274.79	0.86	OK	320.70
		10	255.78	0.68	OK	378.41	241.61	0.64	OK	378.41
		11	235.60	0.73	OK	320.70	310.44	0.97	OK	320.70
		12	269.77	0.46	OK	581.09	323.78	0.56	OK	581.09
	Left sidewall	13	212.48	0.36	OK	587.32	285.50	0.58	OK	489.43
		14	183.15	0.54	OK	340.89	215.26	0.44	OK	284.07
		15	220.47	0.66	OK	334.22	187.16	0.38	OK	334.22
	Left arch	16	146.55	0.66	OK	221.36	125.41	0.57	OK	221.36
		17	76.81	0.35	OK		67.58	0.31	OK	
	Top slab	18	103.70	0.49	OK	254.43	91.82	0.43	OK	254.43
Final judgement			OK				OK			

▭ Member with rebar yielding (structural member coefficient $\gamma_b = 1.56$)

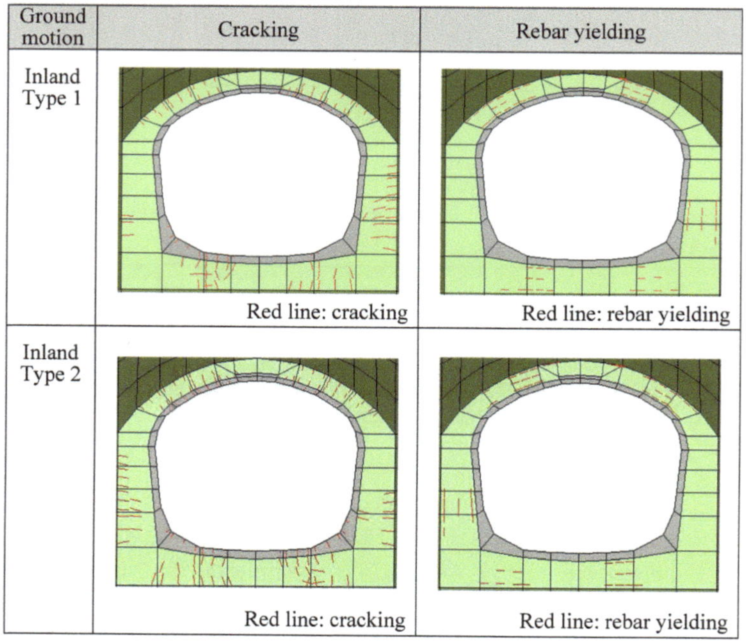

Figure 8.36: Cracking and rebar yielding in new cross section 1 with Mortar No. 2 (Level 2 earthquake ground motion)

(3) Renovation with Mortar No. 3

New renovation cross section 2: The response value of the standard renovation cross section with Mortar No. 2 exceeded the limit value by a factor of up to 2.17 times under Level 2 earthquake force. It was judged that the standard renovation cross section would not meet the performance requirements even if Mortar No. 3 was used. New cross section 2, therefore, was defined by keeping the inner diameter of the standard renovation cross section and raising the centre of the renovated sewer by 50 mm, thus increasing the liner thickness of the bottom slab by 50 mm and reducing the liner thickness of the top plate by 50 mm (internal dimensions 2,680 × 2,290 mm). Figure 8.37 shows the cracking and rebar yielding of new cross section 2 under Level 2 earthquake ground motion. Table 8.25 shows the verification results.

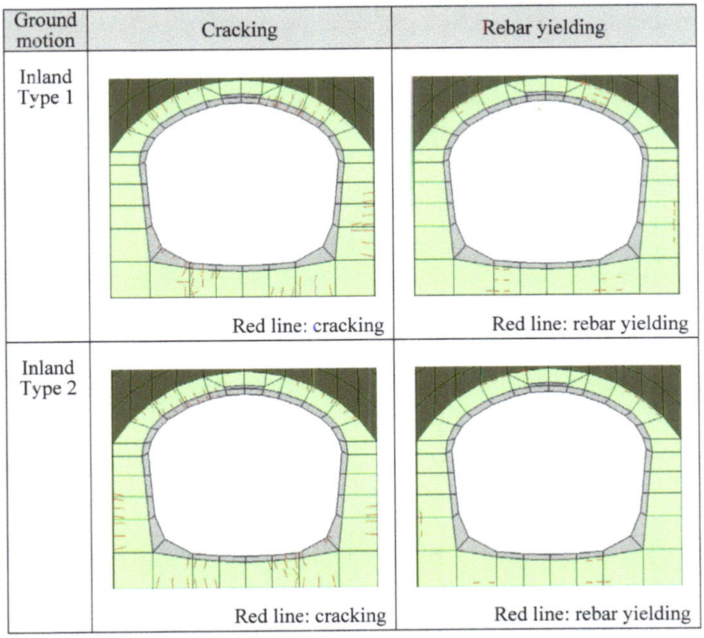

Figure 8.37: Cracking and rebar yielding in new cross section 2 with Mortar No. 3 (Level 2 earthquake ground motion)

8.2.7 Verification of safety under earthquake loading

The results of seismic safety verifications on the existing sewer and the other four types of renovation conditions are shown in Table 8.26. Based on these results, the following comments are made:

- The verification results indicate that under Level 1 earthquake ground motion, the existing sewer meets the earthquake resistance requirements.
- The verification results indicate that under Level 2 earthquake ground motion, the existing sewer does not meet the earthquake resistance requirements, but new cross section 1 with Mortar No. 2 and new cross section 2 with Mortar No. 3 meet the earthquake resistance requirements.
- With new cross section 1 and Mortar No. 2, the load coefficient for cracking is 1.50 and the maximum load coefficient is 5.75, satisfying the performance requirements under normal load conditions.
- With new cross section 2 and Mortar No. 3, the load coefficient for cracking is 2.30 and the maximum load coefficient is 8.82, satisfying the performance requirements under normal load conditions.

Table 8.25: Verification results for new cross section 2 with Mortar No. 3 under Level 2 earthquake ground motion

Cross section			**New cross section 2 (2,680 × 2,290)**							
Renovation conditions			**Increasing bottom slab thickness by 50 mm, and reducing top plate thickness by 50 mm**							
Mortar			**Mortar No. 3**							
Input ground motion			Level 2 earthquake ground motion (Inland Type 1)				Level 2 earthquake ground motion (Inland Type 2)			
Verification item			Response value	Response value/ limit value	Judge-ment	Limit value	Response value	Response value/ limit value	Judge-ment	Limit value
Maximum compressive strain			0.0003		OK	0.0031	0.0003		OK	0.0031
Shear force at check point (kN)	Top slab	1	119.74	0.40	OK	300.44	128.87	0.36	OK	360.52
		2	92.75	0.31	OK		133.10	0.37	OK	
	Right arch	3	70.66	0.24	OK	299.89	95.35	0.32	OK	299.89
		4	148.21	0.49	OK		145.32	0.48	OK	
	Right sidewall	5	177.03	0.57	OK	311.21	212.40	0.57	OK	373.45
		6	233.24	0.75	OK		178.14	0.48	OK	
		7	295.59	0.95	OK		249.19	0.67	OK	
	Bottom slab	8	317.05	0.52	OK	610.92	298.12	0.41	OK	733.10
		9	304.05	0.99	OK	307.45	270.95	0.88	OK	307.45
		10	250.93	0.69	OK	362.51	234.85	0.65	OK	362.51
		11	232.23	0.76	OK	307.45	301.16	0.98	OK	307.45
		12	266.14	0.44	OK	610.92	319.01	0.44	OK	733.10
	Left sidewall	13	212.98	0.57	OK	373.45	294.20	0.95	OK	311.21
		14	181.33	0.49	OK		228.77	0.74	OK	
		15	216.40	0.58	OK		187.21	0.60	OK	
	Left arch	16	141.70	0.47	OK	299.89	122.79	0.41	OK	299.89
		17	100.85	0.34	OK		74.90	0.25	OK	
	Top slab	18	130.06	0.36	OK	360.52	105.07	0.35	OK	300.44
Final judgement			OK				OK			

(Shaded) Member with rebar yielding (structural member coefficient $\gamma_b = 1.56$)

Table 8.26: Results of safety verification under earthquake loading

Renovation condition		Level 1 earthquake ground motion	Level 2 earthquake ground motion			Remarks
Cross section	Mortar		Ultimate displace-ment	Shear force	Judge-ment	
Existing cross section		OK	OK	NG	NG	
Standard renovation cross section	No. 2		OK	NG	NG	
New cross section 1	No. 2		OK	OK	OK	Increasing bottom slab thickness by 70 mm, and reducing top plate thickness by 70 mm
New cross section 2	No. 3		OK	OK	OK	Increasing bottom slab thickness by 50 mm, and reducing top plate thickness by 50 mm

Note: Ultimate displacement under Level 2 earthquake ground motion is verified by checking the maximum compressive strain.

8.2.8 Safety verification of local buckling of the bottom slab

The possibility of local buckling of the renovation layer at the bottom slab under groundwater pressure is checked for three renovation conditions: the standard renovation cross section with Mortar No. 2, new cross section 1 with Mortar No. 2 and new cross section 2 with Mortar No. 3. For details of buckling strength computation, refer to Example I.

Buckling analyses on the three renovation conditions show that, while the possibility of local buckling exists for the case of standard renovation with Mortar No. 2, the other two renovation conditions have sufficient buckling strength against the maximum groundwater pressure. Details of the verification study on local buckling are presented in Table 8.27.

Table 8.27: Safety verification of local buckling of bottom slab

Item	Variable	Unit	Renovation condition		
			Standard renovation cross section	New cross section 1	New cross section 2
			Mortar No. 2	Mortar No. 2	Mortar No. 3
Maximum overburden thickness	h	m	4.990		
Height of sewer	H_o	m	3.066		
Poisson's ratio of liner	v		0.30		
Liner thickness at bottom slab	t	mm	31.7	101.7	81.7
Invert radius	R	mm	7,410		
Invert length of liner	L	mm	1,440		
Moment of inertia of area	I	mm³/mm	2.655	87.656	45.445
Cross-sectional area of liner	A	mm²/mm	31.7	101.7	81.7
Equivalent stiffness of liner	E	N/mm²	12.71×10^3	8.51×10^3	21.62×10^3
Buckling strength	P_{cr}	N/mm²	0.075	0.655	1.028
Maximum groundwater pressure	P_w	N/mm²	0.079		
Safety factor	γ		1.5		
Safety verification			$P_{cr} < \gamma P_w$	$P_{cr} > \gamma P_w$	$P_{cr} > \gamma P_w$
Judgement			NG	OK	OK

8.2.9 Determination of renovation methods

(1) Structural strength

Though the existing sewer can withstand Level 1 earthquake force, it is not strong enough to withstand Level 2 earthquake ground motion, and it fails to meet the performance requirements under normal load conditions. Therefore, structural renovation has to be performed on this sewer line. In cases where Mortar No. 2 is used, the required structural strength is achieved by increasing the bottom slab thickness by 70 mm and reducing the top plate thickness by 70 mm from the standard renovation cross section. In cases where Mortar No. 3 is used, the required structural strength is achieved by increasing the bottom slab thickness by 50 mm and reducing the top plate thickness by 50 mm from the standard renovation cross section.

(2) Achieving the required discharge capacity

The discharge capacity at each section of the renovated sewer is calculated and shown in Table 8.28. Due to the significant reduction of the cross-sectional area in an effort to eliminate the adverse slope in this sewer line, the flow ratios at most of the sections are less than 1.0. However, with the fast reconstruction of the sewer system in this region, the future drainage area of this sewer line will be much reduced and the corresponding flow ratios will then become adequate, as shown in the same table. For details of flow ratio computation, refer to Example I.

Table 8.28: List of renovation plans (Example II)

Sewer No.		**10**	**11**	**12**	**13**	**14**	**15**	**16**	**17**
Sewer length (m)		98.90	61.30	104.70	21.90	64.90	59.90	102.35	18.50
Shape	Existing sewer	2,820 × 2,140	2,850 × 2,160	2,800 × 2,170	2,730 × 2,460	2,720 × 2,450	2,720 × 2,450	2,730 × 2,440	2,730 × 2,470
	After renovation	2,630 × 1,670	2,660 × 1,810	2,600 × 1,730	2,560 × 2,030	2,540 × 2,070	2,540 × 2,160	2,550 × 2,110	2,550 × 2,170
Slope (‰)	Existing sewer	-1.8	0.8	1.2	-3.0	-1.0	1.1	0.6	1.6
	After renovation	0.2	0.2	0.3	0.3	0.3	1.1	0.6	1.6
Flow ratio	Existing sewer		1.18	1.40			1.32	0.93	1.44
	After renovation	0.49	0.53	0.58	0.65	0.64	1.31	0.90	1.42
	Renovated sewer (based on future drainage area)	1.54	1.38	1.40	1.26	1.20	2.40	1.56	2.26

(3) Conclusion

Based on the various analysis results discussed above, sewer renovation will be carried out for this sewer line to strengthen its earthquake resistance, and to mitigate flooding damage by keeping its discharge capacity adequate. Since the existing sewer line contains adverse slope sections, the renovation approach adopted is to use new cross section 2 with Mortar No. 3 because that allows a more flexible adjustment of bottom slab thickness. The renovation plan for each section of this sewer line is listed in Table 8.28.

8.3 Example III: Circular Sewer (Concrete Foundation)

8.3.1 Internal investigation of sewer

(1) Investigation of internal cross section and damage condition
The sewer line investigated here is a circular pipe, and its internal diameter and damage condition at each section are shown in Table 8.29. Based on preliminary inspections, it was decided to conduct a structural investigation on Section No. 8, where the structural deterioration was found to be most severe.

Table 8.29: Structural conditions of inspected sewer line (Example III)

Section No.	Results of inspection (measurements)		Sewer conditions
	Diameter (mm)	Length (m)	Observed damage and judgement
4	ϕ1,210	10.39	No damage
5	ϕ1,210	7.39	No damage
6	ϕ1,210	24.21	No damage
7	ϕ1,210	60.63	Surface deterioration
8	ϕ1,210	80.36	Surface deterioration
9	ϕ1,210	37.36	Surface deterioration
10	ϕ1,210	11.30	No damage

(2) Investigation items and results
The items and results of sewer investigation are shown in Table 8.30.

Table 8.30: Investigation items and results

		Results
Item		Investigation results
Cross-sectional shape		Circular
Internal dimensions (mm)	Diameter	ϕ1,210
Overburden thickness (m)		3.95–5.40
Compressive strength of concrete (N/mm^2)		21.2–37.5
Member thickness (mm)		96.0
Surface corrosion thickness (mm)		10.0
Concrete cover (mm)		40.0 * Including surface corrosion thickness
Rebar diameter and spacing (mm)		ϕ7.0@81
Rebar strength (N/mm^2)	Yield strength	
	Tensile strength	
	Based on JIS standard	SR235

8.3.2 Original design documents and determination of cross sections for structural analysis

As the design drawings used for the construction of the target sewer are available, details of the design cross section are determined based on both the results of investigation and data of the original design documents for obtaining the critical conditions. Table 8.31 compares the results of structural investigation and data obtained from the original drawings and shows the cross-sectional conditions used for structural analysis.

Table 8.31: Structural details of design cross section

		Classification		
Item		**Structural investigation results**	**Standard drawing**	**Design cross section**
Internal dimensions (mm)		ϕ1,230	ϕ1,210	ϕ1,230
Overburden thickness (m)		3.95–5.40		3.95–5.40
Compressive strength of concrete (N/mm²)		21.2		21.2
Member thickness (mm)		86.0	103	86.0
Concrete cover (mm)	Inside	30.0	24.0	30.0
	Outside		24.0	24.0
Rebar diameter and spacing (mm)	Inside	ϕ7.0@81	Reinforcing bars of unknown diameter at 81 mm spacing	ϕ7.0@81
	Outside		Reinforcing bars of unknown diameter at 81 mm spacing	ϕ7.0@81
Rebar strength (N/mm²)	Yield strength			235
	Tensile strength			380
	Based on JIS standard	Steel wire: SR235 or equivalent		SR235

8.3.3 General conditions for structural analysis

(1) Burial conditions, cross-sectional dimensions and rebar arrangements
Figure 8.38 shows the burial conditions of the existing sewer, its general structural dimensions, rebar arrangements and other structural details. The cross section of a sewer with standard renovation is also shown.

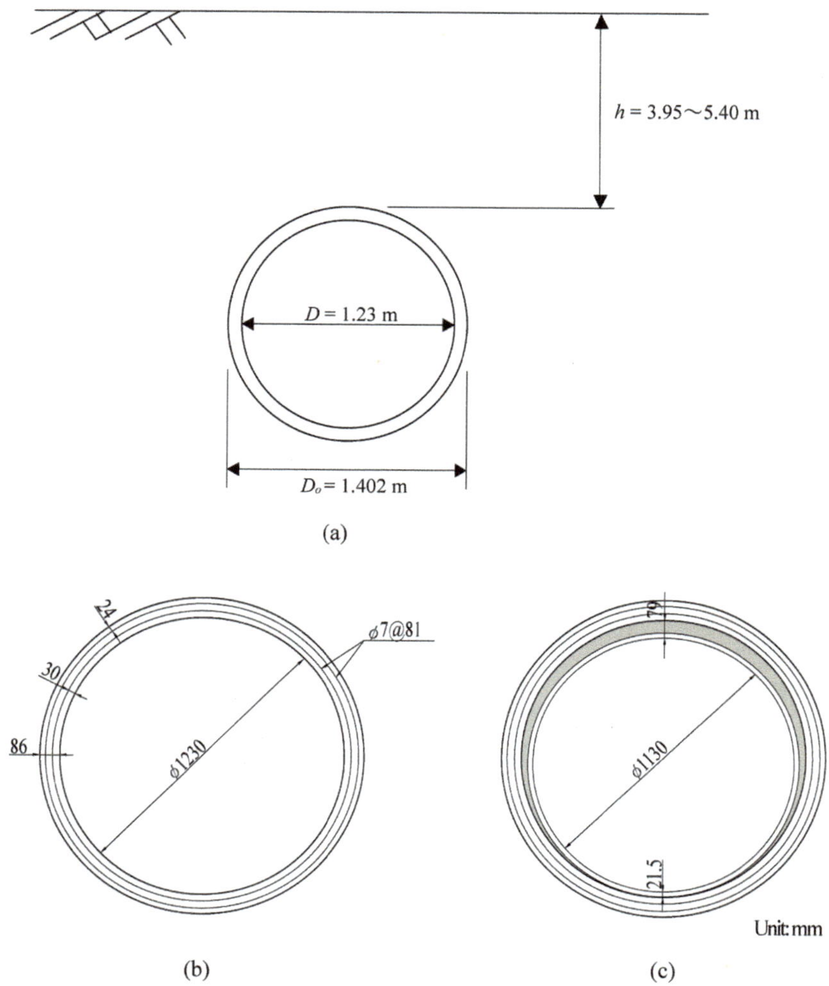

Figure 8.38: Cross section of sewer for structural analysis and renovation design: (a) general structural dimensions and burial conditions; (b) structural details of existing sewer; (c) cross section of sewer with standard renovation

(2) Support conditions

In view of the fact that typical drawings used for other projects around the time of completion (1929) specified 90-degree concrete foundations, the foundation was assumed to be a 90-degree fixed support structure; refer to Fig. 8.39.

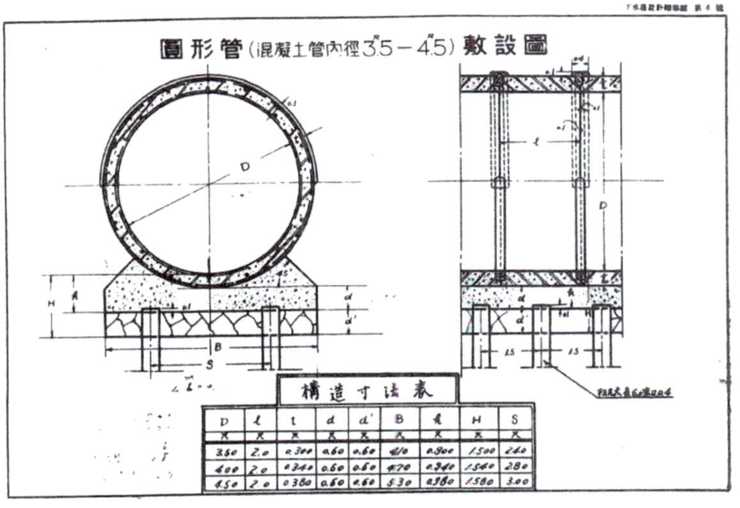

Figure 8.39: Standard drawing (1929)

(3) Material properties

The material properties of the concrete, grouting materials, rebars and profiled steel sheets of the renovation layer are summarised in Table 8.32.

Table 8.32: Material properties

Material	Item	Specification	Unit	Remarks
Concrete members	Compressive strength	21.2	N/mm²	Minimum test value
	Tensile strength	1.76	N/mm²	JSCE (2013)
	Modulus of elasticity	23.6	kN/mm²	〃
	Poisson's ratio	0.2		〃
	Fracture energy	80.9	N/m	〃
Rebar	Yield strength	235	N/mm²	Test
	Modulus of elasticity	200	kN/mm²	JSCE (2013)
Grout (Mortar No. 2)	Compressive strength	21.0	N/mm²	Design value
	Tensile strength	1.83	N/mm²	〃
	Modulus of elasticity	6.60	kN/mm²	〃
	Poisson's ratio	0.25		〃
	Fracture energy	17.6	N/m	〃
Grout (Mortar No. 3)	Compressive strength	35.0	N/mm²	Design value
	Tensile strength	2.92	N/mm²	〃
	Modulus of elasticity	22.0	kN/mm²	〃
	Poisson's ratio	0.22		〃
	Fracture energy	70.8	N/m	〃
Profiled steel sheet #79SW	Yield strength	210	N/mm²	Design value
	Modulus of elasticity	170	kN/mm²	〃
	Cross-sectional area	1571	mm²/m	〃

(4) Soil conditions

In view of the boring data obtained from a nearby location, soil conditions are assumed as shown in Table 8.33.

Table 8.33: Soil conditions

Boring data	Soil No.	Soil type	Depth (m)	Thickness, H_i (m)	Average N-value	Average shear wave velocity, V_{si} (m/s)	H_i/V_{si} (s)
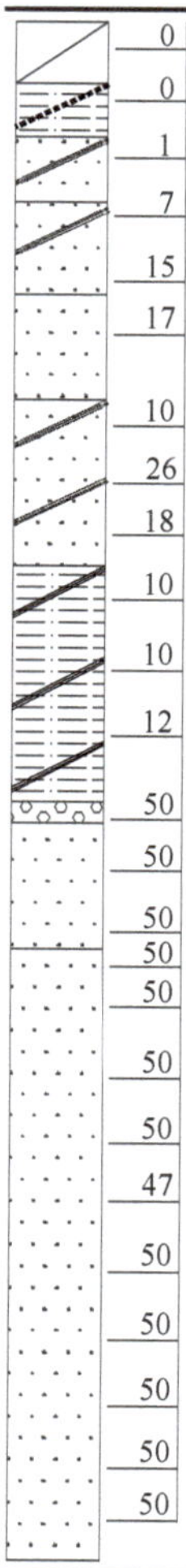	1	Sandy silt	2.50	2.50	1	90.73	0.028
	2	Silty sand	4.00	1.50	1	82.34	0.018
	3	Silty sand	6.20	2.20	11	214.86	0.010
	4	Sand	8.50	2.30	17	255.72	0.009
	5	Silty sand	12.40	3.90	18	261.64	0.015
	6	Clayey silt	17.80	5.40	11	236.76	0.023
	Basement layer	Sand and gravel			> 50	400.00	

Evaluation of ground property: $T_G = 4\sum_{i=1}^{n}\frac{H_i}{V_{si}} = 0.411$ (s), Class II

Classification of ground: $T_G < 0.2$, Class I; $0.2 \le T_G < 0.6$, Class II; $0.6 \le T_G$, Class III

(5) Load conditions

Since the overburden thickness along this sewer line exceeds $2D_o$ (D_o: outside diameter of pipe) in some sections, it was judged that the most critical load conditions would occur when the earth pressure from the loosening zone is taken into consideration under the maximum overburden thickness (5.40 m). Live loads were taken into account in the perpendicular direction. Load conditions are calculated based on relevant codes (JRA, 2013a-b).

1) Static earth pressure (Fig. 8.40)

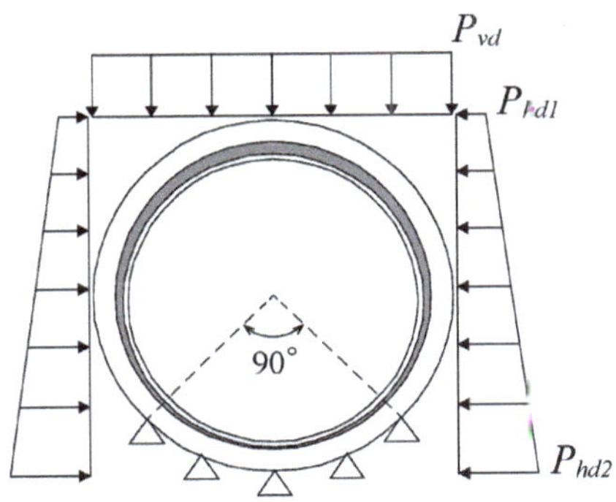

Static earth pressure (kN/m²)		
Vertical load	Horizontal load	
P_{vd}	P_{hd1}	P_{hd2}
65.45	19.64	26.20

Figure 8.40: Static earth pressure

$$P_{vd} = \frac{B1 \cdot (\gamma - c / B1)}{K_0 \cdot \tan\phi}(1 - e^{-K_0 \cdot \tan\phi \cdot h / B1}) \text{ (kN/m}^2\text{)} \quad (8.8)$$

where $B1 = R_0 \cdot \cot\left(\frac{45^\circ + \phi/2}{2}\right)$;

P_{vd}: vertical earth pressure (loosening earth pressure); R_0: excavation radius reflecting loosening width (= 1.482 m); γ: unit weight of soil (= 15.6 kN/m³; weighted average); c: cohesion of soil (= 2.7 kN/m²; weighted average); ϕ: angle of shear resistance of soil (= 7°; weighted average); K_0: coefficient of static earth pressure (1.0 for calculation of loosening earth pressure); h: overburden thickness (5.4 m).

$$P_{hd1} = K_0 \cdot P_{vd} \text{ (kN/m}^2\text{)} \tag{8.9}$$

$$P_{hd2} = K_0 \cdot (P_{vd} + \gamma \cdot D_o) \text{ (kN/m}^2\text{)} \tag{8.10}$$

where P_{hd1}, P_{hd2}: horizontal earth pressure; K_0: coefficient of static earth pressure (= 0.3); D_o: outside diameter of pipe (= 1.402 m).

2) Live loads (Fig. 8.41)

$$P_{vl} = \frac{2Pl \cdot (1+i) \cdot \beta}{C \cdot (a + 2h \cdot \tan\theta)} \text{ (kN/m}^2\text{)} \tag{8.11}$$

where P_{vl}: vertical live load; Pl: rear wheel load (= 100 kN); i: impact factor (= 0.110); β: reduction factor (= 0.9); C: overall vehicle width (= 2.75 m); a: ground contact width in the longitudinal direction (= 0.2 m); θ: load distribution angle (= 45°).

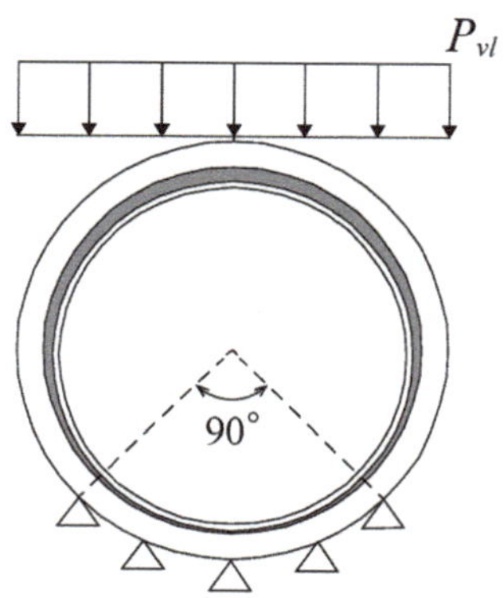

Live load (kN/m²)
P_{vl}
6.61

Figure 8.41: Live loads

(6) Analysis model

Figures 8.42 and 8.43 show FE models for structural analysis under normal load conditions and under seismic load conditions, respectively. In view of the symmetry, a half cross-sectional model is used for structural analysis under normal loads. In the seismic performance evaluation, the ground is modelled with finite elements having sufficient coverage for pipe–soil coupled analysis.

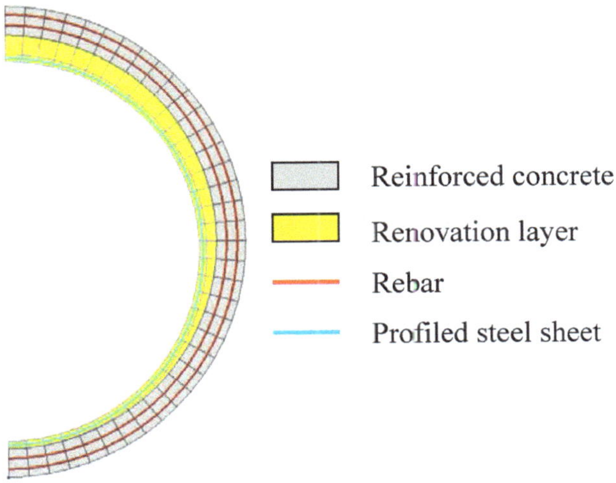

Figure 8.42: FE model for structural analysis under normal loads

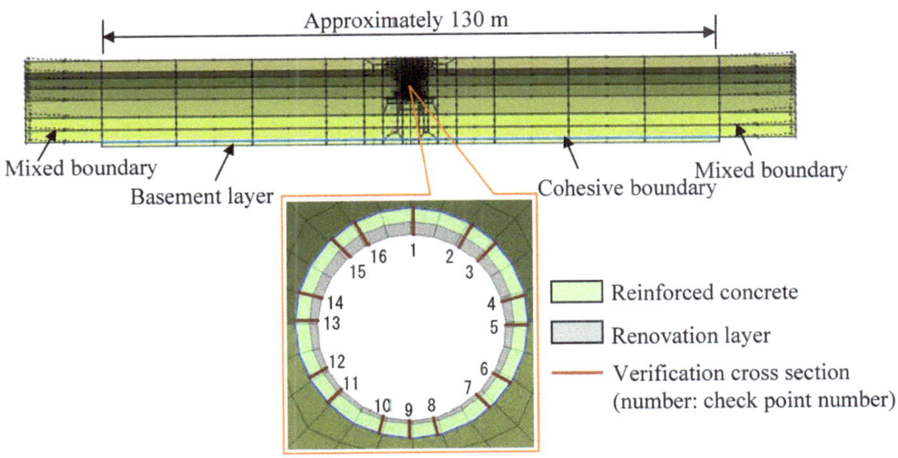

Figure 8.43: FE model for seismic performance analysis

8.3.4 Numerical results under normal load conditions

(1) Existing cross section

Figures 8.44 and 8.45 show the cracking patterns, sectional forces and structural deformations based on the FE model of the existing cross section.

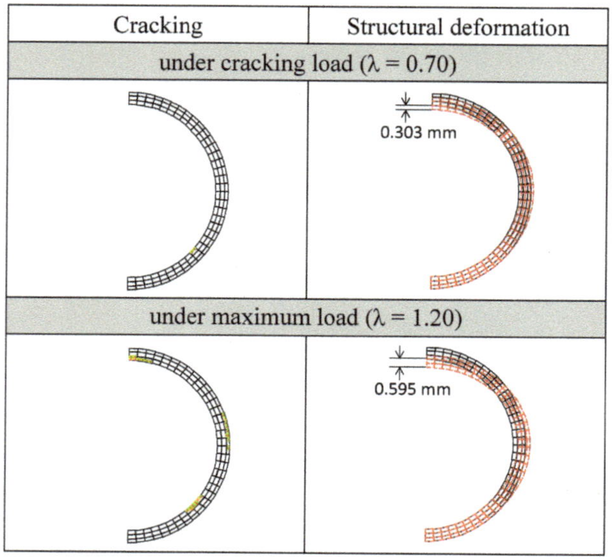

Figure 8.44: Cracking pattern and structural deformation (existing cross section)

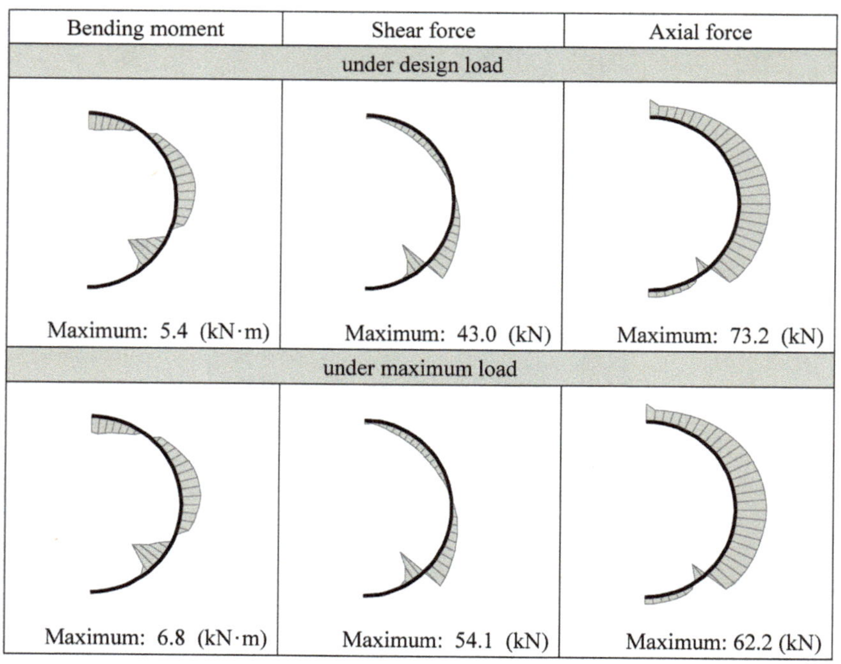

Figure 8.45: Sectional forces (existing cross section)

(2) Standard renovation cross section with Mortar No. 2

Figures 8.46 and 8.47 show the cracking patterns, sectional forces and structural deformations based on the FE model of the standard renovation cross section with Mortar No. 2.

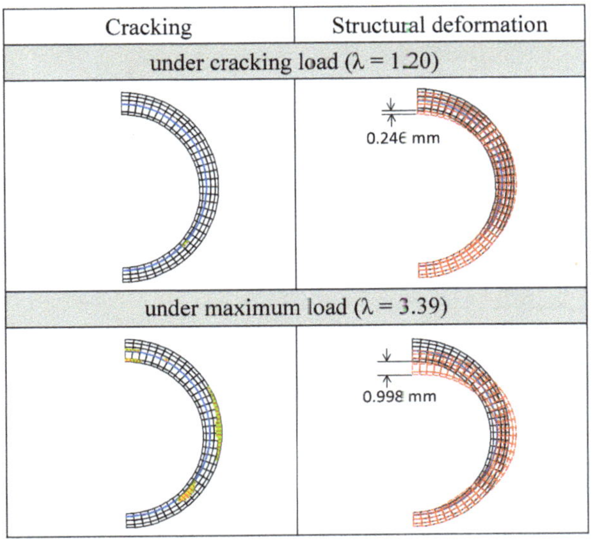

Figure 8.46: Cracking pattern and structural deformation (standard renovation cross section, Mortar No. 2)

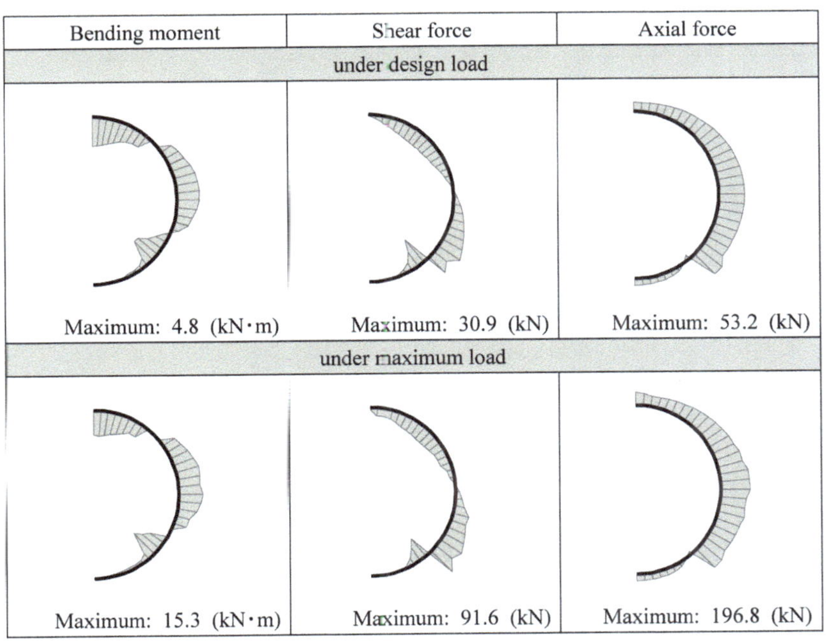

Figure 8.47: Sectional forces (standard renovation cross section, Mortar No. 2)

(3) Standard renovation cross section with Mortar No. 3

Figures 8.48 and 8.49 show the cracking patterns, sectional forces and structural deformations based on the FE model of the standard renovation cross section with Mortar No. 3.

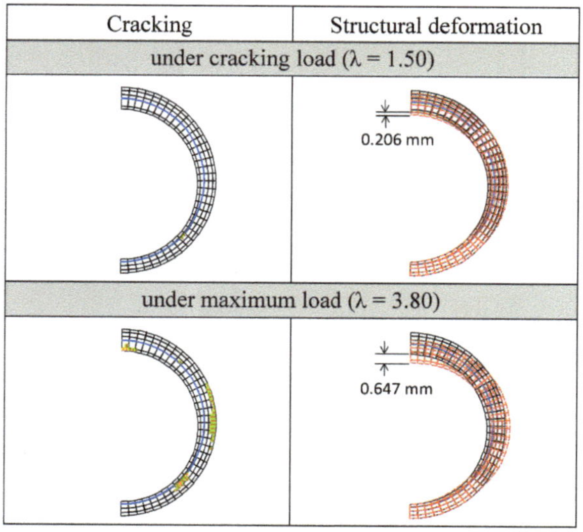

Figure 8.48: Cracking pattern and structural deformation (standard renovation cross section, Mortar No. 3)

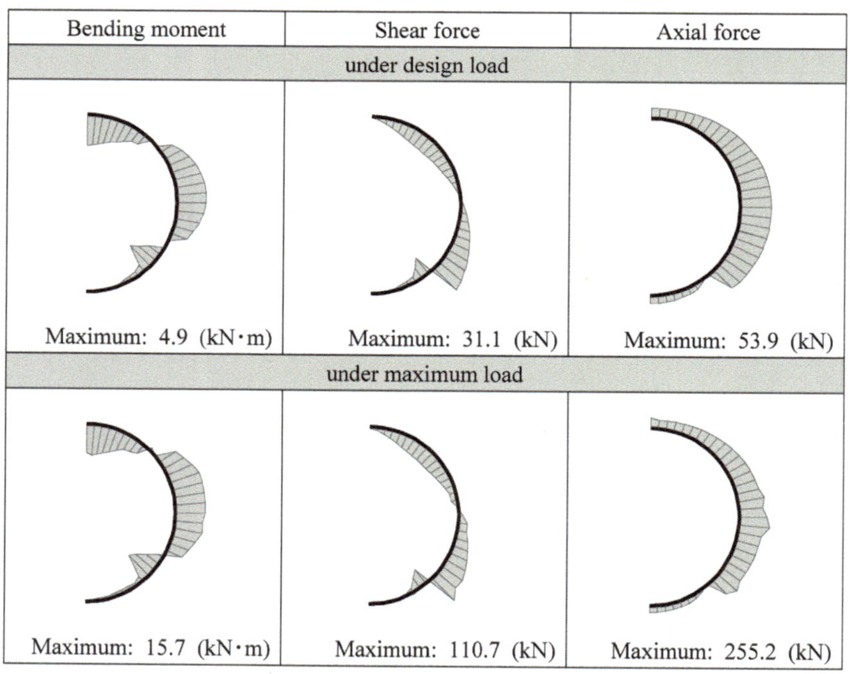

Figure 8.49: Sectional forces (standard renovation cross section, Mortar No. 3)

8.3.5 Verification of safety under normal load conditions

The results of safety verification on the existing sewer and the other two renovation conditions are shown in Table 8.34. Based on these results, the following comments are made:

- In the serviceability limit state, the existing sewer will not meet the performance requirements because the load coefficient for cracking is less than 1.0. The standard renovation cross section with Mortar No. 2 or No. 3 will satisfy the performance requirements because the load coefficient for cracking is greater than 1.0.
- In the ultimate limit state, the existing sewer will not meet the performance requirements because its maximum load coefficient is less than 2.5. The standard renovation cross section with Mortar No. 2 or No. 3 will satisfy the performance requirements because the maximum load coefficient exceeds 2.5.

Table 8.34: Results of safety verification based on load coefficients

Renovation condition		Perpendicular live load		Judgement	Remarks
Cross section	Mortar	Cracking load coefficients	Maximum load coefficients		
Existing cross section		0.70	1.20	NG	
Standard renovation	No. 2	1.20	3.39	OK	
	No. 3	1.50	3.80	OK	
Required load coefficients		> 1.0	≥ 2.5		

8.3.6 Results of seismic performance analysis

(1) Existing cross section

1) Level 1 earthquake ground motion

Figure 8.50 shows the cracking and rebar yielding of the existing cross section under Level 1 earthquake ground motion.

2) Level 2 earthquake ground motion

Figure 8.51 shows the cracking and rebar yielding of the existing cross section under Level 2 earthquake ground motion. Table 8.35 shows the verification results under Level 2 ground motion.

Table 8.35: Verification results for existing cross section under Level 2 earthquake ground motion

Cross section			Existing cross section (ϕ1,230)							
Input ground motion			Level 2 earthquake ground motion (Inland Type 1)				Level 2 earthquake ground motion (Ocean Type 2)			
Verification item			Response value	Response value/ limit value	Judge-ment	Limit value	Response value	Response value/ limit value	Judge-ment	Limit value
Maximum compressive strain			Less than 0.0031		OK	0.0031	Less than 0.0031		OK	0.0031
Shear force at check point (kN)	Top	1	30.41	0.96	OK	31.56	19.51	0.62	OK	31.56
		2	18.64	0.71	OK	26.30	18.16	0.58	OK	31.56
		3	14.84	0.56	OK	26.30	15.37	0.49	OK	31.56
	Side	4	41.39	1.31	NG	31.56	19.62	0.62	OK	31.56
		5	37.75	1.20	NG	31.56	14.00	0.44	OK	31.56
		6	12.92	0.49	OK	26.30	12.36	0.39	OK	31.56
	Bottom	7	13.88	0.53	OK	26.30	13.16	0.42	OK	31.56
		8	27.74	0.88	OK	31.56	10.17	0.32	OK	31.56
		9	27.76	0.88	OK	31.56	17.90	0.57	OK	31.56
		10	22.07	0.70	OK	31.56	21.39	0.68	OK	31.56
		11	16.26	0.62	OK	26.30	16.75	0.53	OK	31.56
	Side	12	19.39	0.61	OK	31.56	17.54	0.56	OK	31.56
		13	32.26	1.02	NG	31.56	23.73	0.75	OK	31.56
		14	34.49	1.09	NG	31.56	17.55	0.56	OK	31.56
	Top	15	9.52	0.36	OK	26.30	11.04	0.35	OK	31.56
		16	17.91	0.57	OK	31.56	13.19	0.42	OK	31.56
Final judgement			NG				OK			

▒ Member with rebar yielding (structural member coefficient $\gamma_b = 1.56$)

Rebar yielding	None
Judgement	OK

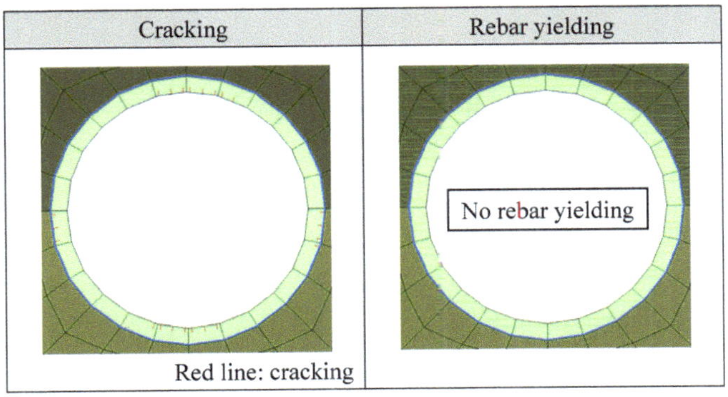

Figure 8.50: Cracking and rebar yielding in existing cross section (Level 1 earthquake ground motion)

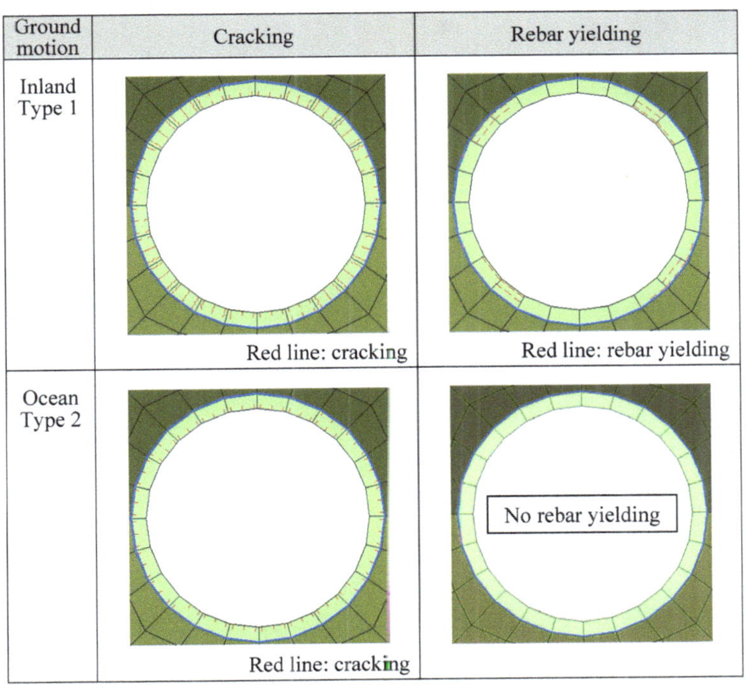

Figure 8.51: Cracking and rebar yielding in existing cross section (Level 2 earthquake ground motion)

(2) Renovation with Mortar No. 2

Figure 8.52 shows the cracking and rebar yielding under Level 2 earthquake ground motion of a standard renovation cross section with Mortar No. 2. Table 8.36 shows the verification results.

Table 8.36: Verification results for standard renovation cross section with Mortar No. 2 under Level 2 earthquake ground motion

Cross section			**Standard renovation cross section (ϕ1,130)**							
Mortar			**Mortar No. 2**							
Input ground motion			Level 2 earthquake ground motion (Inland Type 1)				Level 2 earthquake ground motion (Inland Type 2)			
Verification item			Response value	Response value/ limit value	Judge-ment	Limit value	Response value	Response value/ limit value	Judge-ment	Limit value
Maximum compressive strain			Less than 0.0031		OK	0.0031	Less than 0.0031		OK	0.0031
Shear force at check point (kN)	Top	1	59.40	0.25	OK	241.98	45.73	0.19	OK	241.98
		2	46.30	0.24	OK	193.43	32.74	0.14	OK	232.11
		3	23.97	0.11	OK	220.72	24.57	0.11	OK	220.72
	Side	4	50.67	0.27	OK	187.00	41.43	0.22	OK	187.00
		5	51.46	0.31	OK	167.46	36.12	0.22	OK	167.46
		6	23.38	0.22	OK	107.22	23.25	0.18	OK	128.67
	Bottom	7	16.11	0.31	OK	51.82	18.98	0.31	OK	62.19
		8	36.85	0.68	OK	54.29	25.48	0.47	OK	54.29
		9	35.16	0.66	OK	53.32	30.64	0.57	OK	53.32
		10	29.81	0.55	OK	54.29	35.00	0.64	OK	54.29
		11	25.51	0.41	OK	62.19	22.68	0.36	OK	62.19
	Side	12	35.14	0.27	OK	128.67	30.53	0.24	OK	128.67
		13	71.56	0.43	OK	167.46	36.56	0.22	OK	167.46
		14	51.16	0.33	OK	155.83	32.46	0.17	OK	187.00
	Top	15	18.43	0.10	OK	183.93	19.47	0.11	OK	183.93
		16	32.04	0.14	OK	232.11	31.24	0.13	OK	232.11
Final judgement			OK				OK			

▒ Member with rebar yielding (structural member coefficient $\gamma_b = 1.56$)

Ground motion	Cracking	Rebar yielding
Inland Type 1	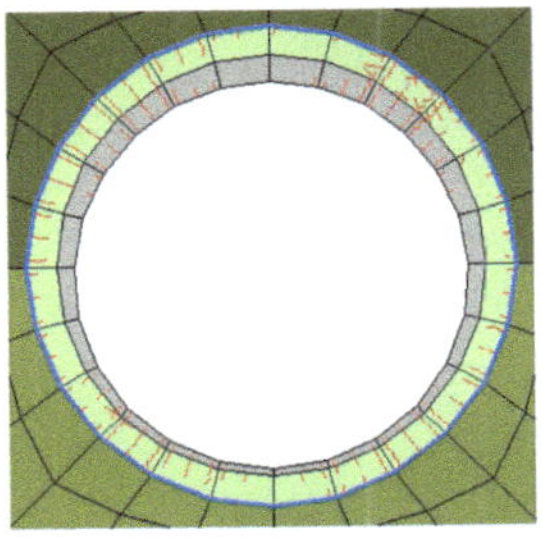 Red line: cracking	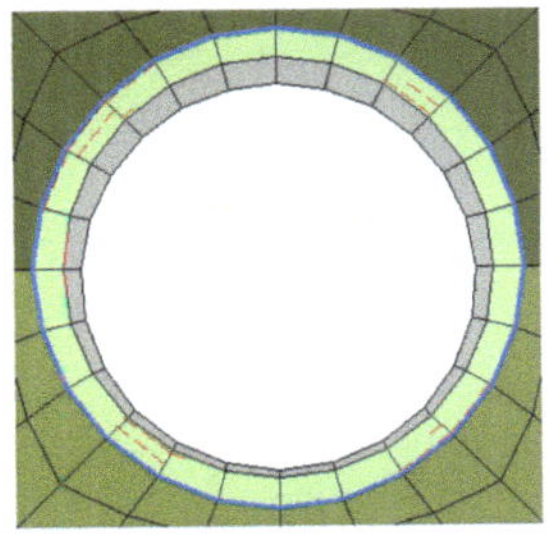 Red line: rebar yielding
Inland Type 2	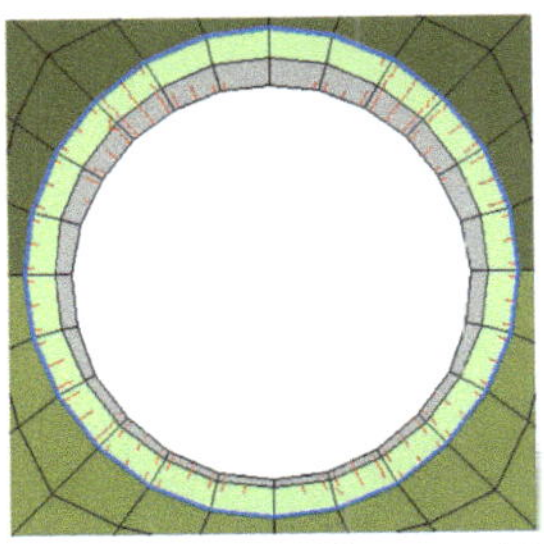 Red line: cracking	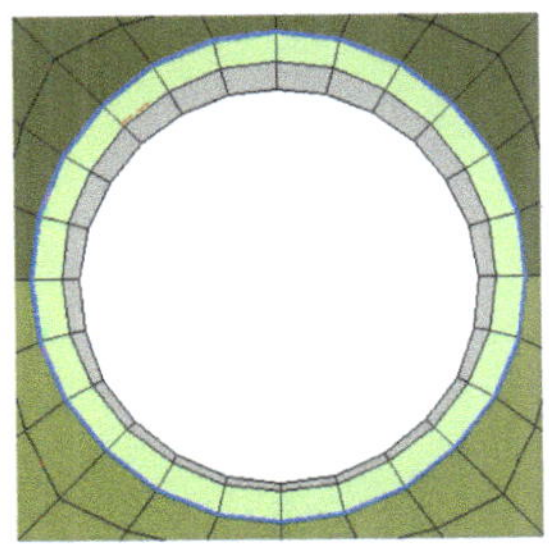 Red line: rebar yielding

Figure 8.52: Cracking and rebar yielding in standard renovation cross section with Mortar No. 2 (Level 2 earthquake ground motion)

(3) Renovation with Mortar No. 3

Figure 8.53 shows the cracking and rebar yielding under Level 2 earthquake ground motion of a standard renovation cross section with Mortar No. 3. Table 8.37 shows the verification results.

Table 8.37: Verification results for standard renovation cross section with Mortar No. 3 under Level 2 earthquake ground motion

Cross section			**Standard renovation cross section (ϕ1,130)**							
Mortar			**Mortar No. 3**							
Input ground motion			Level 2 earthquake ground motion (Inland Type 1)				Level 2 earthquake ground motion (Inland Type 2)			
Verification item			Response value	Response value/ limit value	Judge-ment	Limit value	Response value	Response value/ limit value	Judge-ment	Limit value
Maximum compressive strain			0.0005		OK	0.0031	Less than 0.0031		OK	0.0031
Shear force at check point (kN)	Top	1	63.02	0.23	OK	278.21	56.96	0.20	OK	278.21
		2	37.31	0.17	OK	222.21	40.90	0.15	OK	266.65
		3	26.35	0.12	OK	211.08	26.74	0.11	OK	253.29
	Side	4	53.37	0.25	OK	213.76	47.83	0.22	OK	213.76
		5	53.28	0.28	OK	190.86	40.98	0.21	OK	190.86
		6	26.13	0.22	OK	121.16	29.06	0.20	OK	145.39
	Bottom	7	18.25	0.17	OK	105.54	19.34	0.15	OK	126.65
		8	39.62	0.39	OK	102.14	29.12	0.29	OK	102.14
		9	37.63	0.38	OK	99.12	36.99	0.37	OK	99.12
		10	32.11	0.33	OK	85.12	40.47	0.40	OK	102.14
		11	23.19	0.22	OK	105.54	25.83	0.20	OK	126.65
	Side	12	38.81	0.27	OK	145.39	35.51	0.24	OK	145.39
		13	74.40	0.39	OK	190.86	43.35	0.23	OK	190.86
		14	51.20	0.29	OK	178.13	43.82	0.20	OK	213.76
	Top	15	20.56	0.10	OK	211.08	22.62	0.09	OK	253.29
		16	39.41	0.18	OK	222.21	37.03	0.14	OK	266.65
Final judgement			OK				OK			

(Shaded cells: 222.21, 211.08, 121.16, 105.54, 85.12, 105.54, 178.13, 211.08, 222.21 in the Type 1 limit value column.)

Member with rebar yielding (structural member coefficient $\gamma_b = 1.56$)

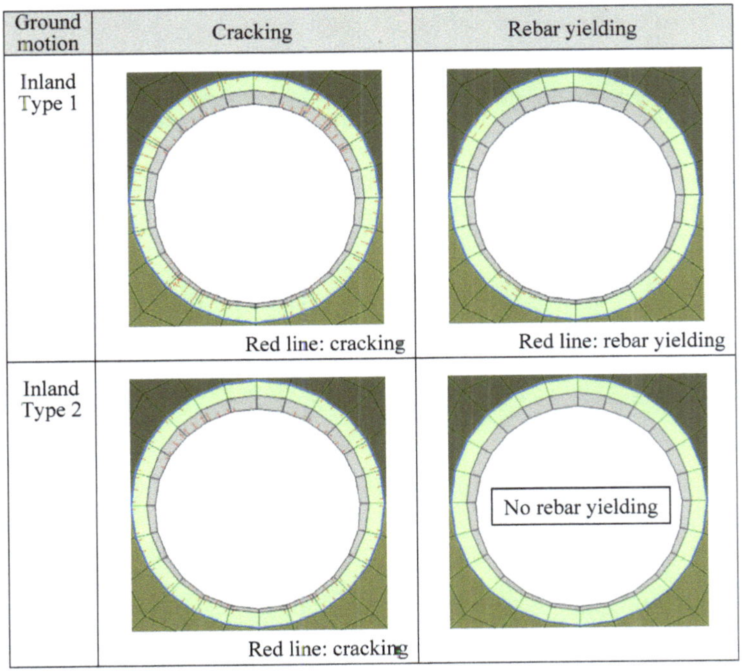

Figure 8.53: Cracking and rebar yielding in standard renovation cross section with Mortar No. 3 (Level 2 earthquake ground motion)

8.3.7 Verification of safety under earthquake loading

The results of seismic safety verification on the existing sewer and the other two types of renovation conditions are shown in Table 8.38. Based on these results, the following comments are made:

- The verification results indicate that under Level 1 earthquake ground motion, the existing sewer meets the earthquake resistance requirements.
- The verification results indicate that under Level 2 earthquake ground motion, the existing sewer does not meet the earthquake resistance requirements, but the standard renovation cross section with Mortar No. 2 or Mortar No. 3 satisfies the earthquake resistance requirements.

Table 8.38: Results of safety verification under earthquake loading

Renovation condition		Level 1 earthquake ground motion	Level 2 earthquake ground motion			Remarks
Cross section	Mortar		Ultimate displacement	Shear force	Judgement	
Existing cross section		OK	OK	NG	NG	
Standard renovation cross section	No. 2		OK	OK	OK	
	No. 3		OK	OK	OK	

Note: Ultimate displacement under Level 2 earthquake ground motion is verified by checking the maximum compressive strain.

8.3.8 Safety verification of local buckling at the bottom of the pipe

Safety verification on the local buckling of the renovation layer at the bottom of the pipe is deemed unnecessary; refer to Section 5.4 for an explanation.

8.3.9 Determination of renovation methods

(1) Structural strength
Though the existing sewer can withstand Level 1 earthquake force, it is not strong enough to sustain Level 2 earthquake ground motion, and it fails to meet the performance requirements under normal load conditions. Therefore, structural renovation has to be performed on this sewer line. The required structural strength can be achieved by employing the standard renovation cross section with Mortar No. 2 or Mortar No. 3.

(2) Achieving the required discharge capacity
The discharge capacity at each section of the renovated sewer is calculated and shown in Table 8.39. For details of flow ratio computation, refer to Example I.

(3) Conclusion
Based on the various analysis results discussed above, sewer renovation will be carried out for this sewer line to strengthen its earthquake resistance, and to mitigate flooding damage by keeping its discharge capacity adequate. The renovation method adopted for this sewer line is to use the standard renovation cross section with Mortar No. 2, so as to achieve the required structural strength against Level 2 earthquake ground motion economically. The renovation plan for each section of this sewer line is listed in Table 8.39.

Table 8.39: List of renovation plans (Example III)

Section No.		4	5	6	7	8	9	10
Sewer length (m)		10.39	7.39	24.21	60.63	80.36	37.36	11.30
Dimensions	Existing sewer	ϕ1,210	ϕ1,210	ϕ1,210	ϕ1,210	ϕ1,210	ϕ1,210	ϕ1,210
	After renovation	ϕ1,110	ϕ1,110	ϕ1,110	ϕ1,110	ϕ1,110	ϕ1,110	ϕ1,110
Slope (‰)	Existing sewer	0.55	0.55	0.55	0.55	0.55	0.91	0.91
	After renovation	0.55	0.55	0.55	0.55	0.55	0.91	0.91
Flow ratio	Existing sewer	1.73	1.73	1.73	1.11	0.99	1.65	0.92
	After renovation	1.79	1.79	1.79	1.15	1.03	1.70	0.95

8.4 Example IV: Circular Sewer (Sand-filled Foundation)

8.4.1 Internal investigation of sewer

(1) Investigation of internal cross section and damage condition
The sewer line investigated here is a circular pipe, and its internal diameter and damage condition at each section are shown in Table 8.40. Based on preliminary inspections, it was decided to conduct a structural investigation on Section No. 1, where the structural deterioration was found to be most severe.

Table 8.40: Structural conditions of inspected sewer line (Example IV)

Section No.	Results of inspection (measurements)		Sewer conditions
	Diameter (mm)	Length (m)	Observed damage and judgement
1	ϕ1,500	76.60	Longitudinal cracking, rebar exposure
2	ϕ1,500	67.65	Rebar exposure
3	ϕ1,500	30.85	Rebar exposure
4	ϕ1,500	21.20	No damage
5	ϕ1,500	88.15	No damage
6	ϕ1,500	62.55	Rebar exposure
7	ϕ1,500	64.50	Rebar exposure
8	ϕ1,500	27.00	No damage
9	ϕ1,500	41.30	No damage
10	ϕ1,500	32.35	No damage
11	ϕ1,500	34.95	No damage
12	ϕ1,500	14.50	No damage

(2) Investigation items and results
The items and results of sewer investigation are shown in Table 8.41.

Table 8.41: Investigation items and results

Item		Results
		Investigation results
Cross-sectional shape		Circular
Internal dimensions (mm)	Diameter	ϕ1,500
Overburden thickness (m)		0.93–2.35
Compressive strength of concrete (N/mm²)		46.1–65.2
Member thickness (mm)		95.0
Concrete cover (mm)		31.0
Rebar diameter and spacing (mm)		ϕ5.0@73
Rebar strength (N/mm²)	Yield strength	
	Tensile strength	
	Based on JIS standard	SR235

8.4.2 Original design documents and determination of cross sections for structural analysis

As the design drawings used for the construction of the target sewer are available, details of the design cross section are determined based on both the results of investigation and data of the original design documents for obtaining the critical conditions. Table 8.42 compares the results of structural investigation and data obtained from the original drawings and shows the cross-sectional conditions used for structural analysis.

Table 8.42: Structural details of design cross section

Item		Classification		
		Structural investigation results	**Standard drawing**	**Design cross section**
Internal dimensions (mm)		ϕ1,500	ϕ1,500	ϕ1,500
Overburden thickness (m)		0.93–2.35		0.93–2.35
Compressive strength of concrete (N/mm²)		46.1		46.1
Member thickness (mm)		95.0	112.0	95.0
Concrete cover (mm)	Inside	31.0		31.0
	Outside			31.0
Rebar diameter and spacing (mm)	Inside	ϕ5.0@73		ϕ5.0@73
	Outside			ϕ5.0@73
Rebar strength (N/mm²)	Yield strength			235
	Tensile strength			380
	Based on JIS standard	SR235		SR235

8.4.3 General conditions for structural analysis

(1) Burial conditions, cross-sectional dimensions and rebar arrangements
Figure 8.54 shows the burial conditions of the existing sewer, its general structural dimensions, rebar arrangements and other structural details. The cross section of a sewer with standard renovation is also shown.

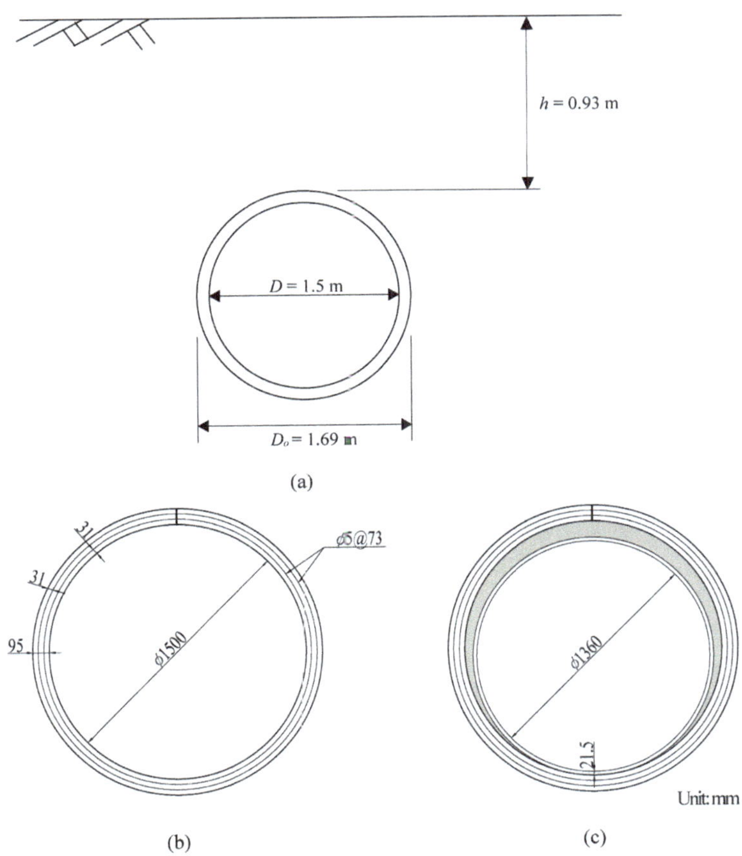

Figure 8.54: Cross section of sewer for structural analysis and renovation design: (a) general structural dimensions and burial conditions; (b) structural details of existing sewer; (c) cross section of sewer with standard renovation

(2) Support conditions
As shown in Fig. 8.55, the as-built drawing shows a sand-filled foundation and specifies the compacted thickness as being up to the level of the top of the sewer. Therefore, the foundation structure is assumed to have a 120-degree range of support area for reaction.

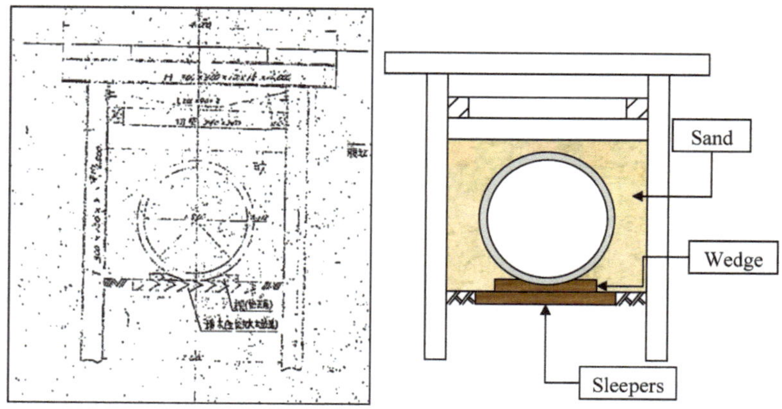

Figure 8.55: As-built drawing (1970)

(3) Material properties

The material properties of the concrete, grouting materials, rebars and profiled steel sheets of the renovation layer are summarised in Table 8.43.

Table 8.43: Material properties

Material	Item	Specification	Unit	Remarks
Concrete members	Compressive strength	46.1	N/mm²	Minimum test value
	Tensile strength	2.96	N/mm²	JSCE (2013)
	Modulus of elasticity	32.2	kN/mm²	〃
	Poisson's ratio	0.2		〃
	Fracture energy	104.8	N/m	〃
Rebar	Yield strength	235	N/mm²	Test
	Modulus of elasticity	200	kN/mm²	JSCE (2013)
Grout (Mortar No. 2)	Compressive strength	21.0	N/mm²	Design value
	Tensile strength	1.83	N/mm²	〃
	Modulus of elasticity	6.60	kN/mm²	〃
	Poisson's ratio	0.25		〃
	Fracture energy	17.6	N/m	〃
Grout (Mortar No. 3)	Compressive strength	35.0	N/mm²	Design value
	Tensile strength	2.92	N/mm²	〃
	Modulus of elasticity	22.0	kN/mm²	〃
	Poisson's ratio	0.22		〃
	Fracture energy	70.8	N/m	〃
Profiled steel sheet #79SW	Yield point	210	N/mm²	Design value
	Modulus of elasticity	170	kN/mm²	〃
	Cross-sectional area	1571	mm²/m	〃

(4) Soil conditions

In view of the boring data obtained from a nearby location, soil conditions are assumed as shown in Table 8.44.

Table 8.44: Soil conditions

Boring data	Soil No.	Soil type	Depth (m)	Thickness, H_i (m)	Average N-value	Average shear wave velocity, V_{si} (m/s)	H_i/V_{si} (s)
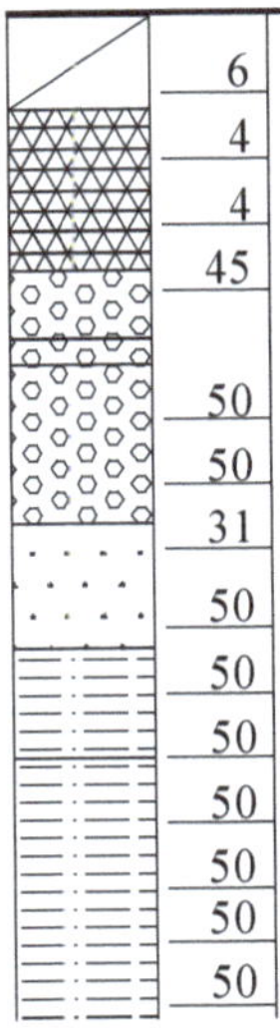	1	Surface soil	1.50	1.50	6	168.60	0.009
	2	Loam	4.90	3.40	4	157.97	0.022
	3	Sand and gravel	7.80	2.90	48	376.42	0.008
	4	Sand	9.70	1.90	31	325.19	0.006
	Basement layer	Hard pan			> 50	400.00	

Evaluation of ground property: $T_G = 4\sum_{i=1}^{n}\frac{H_i}{V_{si}} = 0.176$ (s), Class I

Classification of ground: $T_G < 0.2$, Class I; $0.2 \le T_G < 0.6$, Class II; $0.6 \le T_G$, Class III

(5) Load conditions

Since the overburden thickness along this sewer line is smaller than $2D_o$ (D_o: outside diameter of pipe), it was judged that the most critical load conditions would occur when normal earth pressure is taken into consideration under the minimum overburden thickness (0.93 m). Live loads were taken into account in the perpendicular direction. Load conditions are calculated based on relevant codes (JRA, 2013a-b).

1) Static earth pressure (Fig. 8.56)

Applied static earth pressures are shown in Fig. 8.56. The reaction due to the vertical earth pressure is obtained as

$$P_{rd} = \frac{P_{vd}}{\sin\left(120^\circ/2\right)} \quad (\text{kN/m}^2) \tag{8.12}$$

where P_{vd}: vertical earth pressure; P_{rd}: reaction due to vertical earth pressure. For details of earth pressure computation, refer to Example I. Note that the coefficient of horizontal earth pressure, K_o, is 0.3 for the present case.

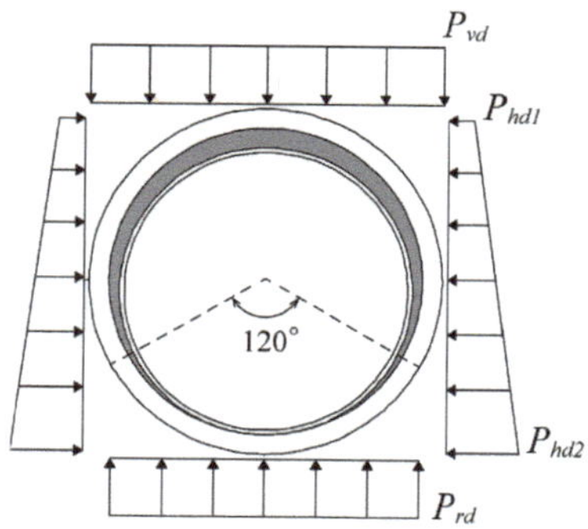

Static earth pressure (kN/m²)			
Vertical load		Horizontal load	
P_{vd}	P_{rd}	P_{hd1}	P_{hd2}
16.74	19.33	5.02	14.15

Figure 8.56: Static earth pressure

2) Live loads (Fig. 8.57)

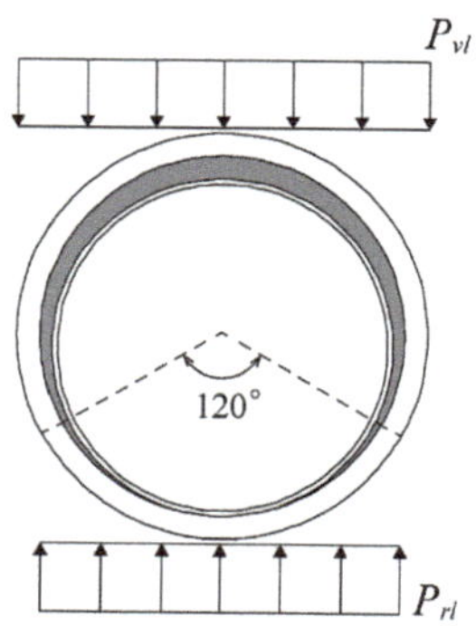

Live load (kN/m²)	
P_{vl}	P_{rl}
47.66	55.04

Figure 8.57: Live loads

Applied live loads are shown in Fig. 8.57. The reaction due to the vertical live load is obtained as

$$P_{rl} = \frac{P_{vl}}{\sin\left(120^\circ/2\right)} \ (\text{kN/m}^2) \tag{8.13}$$

where P_{vl}: vertical live load; P_{rl}: reaction due to vertical live load. For details of live load computation, refer to Example III. Note that the impact factor, i, is 0.5 for the present case.

(6) Analysis model

Figures 8.58 and 8.59 show FE models for structural analysis under normal load conditions and under seismic load conditions, respectively. In view of the symmetry, a half cross-sectional model is used for structural analysis under normal loads. In the seismic performance evaluation, the ground is modelled with finite elements having sufficient coverage for pipe–soil coupled analysis. Regions where longitudinal cracking has occurred are modelled assuming full-depth cracking.

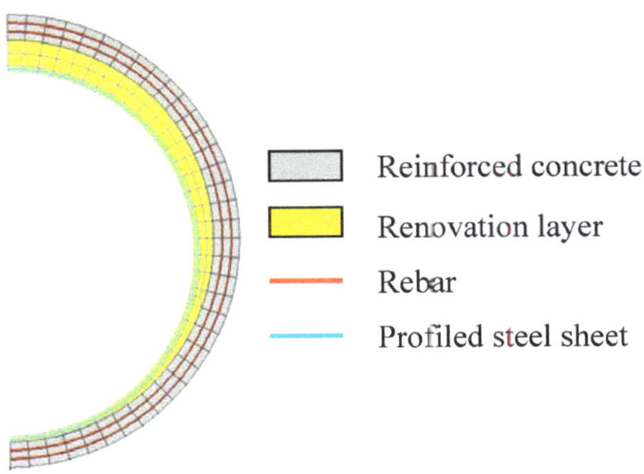

Figure 8.58: FE model for structural analysis under normal loads

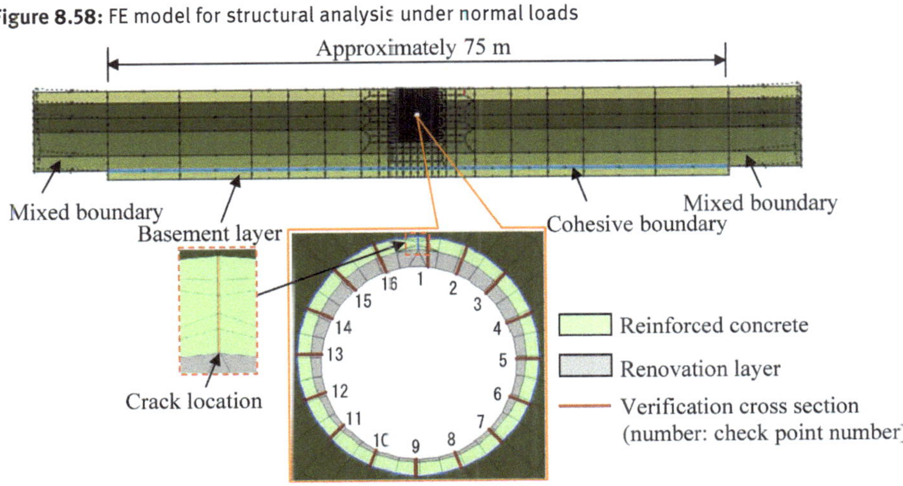

Figure 8.59: FE model for seismic performance analysis

8.4.4 Numerical results under normal load conditions

(1) Existing cross section

Figures 8.60 and 8.61 show the cracking patterns, sectional forces and structural deformations based on the FE model of the existing cross section.

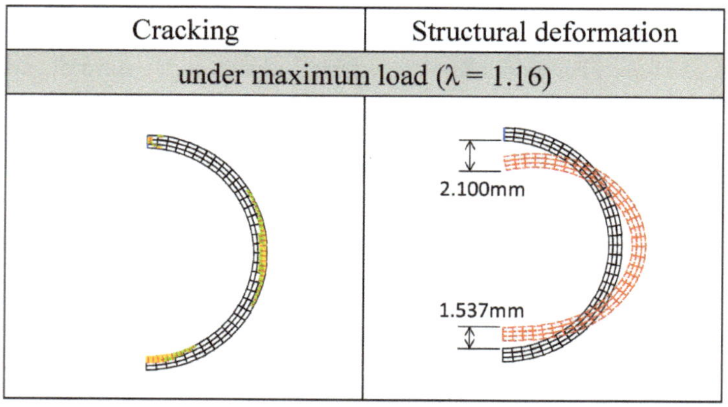

Figure 8.60: Cracking pattern and structural deformation (existing cross section)

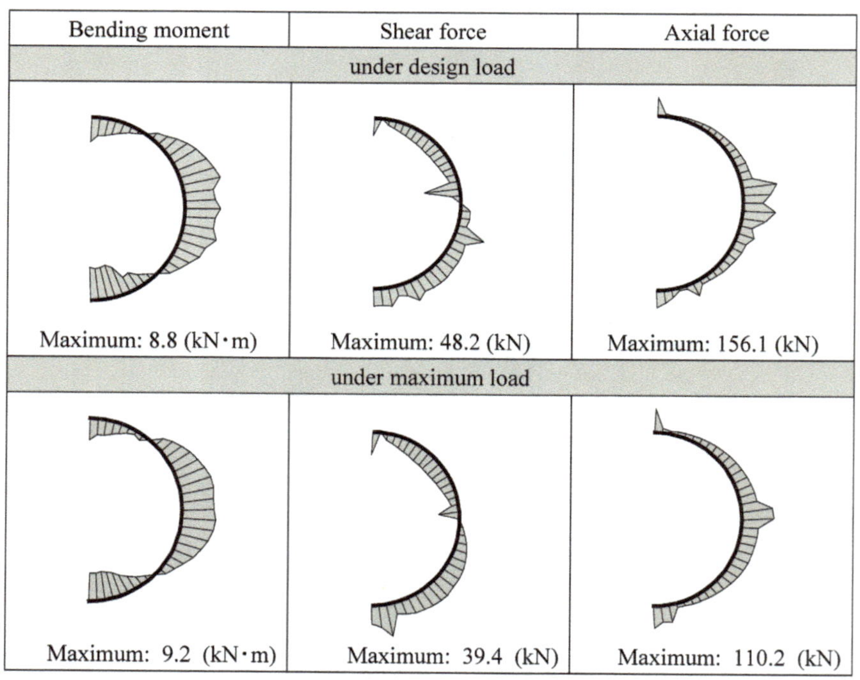

Figure 8.61: Sectional forces (existing cross section)

(2) Standard renovation cross section with Mortar No. 2
Figures 8.62 and 8.63 show the cracking patterns, sectional forces and structural deformations based on the FE model of the standard renovation cross section with Mortar No. 2.

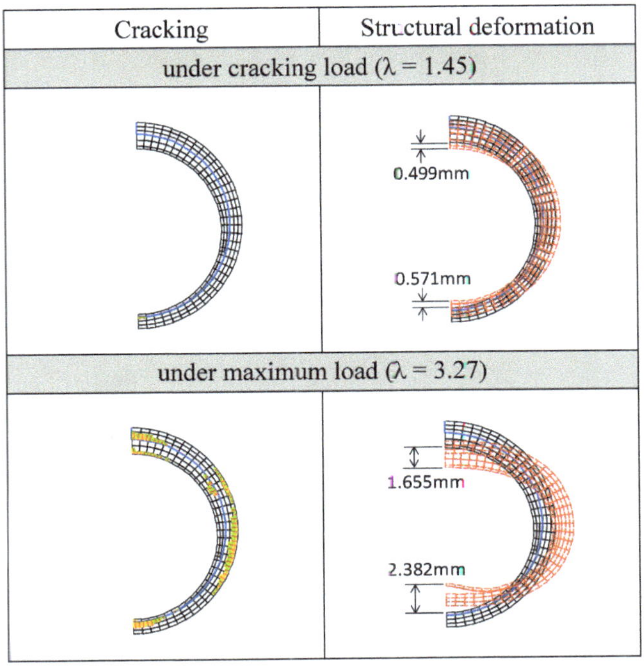

Figure 8.62: Cracking pattern and structural deformation (standard renovation cross section, Mortar No. 2)

(3) Standard renovation cross section with Mortar No. 3
Figures 8.64 and 8.65 show the cracking patterns, sectional forces and structural deformations based on the FE model of the standard renovation cross section with Mortar No. 3.

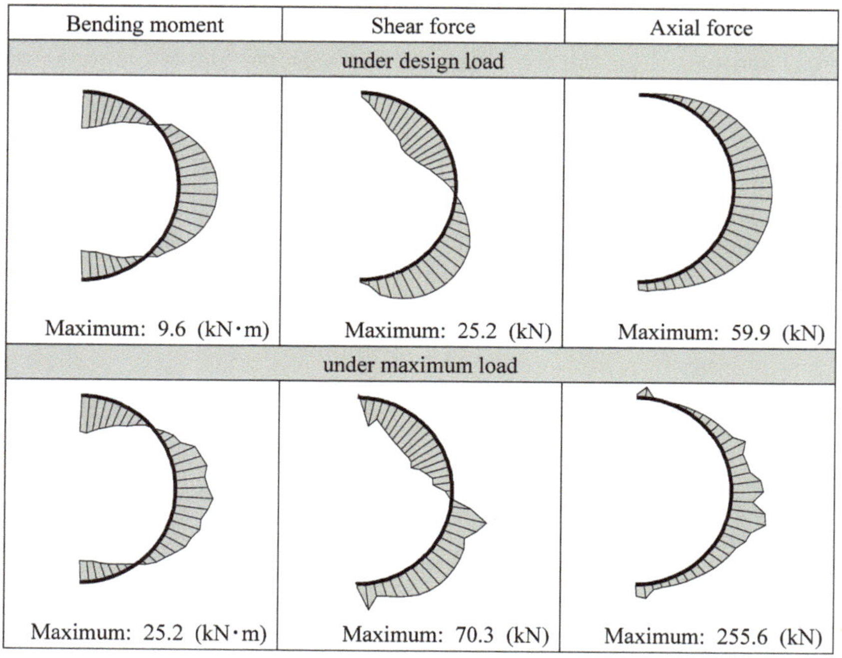

Figure 8.63: Sectional forces (standard renovation cross section, Mortar No. 2)

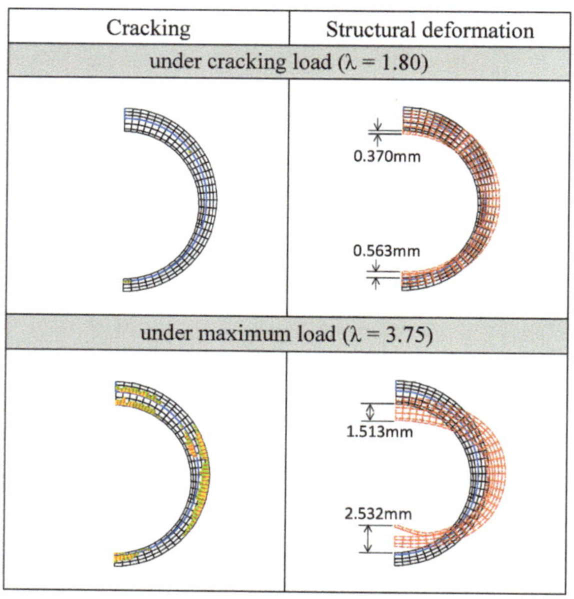

Figure 8.64: Cracking pattern and structural deformation (standard renovation cross section, Mortar No. 3)

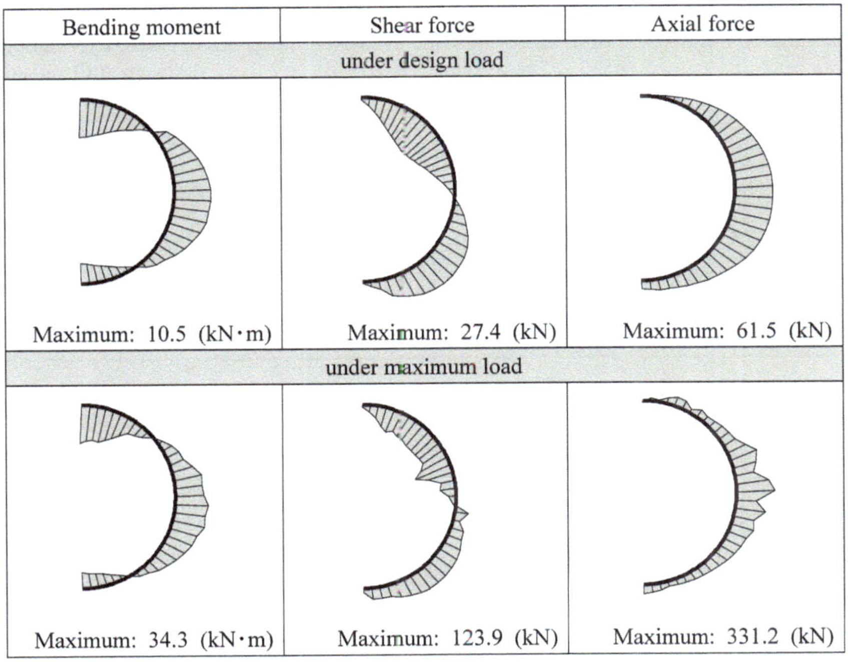

Figure 8.65: Sectional forces (standard renovation cross section, Mortar No. 3)

8.4.5 Verification of safety under normal load conditions

The results of safety verification on the existing sewer and the other two renovation conditions are shown in Table 8.45. Based on these results, the following comments are made:

- In the serviceability limit state, the existing sewer will not meet the performance requirements because it has already sustained severe cracking. The standard renovation cross section with Mortar No. 2 or No. 3 will satisfy the performance requirements because the load coefficient for cracking is greater than 1.0.
- In the ultimate limit state, the existing sewer will not meet the performance requirements because its maximum load coefficient is less than 2.5. The standard renovation cross section with Mortar No. 2 or No. 3 will satisfy the performance requirements because the maximum load coefficient exceeds 2.5.

Table 8.45: Results of safety verification based on load coefficients

Renovation condition		Perpendicular live load		Judgement	Remarks
Cross section	**Mortar**	**Cracking load coefficients**	**Maximum load coefficients**		
Existing cross section			1.16	NG	
Standard renovation	No. 2	1.45	3.27	OK	
	No. 3	1.80	3.75	OK	
Required load coefficients		> 1.0	≥ 2.5		

8.4.6 Results of seismic performance analysis

(1) Existing cross section

1) Level 1 earthquake ground motion

Figure 8.66 shows the cracking and rebar yielding of the existing cross section under Level 1 earthquake ground motion.

Rebar yielding	None
Judgement	OK

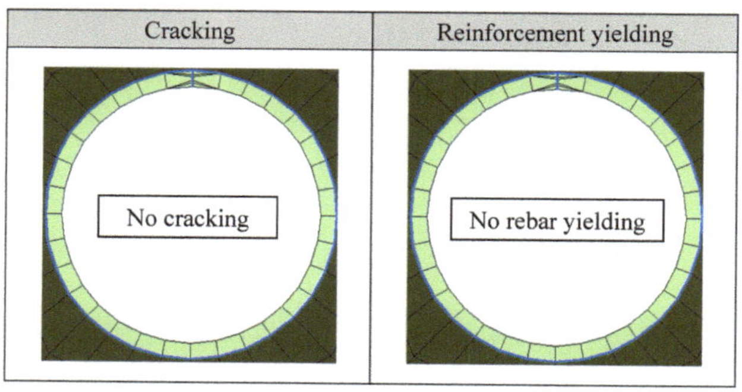

Figure 8.66: Cracking and rebar yielding in existing cross section (Level 1 earthquake ground motion)

2) Level 2 earthquake ground motion

Figure 8.67 shows the cracking and rebar yielding of the existing cross section under Level 2 earthquake ground motion. Table 8.46 shows the verification results under Level 2 ground motion.

Table 8.46: Verification results for existing cross section under Level 2 earthquake ground motion

Cross section			**Existing cross section (ϕ1,500)**							
Input ground motion			Level 2 earthquake ground motion (Inland Type 1)				Level 2 earthquake ground motion (Ocean Type 2)			
Verification item			Response value	Response value/ limit value	Judge-ment	Limit value	Response value	Response value/ limit value	Judge-ment	Limit value
Maximum compressive strain			0.000312		OK	0.0031	Less than 0.0031		OK	0.0031
Shear force at check point (kN)	Top	1	19.86		NG	0.00	11.49		NG	0.00
		2	18.82	0.62	OK	30.51	14.83	0.41	OK	36.61
		3	8.27	0.27	OK	30.51	6.86	0.19	OK	36.61
	Side	4	19.59	0.64	OK	30.51	13.18	0.36	OK	36.61
		5	30.88	0.84	OK	36.61	15.91	0.43	OK	36.61
		6	19.97	0.65	OK	30.51	14.61	0.40	OK	36.61
	Bottom	7	8.86	0.29	OK	30.51	6.53	0.18	OK	36.61
		8	21.46	0.70	OK	30.51	13.64	0.37	OK	36.61
		9	26.57	0.73	OK	36.61	19.99	0.55	OK	36.61
		10	20.87	0.68	OK	30.51	17.85	0.49	OK	36.61
		11	10.07	0.33	OK	30.51	9.3	0.25	OK	36.61
	Side	12	24.52	0.80	OK	30.51	16.36	0.45	OK	36.61
		13	28.31	0.77	OK	36.61	15.57	0.43	OK	36.61
		14	14.09	0.46	OK	30.51	11.21	0.31	OK	36.61
	Top	15	5.72	0.19	OK	30.51	5.03	0.14	OK	36.61
		16	19.93	0.54	OK	36.61	9.98	0.27	OK	36.61
Final judgement			NG				NG			

Member with rebar yielding (structural member coefficient $\gamma_b = 1.56$)

The shear strength of existing members is assumed to be zero because of longitudinal cracking.

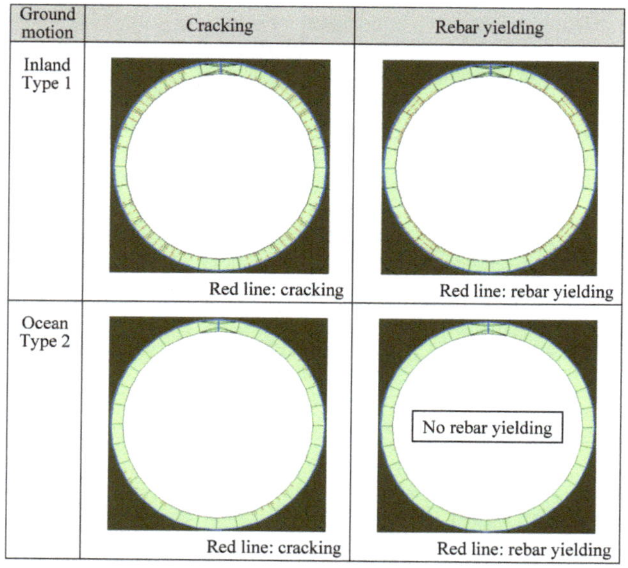

Figure 8.67: Cracking and rebar yielding in existing cross section (Level 2 earthquake ground motion)

(2) Renovation with Mortar No. 2

Figure 8.68 shows the cracking and rebar yielding under Level 2 earthquake ground motion of a standard renovation cross section with Mortar No. 2. Table 8.47 shows the verification results.

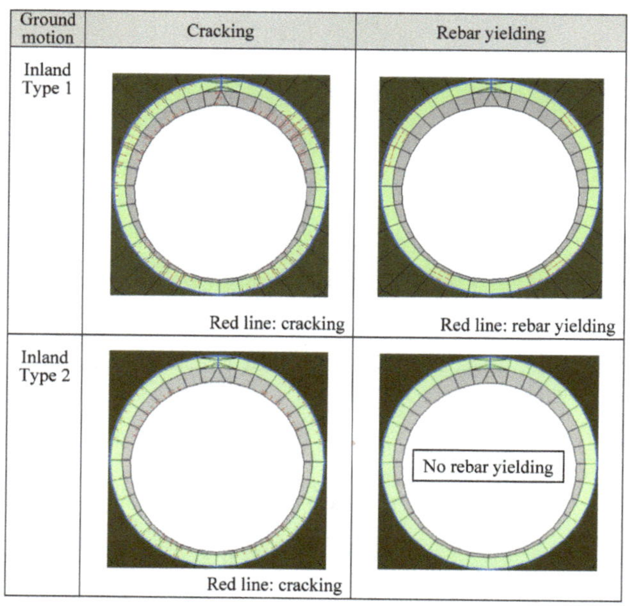

Figure 8.68: Cracking and rebar yielding in standard renovation cross section with Mortar No. 2 (Level 2 earthquake ground motion)

Table 8.47: Verification results for standard renovation cross section with Mortar No. 2 under Level 2 earthquake ground motion

Cross section			**Standard renovation cross section (ϕ1,360)**							
Mortar			**Mortar No. 2**							
Input ground motion			Level 2 earthquake ground motion (Inland Type 1)				Level 2 earthquake ground motion (Inland Type 2)			
Verification item			Response value	Response value/ limit value	Judge-ment	Limit value	Response value	Response value/ limit value	Judge-ment	Limit value
Maximum compressive strain			0.0016		OK	0.0031	Smaller than 0.0031		OK	0.0031
Shear force at check point (kN)	Top	1	53.93	0.17	OK	317.64	50.00	0.16	OK	317.64
		2	35.78	0.10	OK	344.92	34.57	0.10	OK	344.92
		3	23.59	0.09	OK	265.22	12.96	0.04	OK	318.26
	Side	4	32.54	0.12	OK	277.57	27.64	0.10	OK	277.57
		5	44.19	0.19	OK	228.80	38.07	0.17	OK	228.80
		6	30.31	0.17	OK	178.17	33.25	0.19	OK	178.17
	Bottom	7	18.64	0.30	OK	61.27	11.88	0.16	OK	73.53
		8	25.10	0.48	OK	51.93	22.05	0.35	OK	62.32
		9	36.16	0.62	OK	58.36	37.86	0.65	OK	58.36
		10	23.32	0.45	OK	51.93	22.41	0.36	OK	62.32
		11	14.91	0.20	OK	73.53	14.10	0.19	OK	73.53
	Side	12	51.65	0.29	OK	178.17	35.26	0.20	OK	178.17
		13	56.45	0.30	OK	190.67	37.02	0.16	OK	228.80
		14	28.70	0.12	OK	231.31	24.51	0.09	OK	277.57
	Top	15	11.08	0.04	OK	265.22	11.74	0.04	OK	318.26
		16	36.10	0.10	OK	344.92	33.00	0.10	OK	344.92
Final judgement			OK				OK			

Member with rebar yielding (structural member coefficient $\gamma_b = 1.56$)

The shear strength of existing members is assumed to be zero because of longitudinal cracking.

(3) Renovation with Mortar No. 3

Figure 8.69 shows the cracking and rebar yielding under Level 2 earthquake ground motion of a standard renovation cross section with Mortar No. 3. Table 8.48 shows the verification results.

Table 8.48: Verification results for standard renovation cross section with Mortar No. 3 under Level 2 earthquake ground motion

Cross section			Standard renovation cross section (ϕ1,360)							
Mortar			Mortar No. 3							
Input ground motion			Level 2 earthquake ground motion (Inland Type 1)				Level 2 earthquake ground motion (Inland Type 2)			
Verification item			Response value	Response value/ limit value	Judgement	Limit value	Response value	Response value/ limit value	Judgement	Limit value
Maximum compressive strain			0.0003		OK	0.0031	Less than 0.0031		OK	0.0031
Shear force at check point (kN)	Top	1	66.77	0.18	OK	372.33	52.68	0.14	OK	372.33
		2	49.80	0.13	OK	398.00	35.99	0.09	OK	398.00
		3	14.36	0.04	OK	366.76	14.39	0.04	OK	366.76
	Side	4	44.23	0.14	OK	319.06	28.81	0.09	OK	319.06
		5	53.60	0.20	OK	261.89	39.05	0.15	OK	261.89
		6	34.36	0.17	OK	202.54	35.17	0.17	OK	202.54
	Bottom	7	13.17	0.20	OK	64.73	13.09	0.17	OK	77.67
		8	22.17	0.41	OK	54.34	22.16	0.34	OK	65.20
		9	42.07	0.69	OK	60.81	38.89	0.64	OK	60.81
		10	24.33	0.45	OK	54.34	22.55	0.35	OK	65.20
		11	17.50	0.27	OK	64.73	15.78	0.20	OK	77.67
	Side	12	56.94	0.28	OK	202.54	37.18	0.18	OK	202.54
		13	36.10	0.14	OK	261.89	38.98	0.15	OK	261.89
		14	23.79	0.07	OK	319.06	25.57	0.08	OK	319.06
	Top	15	14.96	0.04	OK	366.76	13.04	0.04	OK	366.76
		16	46.43	0.12	OK	398.00	34.48	0.09	OK	398.00
Final judgement			OK				OK			

Member with rebar yielding (structural member coefficient $\gamma_b = 1.56$)

The shear strength of existing members is assumed to be zero because of longitudinal cracking.

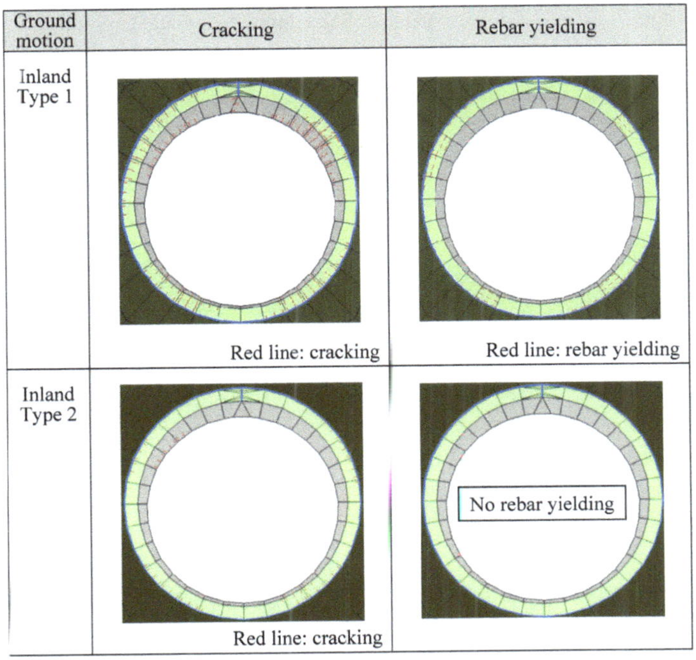

Figure 8.69: Cracking and rebar yielding in standard renovation cross section with Mortar No. 3 (Level 2 earthquake ground motion)

8.4.7 Verification of safety under earthquake loading

The results of seismic safety verification on the existing sewer and the other two types of renovation conditions are shown in Table 8.49. Based on these results, the following comments are made:

- The verification results indicate that under Level 1 earthquake ground motion, the existing sewer meets the earthquake resistance requirements.
- The verification results indicate that under Level 2 earthquake ground motion, the existing sewer does not meet the earthquake resistance requirements, but the standard renovation cross section with Mortar No. 2 or Mortar No. 3 satisfies the earthquake resistance requirements.

Table 8.49: Results of safety verification under earthquake loading

Renovation condition		Level 1 earthquake ground motion	Level 2 earthquake ground motion			Remarks
Cross section	Mortar		Ultimate displace-ment	Shear force	Judge-ment	
Existing cross section		OK	OK	NG	NG	
Standard renovation cross section	No. 2		OK	OK	OK	
	No. 3		OK	OK	OK	

Note: Ultimate displacement under Level 2 earthquake ground motion is verified by checking the maximum compressive strain.

8.4.8 Safety verification of local buckling at the bottom of the pipe

Safety verification on the local buckling of the renovation layer at the bottom of the pipe is deemed unnecessary; refer to Section 5.4 for an explanation.

8.4.9 Determination of renovation methods

(1) Structural strength
Though the existing sewer can withstand Level 1 earthquake force, it is not strong enough to sustain Level 2 earthquake ground motion, and it fails to meet the performance requirements under normal load conditions. Therefore, structural renovation has to be performed on this sewer line. The required structural strength can be achieved by employing the standard renovation cross section with Mortar No. 2 or Mortar No. 3.

(2) Achieving the required discharge capacity
The discharge capacity at each section of the renovated sewer is calculated and shown in Table 8.50. For details of flow ratio computation, refer to Example I.

(3) Conclusion
Based on the various analysis results discussed above, sewer renovation will be carried out for this sewer line to strengthen its earthquake resistance, and to mitigate flooding damage by keeping its discharge capacity adequate. The renovation method adopted for this sewer line is to use the standard renovation cross section with Mortar No. 2, so as to achieve the required structural strength against Level 2 earthquake ground motion economically. The renovation plan for each section of this sewer line is listed in Table 8.50.

Table 8.50: List of renovation plans (Example IV)

Section No.		1	2	3	4	5	6	7
Sewer length (m)		76.60	67.65	30.85	21.20	88.15	62.55	64.50
Dimensions	Existing sewer	ϕ1,500	ϕ1,500	ϕ1,500	ϕ1,500	ϕ1,500	ϕ1,500	ϕ1,500
	After renovation	ϕ1,360	ϕ1,360	ϕ1,360	ϕ1,360	ϕ1,360	ϕ1,360	ϕ1,360
Slope (‰)	Existing sewer	0.4	1.3	1.0	1.2	1.3	1.2	1.7
	After renovation	0.4	1.3	1.0	1.2	1.3	1.2	1.7
Flow ratio	Existing sewer	25.47	19.75	14.83	7.82	11.76	6.73	6.48
	After renovation	25.50	19.77	14.85	7.83	11.77	6.73	6.49

Section No.		8	9	10	11	12
Sewer length (m)		27.00	41.30	32.35	34.95	14.5
Dimensions	Existing sewer	ϕ1,500	ϕ1,500	ϕ1,500	ϕ1,500	ϕ1,500
	After renovation	ϕ1,360	ϕ1,360	ϕ1,360	ϕ1,360	ϕ1,360
Slope (‰)	Existing sewer	0.8	2.0	1.1	0.7	-13.6
	After renovation	0.8	2.0	1.1	0.7	3.4
Flow ratio	Existing sewer	3.09	3.25	15.07	9.76	
	After renovation	3.09	3.25	15.09	9.77	2.01

References

JSCE (2013). Standard Specifications for Concrete Structures: Design. Tokyo, Japan Society of Civil Engineers.

JRA (2013a). Specifications for Highway Bridges: Part I, General Specifications. Tokyo. Japan Road Association.

JRA (2013b). Highway Earthworks: Guidelines for Culverts. Tokyo, Japan Road Association.

List of Figures

List of Photos

List of Tables

Index

Kolibri 360 GmbH
Ferdinand-Jühlke-Straße 7
99095 Erfurt, Deutschland
produktsicherheit@kolibri360.de